Modern Mathematical Statistics with Applications

Modern Mathematical Statistics with Applications

Edited by Canaan Knight

Larsen & Keller
www.larsen-keller.com

Modern Mathematical Statistics with Applications
Edited by Canaan Knight
ISBN: 979-8-88836-047-7 (Hardback)

Larsen & Keller
Published by Larsen and Keller Education,
5 Penn Plaza,
19th Floor,
New York, NY 10001, USA

Cataloging-in-Publication Data

Modern mathematical statistics with applications / edited by Canaan Knight.
p. cm.
Includes bibliographical references and index.
ISBN 979-8-88836-047-7
1. Mathematical statistics. 2. Mathematical statistics--Problems, exercises, etc. 3. Statistics.
4. Mathematics. I. Knight, Canaan.
QA276 .M63 2023
519.5--dc23

For more information regarding Larsen and Keller Education and its products, please visit the publisher's website www.larsen-keller.com

Table of Contents

	Preface	**VII**
Chapter 1	**An Estimation of Sensitive Attribute Applying Geometric Distribution under Probability Proportional to Size Sampling** Gi-Sung Lee, Ki-Hak Hong and Chang-Kyoon Son	**1**
Chapter 2	**A Functional Data Analysis Approach for the Detection of Air Pollution Episodes and Outliers** Javier Martínez Torres, Jorge Pastor Pérez, Joaquín Sancho Val, Aonghus McNabola, Miguel Martínez Comesaña and John Gallagher	**17**
Chapter 3	**Skew-Reflected-Gompertz Information Quantifiers with Application to Sea Surface Temperature Records** Javier E. Contreras-Reyes, Mohsen Maleki and Daniel Devia Cortés	**36**
Chapter 4	**Classical Lagrange Interpolation based on General Nodal Systems at Perturbed Roots of Unity** Elías Berriochoa, Alicia Cachafeiro, Alberto Castejón and José Manuel García-Amor	**50**
Chapter 5	**On the Performance of Variable Selection and Classification via Rank-Based Classifier** Md Showaib Rahman Sarker, Michael Pokojovy and Sangjin Kim	**66**
Chapter 6	**A Functional Data Analysis for Assessing the Impact of a Retrofitting in the Energy Performance of a Building** Miguel Martínez Comesaña, Sandra Martínez Mariño, Pablo Eguía Oller, Enrique Granada Álvarez and Aitor Erkoreka González	**82**
Chapter 7	**Combination of Ensembles of Regularized Regression Models with Resampling-Based Lasso Feature Selection in High Dimensional Data** Abhijeet R Patil and Sangjin Kim	**102**
Chapter 8	**Outliers Detection Models in Shewhart Control Charts; An Application in Photolithography: A Semiconductor Manufacturing Industry** Ishaq Adeyanju Raji, Muhammad Hisyam Lee, Muhammad Riaz, Mu'azu Ramat Abujiya and Nasir Abbas	**125**
Chapter 9	**Comparing Groups of Decision-Making Units in Efficiency based on Semiparametric Regression** Hohsuk Noh and Seong J. Yang	**142**
Chapter 10	**Functional Location-Scale Model to Forecast Bivariate Pollution Episodes** Manuel Oviedo-de La Fuente, Celestino Ordóñez and Javier Roca-Pardiñas	**158**
Chapter 11	**Two-Stage Classification with SIS using a New Filter Ranking Method in High Throughput Data** Sangjin Kim and Jong-Min Kim	**170**

Chapter 12 **Constructing a Control Chart using Functional Data** **186**
Miguel Flores, Salvador Naya, Rubén Fernández-Casal, Sonia Zaragoza, Paula Raña and Javier Tarrío-Saavedra

Chapter 13 **Robust Linear Trend Test for Low-Coverage Next-Generation Sequence Data Controlling for Covariates** **212**
Jung Yeon Lee, Myeong-Kyu Kim and Wonkuk Kim

Chapter 14 **Application of Functional Data Analysis and FTIR-ATR Spectroscopy to Discriminate Wine Spirits Ageing Technologies** **226**
Ofélia Anjos, Miguel Martínez Comesaña, Ilda Caldeira, Soraia Inês Pedro, Pablo Eguía Oller and Sara Canas

Permissions

List of Contributors

Index

Preface

Statistics is a branch of applied mathematics that deals with collecting, describing, presenting and analyzing data. It also involves making inferences or conclusions from the given quantitative data. There are two major areas of statistics, namely, descriptive statistics and inferential statistics. Descriptive statistics is focused on describing the properties associated with the sample and population data. In inferential statistics, sample data is analyzed to test hypotheses and draw conclusions. Some of the common and widely used statistical tools and procedures are variance, skewness, linear regression analysis, null hypothesis testing, probit models, ANOVA, and mean. Statistics and statistical techniques draw heavily on various mathematical theories such as differential and integral calculus, linear algebra, and probability theory. Statistics finds applications in a variety of disciplines and professions including economics and finance, accounting, academic research, and investment analysis. The book studies, and analyzes mathematical statistics and its applications in modern times. It is an essential guide for both academicians and those who wish to pursue this discipline further.

The information shared in this book is based on empirical researches made by veterans in this field of study. The elaborative information provided in this book will help the readers further their scope of knowledge leading to advancements in this field.

Finally, I would like to thank my fellow researchers who gave constructive feedback and my family members who supported me at every step of my research.

Editor

1

An Estimation of Sensitive Attribute Applying Geometric Distribution under Probability Proportional to Size Sampling

Gi-Sung Lee [1], Ki-Hak Hong [2] and Chang-Kyoon Son [3,*]

1 Department of Children Welfare, Woosuk University, Wanju Jeonbuk 55338, Korea; gisung@woosuk.ac.kr
2 Department of Computer Science, Dongshin University, Naju Jeonnam 58245, Korea; khhong@dsu.ac.kr
3 Department of Applied Statistics, Dongguk University, Gyeongju Gyeongbuk 38066, Korea
* Correspondence: sonchangkyoon@gmail.com

Abstract: In this paper, we extended Yennum et al.'s model, in which geometric distribution is used as a randomization device for a population that consists of different-sized clusters, and clusters are obtained by probability proportional to size (PPS) sampling. Estimators of a sensitive parameter, their variances, and their variance estimators are derived under PPS sampling and equal probability two-stage sampling, respectively. We also applied these sampling schemes to Yennum et al.'s generalized model. Numerical studies were carried out to compare the efficiencies of the proposed sampling methods for each case of Yennum et al.'s model and Yennum et al.'s generalized model.

Keywords: probability proportional to size (PPS) sampling; geometric distribution; sensitive attribute; randomization device; Yennum et al.'s model

1. Introduction

The randomized response model (RRM) was suggested by [1] to estimate the true population proportion of sensitive characteristics, such as illegal gambling, drug-abuse, tax evasion, the extent of illegal income, and the experience of abortion, among others [2–4].

Since Warner's work, many scholars have developed the RRM in various ways. In [5,6], they arranged, summarized, and systemized various RRMs and emphasized their importance. In [7], sampling survey of sensitive attributes applied two-stage cluster sampling to RRM for a population consisting of equal-sized clusters, and [8] considered the cluster RRM for a population consisting of different-sized clusters, where the clusters are selected by probability proportional to size (PPS) sampling.

Recently, Yennum et al. [9] suggested a new randomization device to gather sensitive data in two-stages under the assumption of geometric distribution and made a generalization of their model encompassing generalized geometric distribution using [10] model.

Based on Yennum et al.'s work, it is assumed that the respondents are selected by simple random sampling with replacements, but a real survey selects respondents from various sampling schemes.

Now, we can consider a large sample of clusters. For example, to estimate the true population proportion of drug-abuse among high school students, it is possible to use a randomization device like Yennum et al.'s model via proportional sampling by considering the primary sampling unit as the school and the secondary sampling unit as the students.

From this point of view, we extend Yennum et al.'s model, in which geometric distribution is used as a randomization device based on a population that consists of different-sized clusters, and the clusters are selected by PPS sampling. Estimators of a sensitive parameter, their variances, and their variance estimators are derived by PPS sampling and equal probability two-stage sampling, respectively.

We also apply these methods to the case of Yennum et al.'s generalized model. Numerical studies are carried out to compare the efficiencies of the suggested methods in each case of Yennum et al.'s model and Yennum et al.'s generalized model.

2. An Estimation of Sensitive Attributes with Probability Proportional to Size Sampling under Yennum et al.'s Model

In Section 2, we consider a new sampling scheme to estimate sensitive attributes using Yennum et al.'s model, in which geometric distribution is used as a randomization device when n clusters are selected with proportional to size (PPS) sampling or equal probability sampling from a population that consists of N clusters of size, $M_i(i = 1, 2, \cdots, N)$ and $m_i(i = 1, 2, \cdots, n)$ units are selected by simple random sampling from each sampled cluster.

In Section 2.1, we consider the sampling method for the clusters via PPS sampling with replacements. Clusters by PPS sampling without replacement are considered in Section 2.2, and clusters by equal probability sampling are examined in Section 2.3.

2.1. PPS Sampling with Replacement

Let the population be composed of N clusters. In the first stage, the size of the n sample of the first sampling units (FSU) is selected with replacement by the selection probability p_i for the ith cluster. In the second stage, m_i second sampling units (SSU) are drawn by simple random sampling with replacement (SRSWR) from each FSU and are guided to carry out Yennum et al.'s randomization device.

First of all, the randomization device consists of two elements. The first randomization device for the ith cluster consists of two kinds of urns with white and black balls, where the selection probability of a white ball is W_i, and the selection probability of a black ball is $1 - W_i$.

On the other hand, the second randomization device is composed of two kinds of urns with balls. The first device with balls contains a slip of paper including two statements, such as "I have a sensitive attribute" with selection probability P_i, and the other balls includes a statement such as "I do not have a sensitive attribute" with selection probability $1 - P_i$. The second device with balls contains a slip of paper with the statement "I do not have a sensitive attribute" with selection probability T_i and balls with the statement "I have a sensitive attribute" with selection probability $1 - T_i$.

In the first stage, for the ith cluster, each interviewee draws a ball from the first randomization device, such as the urn with the white and black balls. When he or she selects a white ball, he or she is guided to pick balls from the first urn of the second randomization device, one after another, with replacement, until the first ball containing a statement matching his or her own status appears.

We assume that X_{i1} is the total number of balls drawn before he or she obtains the first ball including his or her own status in the ith cluster, and X_{i2} is the total number of balls drawn before he or she obtains the first ball with his or her own status of not having a sensitive attribute in the ith cluster. Similarly, when he or she draws a black ball, he or she is guided to pick balls from the second urn of the second randomization device, one after another, with replacement, until the first ball containing a statement matching his or her own status appears.

For the ith cluster, using the randomization device in Figure 1, the total number of balls taken by interviewees $X_{i1}, X_{i2}, Y_{i1}, Y_{i2}$ are distributed via generalized geometric distribution. Let π_i and $1 - \pi_i$ be the true population proportion of persons who have a sensitive attribute A_i and A_i^c for the ith cluster. Assume that each interviewee in the ith cluster is drawn by SRSWR.

For the ith cluster, the total number for each ball selected by interviewees through the proposed two-stage device distributes one of the following random variables: $X_{i1} \sim Ge(P_i)$, $X_{i2} \sim Ge(1 - P_i)$, $Y_{i1} \sim Ge(T_i)$ and $Y_{i2} \sim Ge(1 - T_i)$, where $Ge(\cdot)$ represents the geometric distribution with a success probability. Let π_i and $1 - \pi_i$ be the true population proportions of persons who have a sensitive attribute (A_i and A_i^c, respectively) for the ith cluster. Assume that each interviewee in the ith cluster is drawn by SRSWR.

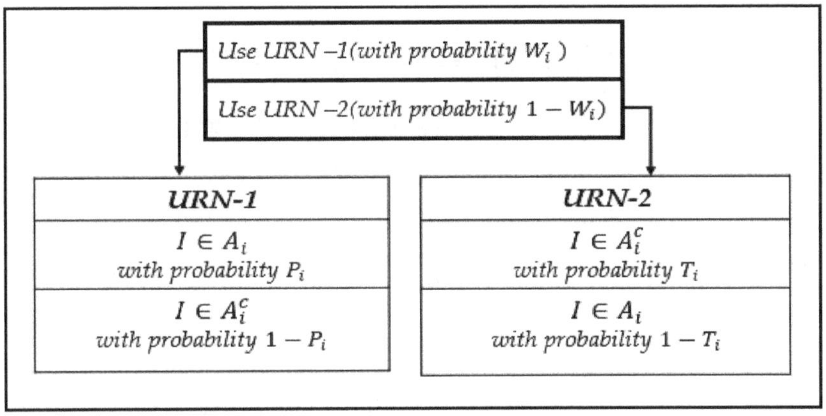

Figure 1. Randomization device for the ith cluster.

Let Z_{ij} be the jth observed answer in the ith cluster; Z_{ij} can be expressed as

$$Z_{ij} = \begin{cases} X_{i1}, \text{with probability } W_i\pi_i \\ Y_{i2}, \text{with probability } (1-W_i)\pi_i \\ X_{i2}, \text{with probability } W_i(1-\pi_i) \\ Y_{i1}, \text{with probability } (1-W_i)(1-\pi_i) \end{cases} \tag{1}$$

Then, we can find the expected value of Z_{ij} as follows:

$$\begin{aligned} E(Z_{ij}) &= \pi_i\left[\frac{W_i}{P_i} + \frac{(1-W_i)}{(1-T_i)}\right] + (1-\pi_i)\left[\frac{W_i}{(1-P_i)} + \frac{(1-W_i)}{T_i}\right] \\ &= \pi_i\left[\frac{W_i}{P_i} + \frac{(1-W_i)}{(1-T_i)} - \frac{W_i}{(1-P_i)} - \frac{(1-W_i)}{T_i}\right] + \frac{W_i}{(1-P_i)} + \frac{(1-W_i)}{T_i}. \end{aligned} \tag{2}$$

The expected value (2) can be expressed as follows:

$$\frac{(1-T_i)P_i\{E(Z_{ij})(1-P_i)T_i - W_iT_i - (1-W_i)(1-P_i)\}}{P_iT_i(1-P_i)(1-T_i)} = \frac{\pi_i\psi_i}{P_iT_i(1-P_i)(1-T_i)}, \tag{3}$$

where $\psi_i = W_i(1-2P_i)T_i(1-T_i) + (1-W_i)(2T_i-1)P_i(1-P_i)$.

Now the estimator $\hat{\pi}_i$ for the true population proportion π_i in the ith cluster is given by:

$$\hat{\pi}_i = \frac{1}{\psi_i}\left[P_iT_i(1-P_i)(1-T_i)\frac{1}{m_i}\sum_{i=1}^{m_i} Z_{ij} - W_iT_iP_i(1-T_i) - P_i(1-W_i)(1-P_i)(1-T_i)\right]. \tag{4}$$

When the interviewees are drawn by SRSWR from the ith cluster selected with a replacement by the sampling probability p_i, the estimator $\hat{\pi}_{ppswr}$ of the true population proportion π for a sensitive character is given by:

$$\begin{aligned} \hat{\pi}_{ppswr} &= \frac{1}{nM_0}\sum_{i=1}^{n}\frac{M_i\hat{\pi}_i}{p_i} \\ &= \frac{1}{nM_0}\sum_{i=1}^{n}\frac{M_i}{p_i\psi_i}\left[P_iT_i(1-P_i)(1-T_i)\frac{1}{m_i}\sum_{j=1}^{m_i} Z_{ij} - W_iT_iP_i(1-T_i) - P_i(1-W_i)(1-P_i)(1-T_i)\right], \end{aligned} \tag{5}$$

where $M_0 = \sum_{i=1}^{N} M_i$.

Theorem 1: *The estimator $\hat{\pi}_{ppswr}$ of the true population proportion of a sensitive attribute π under PPS with a replacement sampling scheme is an unbiased estimator.*

Proof:

$$\begin{aligned} E_1E_2\left(\hat{\pi}_{ppswr}\right) &= E_1E_2\left[\frac{1}{nM_0}\sum_{i=1}^{n}\frac{M_i\hat{\pi}_i}{p_i}\right] \\ &= E_1\left[\frac{1}{nM_0}\sum_{i=1}^{n}\frac{M_iE_2(\hat{\pi}_i)}{p_i}\right], \end{aligned}$$

and since:

$$\begin{aligned} E_2(\hat{\pi}_i) &= \frac{1}{\psi_i}\left[P_iT_i(1-P_i)(1-T_i)\frac{1}{m_i}\sum_{i=1}^{m_i}E_2\left(Z_{ij}\right) - W_iT_iP_i(1-T_i) - P_i(1-W_i)(1-P_i)(1-T_i)\right] \\ &= \pi_i. \end{aligned}$$

we can obtain:

$$\begin{aligned} E_1E_2\left(\hat{\pi}_{ppswr}\right) &= E_1\left[\frac{1}{nM_0}\sum_{i=1}^{n}\frac{M_i\pi_i}{p_i}\right] \\ &= \frac{1}{M_0}\sum_{i=1}^{N}p_i\frac{M_i\pi_i}{p_i} \\ &= \pi. \end{aligned}$$

□

Theorem 2: *The variance of $\hat{\pi}_{ppswr}$ is obtained from a two-stage procedure, such that a sample of size n FSU is selected by replacement with sampling probability p_i for the unit i from the population of N clusters with size M_i elements in the ith cluster, and the SSUs with size m_i are drawn by SRSWR from each FSU, as given by:*

$$\begin{aligned} V(\hat{\pi}_{ppswr}) &= \frac{1}{nM_0^2}\sum_{i=1}^{N}p_i\left[\frac{M_i\pi_i}{p_i} - M_0\pi\right]^2 \\ &+ \frac{1}{nM_0^2}\sum_{i=1}^{N}\frac{M_i^2}{m_ip_i}\left[\pi_i(1-\pi_i) - \frac{\pi_i}{\psi_i^2}A_i + \frac{1}{\psi_i^2}B_i\right], \end{aligned} \tag{6}$$

where:

$$\begin{aligned} A_i &= W_i(1-2P_i)(2-P_i+P_i^2)T_i^2(1-T_i)^2 + (1-W_i)(2T_i-1)(2-T_i+T_i^2)P_i^2(1-P_i)^2 \\ &+ W_i^2T_i^2(1-T_i)^2(2P_i-1) + (1-W_i)^2P_i^2(1-P_i)^2(1-2T_i) \\ &+ 2W_i(1-W_i)P_iT_i(1-P_i)(1-T_i)(P_i-T_i), \end{aligned}$$

$$B_i = (1-T_i)^2P_i^2\left\{W_i(1-W_i)(P_i+T_i-1)^2 + W_iP_iT_i^2 + (1-W_i)(1-T_i)(1-P_i)^2\right\}.$$

Proof: Given $X_{i1} \sim G(P_i)$, $X_{i2} \sim G(1-P_i)$, $Y_{i1} \sim G(T_i)$, $Y_{i2} \sim G(1-T_i)$, where G represents the geometric distribution with a success probability. Since the expected values of Z_{ij} and Z_{ij}^2 are

$$E(Z_{ij}) = \pi_i\left[\frac{W_i}{P_i} + \frac{(1-W_i)}{(1-T_i)}\right] + (1-\pi_i)\left[\frac{W_i}{(1-P_i)} + \frac{(1-W_i)}{T_i}\right], \tag{7}$$

then:

$$E(Z_{ij}^2) = \pi_i\left[\frac{W_i(2-P_i)}{P_i^2}\right] + (1-\pi_i)\left[\frac{W_i(1+P_i)}{(1-P_i)^2}\right] + (1-\pi_i)(1-V_i)\left[\frac{(2-T_i)}{T_i^2}\right] + \pi_i(1-W_i)\left[\frac{(1+T_i)}{(1-T_i)^2}\right]. \tag{8}$$

Based on (7) and (8), the variance of Z_{ij} is:

$$\sigma_{iZ}^2 = E(Z_{ij}^2) - \left[E(Z_{ij})\right]^2, \tag{9}$$

and, since Z_{ij} is independent, the variance of $\hat{\pi}_i$ can be expressed by:

$$\begin{aligned} V(\hat{\pi}_i) &= V\left[\frac{1}{\psi_i}\left(\frac{P_iT_i(1-P_i)(1-T_i)}{m_i}\sum_{j=1}^{m_i} Z_{ij} - W_iT_iP_i(1-T_i) - P_i(1-W_i)(1-P_i)(1-T_i)\right)\right] \\ &= \frac{P_i^2T_i^2(1-P_i)^2(1-T_i)^2}{n_h^2\psi_h^2}\sum_{j=1}^{m_i}\sigma_{iZ}^2 \\ &= \frac{\pi_i(1-\pi_i)}{m_i} + \frac{\pi_i}{m_i\psi_i^2}A_i + \frac{1}{m_i\psi_i^2}B_i. \end{aligned} \tag{10}$$

Since $V(\hat{\pi}_{ppswr}) = V_1E_2(\hat{\pi}_{ppswr}) + E_1V_2(\hat{\pi}_{ppswr})$, then the first and second terms are given, respectively, as:

$$\begin{aligned} V_1E_2(\hat{\pi}_{ppswr}) &= V_1E_2\left[\frac{1}{nM_0}\sum_{i=1}^{n}\frac{M_i\hat{\pi}_i}{p_i}\right] \\ &= V_1\left[\frac{1}{nM_0}\sum_{i=1}^{n}\frac{M_i\pi_i}{p_i}\right] \\ &= \frac{1}{nM_0^2}\sum_{i=1}^{N}p_i\left[\frac{M_i\pi_i}{p_i} - M_0\pi\right]^2, \end{aligned}$$

and

$$\begin{aligned} E_1V_2(\hat{\pi}_{ppswr}) &= E_1V_2\left[\frac{1}{nM_0}\sum_{i=1}^{n}\frac{M_i\hat{\pi}_i}{p_i}\right] \\ &= E_1\left[\frac{1}{(nM_0)^2}\sum_{i=1}^{n}\frac{M_i^2}{p_i^2}V_2(\hat{\pi}_i)\right] \\ &= E_1\left[\frac{1}{(nM_0)^2}\sum_{i=1}^{n}\frac{M_i^2}{p_i^2}V_2\left\{\frac{1}{\psi_i}P_iT_i(1-P_i)(1-T_i)\frac{1}{m_i}\sum_{i=1}^{m_i}Z_{ij} - W_iT_iP_i(1-T_i) - P_i(1-W_i)(1-P_i)(1-T_i)\right\}\right] \\ &= E_1\left[\frac{1}{(nM_0)^2}\sum\frac{M_i^2}{m_ip_i^2}\left\{\pi_i(1-\pi_i) + \frac{\pi_i}{\psi_i^2}A_i + \frac{1}{\psi_i^2}B_i\right\}\right]. \end{aligned}$$

Then, we can obtain the variance (10).

Moreover, an unbiased estimator of $V(\hat{\pi}_{ppswr})$ is given by

$$\begin{aligned} \hat{V}(\hat{\pi}_{ppswr}) &= \frac{1}{nM_0^2}\sum_{i=1}^{n}p_i\left[\frac{M_i\hat{\pi}_i}{p_i} - M_0\hat{\pi}_{ppswr}\right]^2 \\ &+ \frac{1}{nM_0^2}\sum_{i=1}^{n}\frac{M_i^2}{p_i(m_i-1)}\left[\hat{\pi}_i(1-\hat{\pi}_i) - \frac{\hat{\pi}_i}{\psi_i^2}A_i + \frac{1}{\psi_i^2}B_i\right]. \end{aligned} \tag{11}$$

□

If the FSUs are selected proportional to size with M_i, then $p_i = M_i/M_0$. For this reason, we call this method "probability proportional to size" (PPS) sampling. When a sample of the FSU is selected by PPS sampling with replacement via sampling probability, $p_i = M_i/M_0$ for the ith cluster, and m_i SSU are selected by SRSWR from each FSU. The estimator $\hat{\pi}_{ppswr}$ of π is given by:

$$\begin{aligned} \hat{\pi}_{ppswr} &= \frac{1}{n}\sum_{i=1}^{n}\hat{\pi}_i \\ &= \frac{1}{n}\sum_{i=1}^{n}\frac{1}{m_i}\left[\pi_i(1-\pi_i) + \frac{\pi_i}{\psi_i^2}A_i + \frac{1}{\psi_i^2}B_i\right], \end{aligned} \tag{12}$$

and the variance of $\hat{\pi}_{ppswr}$ and its estimator are as follows:

$$\begin{aligned} V\left(\hat{\pi}_{ppswr}\right) &= \frac{1}{nM_0}\sum_{i=1}^{N}M_i(\pi_i - \pi)^2 \\ &+ \frac{1}{nM_0}\sum_{i=1}^{N}\frac{M_i}{m_i}\left[\pi_i(1-\pi_i) + \frac{\pi_i}{\psi_i^2}A_i + \frac{1}{\psi_i^2}B_i\right], \end{aligned} \tag{13}$$

$$\begin{aligned}\hat{V}\left(\hat{\pi}_{ppswr}\right) &= \tfrac{1}{nM_0}\sum_{i=1}^{n} M_i(\hat{\pi}_i - \hat{\pi}_{ppswr})^2 \\ &+ \tfrac{1}{nM_0}\sum_{i=1}^{n} \tfrac{M_i}{m_i-1}\left[\hat{\pi}_i(1-\hat{\pi}_i) + \tfrac{\hat{\pi}_i}{\psi_i^2}A_i + \tfrac{1}{\psi_i^2}B_i\right].\end{aligned} \tag{14}$$

2.2. The PPS without Replacement

In this subsection, we consider PPS sampling without replacement to estimate the true population proportion of a sensitive character by applying Yennum et al.'s model, in which n FSUs are drawn by PPS sampling without replacement from the population of N clusters with M_i elementary units for the ith cluster, and m_i SSUs are drawn by SRSWR from each FSU.

From this two-stage sampling, the estimator $\hat{\pi}_{ppswor}$ of π is:

$$\hat{\pi}_{ppswor} = \frac{1}{M_0}\sum_{i=1}^{n}\frac{M_i\hat{\pi}_i}{\theta_i}, \tag{15}$$

where θ_i is the first inclusion probability for the ith cluster.

The variance of $\hat{\pi}_{ppswor}$ is given by:

$$\begin{aligned}V\left(\hat{\pi}_{ppswor}\right) &= \tfrac{1}{M_0^2}\sum_{i=1}^{N}\sum_{j>i}^{N}(\theta_i\theta_j - \theta_{ij})\left[\tfrac{M_i\pi_i}{\theta_i} - \tfrac{M_j\pi_j}{\theta_j}\right]^2 \\ &+ \tfrac{1}{M_0^2}\sum_{i=1}^{N}\tfrac{M_i^2}{m_i\theta_i}\left[\pi_i(1-\pi_i) + \tfrac{\pi_i}{\psi_i^2}A_i + \tfrac{1}{\psi_i^2}B_i\right],\end{aligned} \tag{16}$$

where θ_{ij} is the second inclusion probability of the ith and jth clusters.

Furthermore, the variance estimator of $\hat{\pi}_{ppswor}$ is as follows:

$$\begin{aligned}\hat{V}\left(\hat{\pi}_{ppswor}\right) &= \tfrac{1}{M_0^2}\sum_{i=1}^{n}\sum_{j>i}^{n}\tfrac{(\theta_i\theta_j-\theta_{ij})}{\theta_{ij}}\left[\tfrac{M_i\hat{\pi}_i}{\theta_i} - \tfrac{M_j\hat{\pi}_j}{\theta_j}\right]^2 \\ &+ \tfrac{1}{M_0^2}\sum_{i=1}^{n}\tfrac{M_i^2}{\theta_i(m_i-1)}\left[\hat{\pi}_i(1-\hat{\pi}_i) + \tfrac{\hat{\pi}_i}{\psi_i^2}A_i + \tfrac{1}{\psi_i^2}B_i\right].\end{aligned} \tag{17}$$

2.3. Two-Stage Equal Probability Sampling

In this subsection, we consider a two-stage equal probability sampling design to estimate the true population proportion of a sensitive characteristic by applying Yennum et al.'s model, in which n FSUs are drawn by simple random sampling without replacement (SRSWOR) from a population of N clusters with M_i elementary units for the ith cluster, and m_i SSUs are drawn by SRSWR from each FSU.

From this two-stage sampling, the estimator $\hat{\pi}_{wr}$ of π is given by:

$$\hat{\pi}_{wr} = \frac{N}{nM_0}\sum_{i=1}^{n} M_i\hat{\pi}_i, \tag{18}$$

where $\hat{\pi}_i$ is an estimator of the true population proportion for a sensitive characteristic for the ith cluster, which is the same as (4).

The variance of $\hat{\pi}_{wr}$ and its estimator are given as:

$$\begin{aligned}V(\hat{\pi}_{wr}) &= \tfrac{N^2}{nM_0^2}\tfrac{1}{(N-1)}\sum_{i=1}^{N}\left(M_i\pi_i - \overline{M}\pi\right)^2 \\ &+ \tfrac{N}{nM_0^2}\sum_{i=1}^{N}\tfrac{M_i^2}{m_i}\left[\pi_i(1-\pi_i) + \tfrac{\pi_i}{\psi_i^2}A_i + \tfrac{1}{\psi_i^2}B_i\right],\end{aligned} \tag{19}$$

$$\begin{aligned}\hat{V}(\hat{\pi}_{wr}) &= \frac{N^2}{nM_0^2}\frac{1}{(n-1)}\sum_{i=1}^{n}\left(M_i\hat{\pi}_i - \overline{M}\hat{\pi}_{wr}\right)^2 \\ &+ \frac{N}{nM_0^2}\sum_{i=1}^{n}\frac{M_i^2}{m_i-1}\left[\hat{\pi}_i(1-\hat{\pi}_i) + \frac{\hat{\pi}_i}{\psi_i^2}A_i + \frac{1}{\psi_i^2}B_i\right],\end{aligned} \tag{20}$$

where $\overline{M} = M_0/N$.

3. An Estimation of Sensitive Attributes with Probability Proportional to Size Sampling Under Yennum et al.'s Generalized Model

We consider Yennum et al.'s generalized model, in which generalized geometric distribution is used as a randomization device when n clusters are sampled by PPS sampling or equal probability sampling from the population, which consists of N clusters with size $M_i(i = 1, 2, \cdots, N)$, and $m_i(i = 1, 2, \cdots, n)$ units are drawn by simple random sampling from each sampled cluster.

We develop the sampling schemes for PPS sampling with replacement in Section 3.1 and those for PPS sampling without replacement in Section 3.2. Finally, equal probability sampling is presented in Section 3.3.

3.1. PPS Sampling with Replacement

Let the population be composed of N clusters. In the first stage, a sample of n FSUs is drawn by replacement with the sampling probability p_i for the ith cluster. In the second stage, m_i SSUs are selected by SRSWR from each FSU and guided to apply Yennum et al.'s generalized randomization device.

If the interviewees in the ith cluster choose a white ball during the first stage, and if they have a sensitive attribute A (or A^c), then they are guided to pick replacement balls from the first urn of the second stage device until they take k_{i2} (or k_{i1}) successive balls with their actual status for the first time and are then asked to determine the total number of balls as X_{i1} (or X_{i2}).

If the interviewee in the ith cluster draws a black ball in the first stage, and if they have a sensitive attribute A^c (or A), then they are guided to take replacement balls from the second urn of the second stage device until they take k_{i2} (or k_{i1}) successive balls with their actual status for the first time and are then asked to determine the total number of balls as Y_{i1} (or Y_{i2}).

For the ith cluster, using the randomization device in Figure 1, the total number of balls taken by interviewees X_{i1}, X_{i2}, Y_{i1}, and Y_{i2} are distributed via generalized geometric distribution. Let π_i and $1-\pi_i$ be the true population proportion of persons who have a sensitive attribute A and A^c for the ith cluster. Assume that each interviewee in the ith cluster is drawn by SRSWR.

For the jth surveyed answer in the ith cluster, Z_{ij} can be expressed as:

$$Z_{ij} = \begin{cases} X_{i1} \text{ with probability} & W_i\pi_i, \\ Y_{i2} \text{ with probability} & (1-W_i)\pi_i, \\ X_{i2} \text{ with probability} & W_i(1-\pi_i), \\ Y_{i1} \text{ with probability} & (1-W_i)(1-\pi_i), \end{cases} . \tag{21}$$

The expected value of Z_{ij} is given by:

$$\begin{aligned}E(Z_{ij}) &= W_i\pi_i E(X_{i1}) + \pi_i(1-W_i)E(Y_{i2}) + (1-\pi_i)W_i E(X_{i2}) + (1-W_i)(1-\pi_i)E(Y_{i1}) \\ &= \pi_i\left[W_i\left\{\frac{1-P_i^{k_{i1}}}{(1-P_i)P_i^{k_{i1}}} - \frac{1-(1-P_i)^{k_{i2}}}{P_i(1-P_i)^{k_{i2}}}\right\} + (1-W_i)\left\{\frac{1-(1-T_i)^{k_{i1}}}{T_i(1-T_i)^{k_{i1}}} - \frac{1-T_i^{k_{i2}}}{(1-T_i)T_i^{k_{i2}}}\right\}\right] \\ &+ W_i\left\{\frac{1-(1-P_i)^{k_{i2}}}{P_i(1-P_i)^{k_{i2}}}\right\} + (1-W_i)\left\{\frac{1-T_i^{k_{i2}}}{(1-T_i)T_i^{k_{i2}}}\right\}.\end{aligned} \tag{22}$$

Then, the formula (22) can be expressed as:

$$\begin{aligned} E(Z_{ij}) - W_i\left\{\frac{1-(1-P_i)^{k_{i2}}}{P_i(1-P_i)^{k_{i2}}}\right\} - (1-W_i)\left\{\frac{1-T_i^{k_{i2}}}{(1-T_i)T_i^{k_{i2}}}\right\} \\ = \pi_i\left[W_i\left\{\frac{1-P_i^{k_{i1}}}{(1-P_i)P_i^{k_{i1}}} - \frac{1-(1-P_i)^{k_{i2}}}{P_i(1-P_i)^{k_{i2}}}\right\} + (1-W_i)\left\{\frac{1-(1-T_i)^{k_{i2}}}{T_i(1-T_i)^{k_{i2}}} - \frac{1-T_i^{k_{i1}}}{(1-T_i)T_i^{k_{i1}}}\right\}\right]. \end{aligned} \tag{23}$$

The estimator $\hat{\pi}_{iG}$ of the population proportion π_i for the ith cluster is given by:

$$\hat{\pi}_{iG} = \frac{(1-T_i)^{k_{i1}+1}T_i^{k_{i2}+1}(1-P_i)^{k_{i2}+1}P_i^{k_{i1}+1}}{m_i\varphi_{i2}}\left(\sum_{j=1}^{m_i} Z_{ij} - \varphi_{i1}\right), \tag{24}$$

where:

$$\begin{aligned} \varphi_{i1} &= W_i\left\{1-(1-P_i)^{k_{i2}}\right\}(1-T_i)^{k_{i1}+1}T_i^{k_{i2}+1}(1-P_i)P_i^{k_{i1}} \\ &\quad + (1-W_i)(1-T_i^{k_{i2}})P_i^{k_{i1}+1}(1-P_i)^{k_{i2}+1}T_i(1-T_i)^{k_{i1}+1}, \end{aligned} \tag{25}$$

and:

$$\begin{aligned} \varphi_{i2} &= W_i\left[(1-P_i)^{k_{i1}}P_iT_i^{k_{i2}+1}(1-T_i)^{k_{i1}+1}(1-P_i)^{k_{i2}} - \left\{1-(1-P_i)^{k_{i2}}\right\}P_i^{k_{i1}}(1-P_i)T_i^{k_{i2}+1}(1-T_i)^{k_{i1}+1}\right] \\ &\quad + (1-W_i)\left[\left\{1-(1-T_i)^{k_{i1}}\right\}P_i^{k_{i1}+1}(1-T_i)(1-P_i)^{k_{i2}+1}T_i^{k_{i2}} - (1-T_i^{k_{i2}})P_i^{k_{i1}+1}T_i(1-P_i)^{k_{i2}+1}(1-T_i)^{k_{i1}}\right]. \end{aligned} \tag{26}$$

When the interviewees are sampled by SRSWR for the ith cluster selected with a replacement by sampling probability p_i, the estimator $\hat{\pi}_{Gppswr}$ of the true population proportion π of a sensitive attribute is:

$$\begin{aligned} \hat{\pi}_{Gppswr} &= \frac{1}{nM_0}\sum_{i=1}^{n}\frac{M_i\hat{\pi}_{iG}}{p_i} \\ &= \frac{1}{nM_0}\sum_{i=1}^{n}\frac{M_i}{p_i}\left[\frac{(1-T_i)^{k_{i1}+1}T_i^{k_{i2}+1}(1-P_i)^{k_{i2}+1}P_i^{k_{i1}+1}}{m_i\varphi_{i2}}\left(\sum_{j=1}^{m_i} Z_{ij} - \varphi_{i1}\right)\right], \end{aligned} \tag{27}$$

where $M_0 = \sum_{i=1}^{N} M_i$.

Theorem 3: *The estimator $\hat{\pi}_{Gppswr}$ of the true population proportion π of a sensitive character is an unbiased estimator.*

Proof:

$$E_1E_2\left(\hat{\pi}_{Gppswr}\right) = E_1E_2\left[\frac{1}{nM_0}\sum_{i=1}^{n}\frac{M_i\hat{\pi}_{iG}}{p_i}\right] = E_1\left[\frac{1}{nM_0}\sum_{i=1}^{n}\frac{M_iE_2(\hat{\pi}_{iG})}{p_i}\right],$$

and, since:

$$\begin{aligned} E_2(\hat{\pi}_{iG}) &= E_2\left[\frac{(1-T_i)^{k_{i1}+1}T_i^{k_{i2}+1}(1-P_i)^{k_{i2}+1}P_i^{k_{i1}+1}}{m_i\varphi_{i2}}\left(\sum_{j=1}^{m_i} Z_{ij} - \varphi_{i1}\right)\right] \\ &= \frac{(1-T_i)^{k_{i1}+1}T_i^{k_{i2}+1}(1-P_i)^{k_{i2}+1}P_i^{k_{i1}+1}}{m_i\varphi_{i2}}\left(\sum_{j=1}^{m_i} E_2(Z_{ij}) - \varphi_{i1}\right) \\ &= \pi_i, \end{aligned}$$

we can obtain:

$$E_1E_2\left(\hat{\pi}_{Gppswr}\right) = E_1\left[\frac{1}{nM_0}\sum_{i=1}^{n}\frac{M_i\pi_i}{p_i}\right] = \frac{1}{M_0}\sum_{i=1}^{N}p_i\frac{M_i\pi_i}{p_i} = \pi.$$

□

Theorem 4: *The variance of $\hat{\pi}_{Gppswr}$ is obtained by a two-stage sampling scheme, such that a sample of n FSU is selected with replacement by sampling probability p_i for the ith cluster from the population of N clusters consisting of M_i elements for the ith cluster, and m_i SSUs are drawn by SRSWR from each FSU, as given by:*

$$\begin{aligned} V\left(\hat{\pi}_{Gppswr}\right) &= \frac{1}{nM_0^2}\sum_{i=1}^{N} p_i\left[\frac{M_i\pi_i}{p_i} - M_0\pi\right]^2 \\ &+ \frac{1}{nM_0^2}\sum_{i=1}^{N}\frac{M_i^2}{m_ip_i}\left[\frac{\left\{(1-T_i)^{k_{i1}+1}T_i^{k_{i2}+1}(1-P)^{k_{i2}+1}P_i^{k_{i1}+1}\right\}^2}{\varphi_{i2}^2}\sigma_{iZ}^2\right], \end{aligned} \tag{28}$$

where:

$$\begin{aligned} \sigma_{iZ}^2 &= E(Z_{ij}^2) - (E(Z_{ij}))^2 \\ &= \pi_i\left[W_i\left(\frac{1-(2k_{i1}+1)(1-P_i)P_i^{k_{i1}}-P_i^{2k_{i1}+1}+(1-P_i^{k_{i1}})^2}{(1-P_i)^2P_i^{2k_{i1}}}\right)\right. \\ &\quad\left. + (1-W_i)\left(\frac{1-(2k_{i1}+1)T_i(1-T_i)^{k_{i1}}-(1-T_i)^{2k_{i1}+1}+(1-(1-T_i)^{k_{i1}})^2}{T_i^2(1-T_i)^{2k_{i1}}}\right)\right] \\ &+(1-\pi_i)\left[W_i\left(\frac{1-(2k_{i2}+1)P_i(1-P_i)^{k_{i2}}-(1-P_i)^{2k_{i2}+1}+(1-(1-P_i^{k_{i1}})^2)}{P_i^2(1-P_i)^{2k_{i2}}}\right)\right. \\ &\quad\left. + (1-W_i)\left(\frac{1-(2k_{i2}+1)(1-T_i)T_i^{k_{i2}}-T_i^{2k_{i2}+1}+(1-T_i^{k_{i2}})^2}{(1-T_i)^2T_i^{2k_{i2}}}\right)\right] \\ &- \left[\pi_i\left\{ W_i\left(\frac{1-P_i^{k_{i1}}}{(1-P_i)P_i^{2k_{i1}}} - \frac{1-(1-P_i)^{k_{i2}}}{P_i(1-P_i)^{k_{i2}}}\right) + (1-W_i)\left(\frac{1-(1-T_i)^{k_{i1}}}{T_i(1-T_i)^{k_{i1}}} - \frac{1-T_i^{k_{i2}}}{(1-T_i)T_i^{k_{i2}}}\right)\right\}\right. \\ &\quad\left. +(1-\pi_i)\left\{ W_i\left(\frac{1-(1-P_i)^{k_{i2}}}{P_i(1-P_i)^{k_{i2}}}\right) + (1-W_i)\left(\frac{1-T_i^{k_{i2}}}{(1-T_i)T_i^{k_{i2}}}\right)\right\}\right]^2. \end{aligned} \tag{29}$$

Proof: The total number of balls taken by interviewees for the ith cluster, X_{i1}, X_{i2}, Y_{i1} and Y_{i2}, are random variables with variances:

$$V(X_{i1}) = \frac{1-(2k_{i1}+1)(1-P_i)P_i^{k_{i1}} - P_i^{2k_{i1}+1}}{(1-P_i)^2P_i^{2k_{i1}}}, \tag{30}$$

$$V(X_{i2}) = \frac{1-(2k_{i2}+1)P_i(1-P_i)^{2k_{i2}} - (1-P_i)^{2k_{i2}+1}}{P_i^2(1-P_i)^{2k_{i2}}}, \tag{31}$$

$$V(Y_{i1}) = \frac{1-(2k_{i2}+1)(1-T_i)T_i^{k_{i2}} - T_i^{2k_{i2}+1}}{(1-T_i)^2T_i^{2k_{i2}}}, \tag{32}$$

$$V(Y_{i2}) = \frac{1-(2k_{i1}+1)T_i(1-T_i)^{k_{i1}} - (1-T_i)^{2k_{i1}+1}}{T_i^2(1-T_i)^{2k_{i1}}}. \tag{33}$$

From (21), to drive the variance of $\hat{\pi}_{Gppswr}$ we can obtain the expected values of Z_{ij} and Z_{ij}^2 as follows:

$$\begin{aligned} E(Z_{ij}) &= \pi_i\left[W_i\left(\frac{1-P_i^{k_{i1}}}{(1-P_i)P_i^{k_{i1}}}\right) + (1-W_i)\left(\frac{1-(1-T_i)^{k_{i1}}}{T_i(1-T_i)^{k_{i1}}}\right)\right] \\ &+ (1-\pi_i)\left[W_i\left(\frac{1-(1-P_i)^{k_{i2}}}{P_i(1-P_i)^{k_{i2}}}\right) + (1-W_i)\left(\frac{1-T_i^{k_{i2}}}{(1-T_i)T_i^{k_{i2}}}\right)\right], \end{aligned} \tag{34}$$

$$\begin{aligned} E(Z_{ij}^2) &= \pi_i\Big[W_iE(X_{i1}^2) + (1-W_i)E(Y_{i2}^2)\Big] + (1-\pi_i)\Big[W_iE(X_{i2}^2) + (1-W_i)E(Y_{i2}^2)\Big] \\ &= \pi_i\Bigg[\quad W_i\left(\frac{1-(2k_{i1}+1)(1-P_i)P_i^{k_{i1}} - P_i^{2k_{i1}+1} + (1-P_i^{k_{i1}})^2}{(1-P_i)^2P_i^{2k_{i1}}}\right) \\ &\quad + (1-W_i)\left(\frac{1-(2k_{i1}+1)T_i(1-T_i)^{k_{i1}} - (1-T_i)^{2k_{i1}+1} + \{1-(1-T_i)^{k_{i1}}\}^2}{T_i^2(1-T_i)^{2k_{i1}}}\right) \Bigg] \\ &\quad + (1-\pi_i)\Bigg[\quad W_i\left(\frac{1-(2k_{i2}+1)P_i(1-P_i)^{k_{i2}} - (1-P_i)^{2k_{i2}+1} + \left\{1-(1-P_i^{k_{i2}})^2\right\}}{P_i^2(1-P_i)^{2k_{i2}}}\right) \\ &\quad + (1-W_i)\left(\frac{1-(2k_{i2}+1)(1-T_i)T_i^{k_{i2}} - T_i^{2k_{i2}+1} + (1-T_i^{k_{i2}})^2}{(1-T_i)^2T_i^{2k_{i2}}}\right) \Bigg]. \end{aligned} \tag{35}$$

Since $V(\hat{\pi}_{Gppswr}) = V_1E_2(\hat{\pi}_{Gppswr}) + E_1V_2(\hat{\pi}_{Gppswr})$,

$$\begin{aligned} V_1E_2(\hat{\pi}_{Gppswr}) &= V_1E_2\left[\frac{1}{nM_0}\sum_{i=1}^{n}\frac{M_i\hat{\pi}_{iG}}{p_i}\right] \\ &= V_1\left[\frac{1}{nM_0}\sum_{i=1}^{n}\frac{M_i\pi_i}{p_i}\right] \\ &= \frac{1}{nM_0^2}\sum_{i=1}^{N}p_i\left[\frac{M_i\pi_{iG}}{p_i} - M_0\pi\right]^2, \end{aligned}$$

and:

$$\begin{aligned} E_1V_2(\hat{\pi}_{Gppswr}) &= E_1V_2\left[\frac{1}{nM_0}\sum_{i=1}^{n}\frac{M_i\hat{\pi}_{iG}}{p_i}\right] \\ &= E_1\left[\frac{1}{(nM_0)^2}\sum_{i=1}^{n}\frac{M_i^2}{p_i^2}V_2(\hat{\pi}_{iG})\right] \\ &= E_1\left[\frac{1}{(nM_0)^2}\sum_{i=1}^{n}\frac{M_i^2}{p_i^2}V_2\left\{\frac{(1-T_i)^{k_{i1}+1}T_i^{k_{i2}+1}(1-P_i)^{k_{i2}+1}P_i^{k_{i1}+1}}{m_i\varphi_{i2}}\left(\sum_{j=1}^{m_i}Z_{ij} - \varphi_{i1}\right)\right\}\right] \\ &= E_1\left[\frac{1}{(nM_0)^2}\sum_{i=1}^{n}\frac{M_i^2}{p_i^2}\frac{1}{m_i}\left\{\frac{\left\{(1-T_i)^{k_{i1}+1}T_i^{k_{i2}+1}(1-P_i)^{k_{i2}+1}P_i^{k_{i1}+1}\right\}^2}{\varphi_{i2}^2}\sigma_{iZ}^2\right\}\right] \\ &= \frac{1}{nM_0^2}\sum_{i=1}^{N}\frac{M_i^2}{m_ip_i}\left[\frac{\left\{(1-T_i)^{k_{i1}+1}T_i^{k_{i2}+1}(1-P_i)^{k_{i2}+1}P_i^{k_{i1}+1}\right\}^2}{\varphi_{i2}^2}\sigma_{iZ}^2\right]. \end{aligned}$$

We can then obtain the variance (28). Also, an unbiased estimator of $V(\hat{\pi}_{Gppswr})$ is given by:

$$\begin{aligned} \hat{V}\left(\hat{\pi}_{Gppswr}\right) &= \frac{1}{nM_0^2}\sum_{i=1}^{n}p_i\left[\frac{M_i\hat{\pi}_{iG}}{p_i} - M_0\hat{\pi}_{Gppswr}\right]^2 \\ &\quad + \frac{1}{nM_0^2}\sum_{i=1}^{n}\frac{M_i^2}{p_i(m_i-1)}\left[\frac{\left\{(1-T_i)^{k_{i1}+1}T_i^{k_{i2}+1}(1-P)^{k_{i2}+1}P_i^{k_{i1}+1}\right\}^2}{\varphi_{i2}^2}\hat{\sigma}_{iZ}^2\right]. \end{aligned} \tag{36}$$

□

3.2. PPS Sampling Without Replacement

In this subsection, we consider PPS sampling without replacement to estimate the true population proportion of a sensitive characteristic by applying Yennum et al.'s generalized model, in which n FSUs are drawn by PPS sampling without replacement from a population of N clusters with M_i elementary units for the ith cluster, and m_i SSUs are drawn by SRSWR from each FSU.

From this procedure, the estimator $\hat{\pi}_{Gppswor}$ of π is given by:

$$\hat{\pi}_{Gppswor} = \frac{1}{M_0}\sum_{i=1}^{n}\frac{M_i\hat{\pi}_{iG}}{\theta_i}, \tag{37}$$

where θ_i is the first inclusion probability for the ith cluster.

The variance of $\hat{\pi}_{Gppswor}$ is given by:

$$\begin{aligned} V\left(\hat{\pi}_{Gppswor}\right) &= \frac{1}{M_0^2}\sum_{i=1}^{N}\sum_{j>i}^{N}(\theta_i\theta_j - \theta_{ij})\left[\frac{M_i\pi_i}{\theta_i} - \frac{M_j\pi_j}{\theta_j}\right]^2 \\ &+ \frac{1}{M_0^2}\sum_{i=1}^{N}\frac{M_i^2}{m_i\theta_i}\left[\frac{\left\{(1-T_i)^{k_{i1}+1}T_i^{k_{i2}+1}(1-P)^{k_{i2}+1}P_i^{k_{i1}+1}\right\}^2}{\varphi_{i2}^2}\sigma_{iZ}^2\right], \end{aligned} \tag{38}$$

where θ_{ij} is the second inclusion probability for ith and jth clusters.

Also, the variance estimator of $\hat{\pi}_{Gppswor}$ is:

$$\begin{aligned} \hat{V}\left(\hat{\pi}_{Gppswor}\right) &= \frac{1}{M_0^2}\sum_{i=1}^{n}\sum_{j>i}^{n}\frac{(\theta_i\theta_j - \theta_{ij})}{\theta_{ij}}\left[\frac{M_i\hat{\pi}_{iG}}{\theta_i} - \frac{M_j\hat{\pi}_{jG}}{\theta_j}\right]^2 \\ &+ \frac{1}{M_0^2}\sum_{i=1}^{n}\frac{M_i^2}{\theta_i(m_i-1)}\left[\frac{\left\{(1-T_i)^{k_{i1}+1}T_i^{k_{i2}+1}(1-P)^{k_{i2}+1}P_i^{k_{i1}+1}\right\}^2}{\varphi_{i2}^2}\hat{\sigma}_{iZ}^2\right]. \end{aligned} \tag{39}$$

3.3. *Two-Stage Equal Probability Sampling*

In this subsection, we consider a two-stage equal probability sampling scheme to estimate the true population proportion of a sensitive attribute by applying Yennum et al.'s generalized model, in which n FSUs are drawn by SRSWOR from a population of N clusters consisting of M_i elementary units for the ith cluster, and m_i SSUs are drawn by SRSWR from each FSU.

From this procedure, the estimator $\hat{\pi}_{Gwr}$ of the true population proportion π for a sensitive attribute is given by:

$$\hat{\pi}_{Gwr} = \frac{N}{nM_0}\sum_{i=1}^{n}M_i\hat{\pi}_{iG}, \tag{40}$$

where the estimator $\hat{\pi}_{iG}$ is the estimator of a sensitive characteristic of the ith cluster, which is the same as (24).

The variance and variance estimator of $\hat{\pi}_{Gwr}$ are:

$$\begin{aligned} V(\hat{\pi}_{Gwr}) &= \frac{N^2}{nM_0^2}\sum_{i=1}^{N}\frac{1}{N-1}\left[M_i\pi_i - \overline{M}\pi\right]^2 \\ &+ \frac{N}{nM_0^2}\sum_{i=1}^{N}\frac{M_i^2}{m_i}\left[\frac{\left\{(1-T_i)^{k_{i1}+1}T_i^{k_{i2}+1}(1-P)^{k_{i2}+1}P_i^{k_{i1}+1}\right\}^2}{\varphi_{i2}^2}\sigma_{iZ}^2\right], \end{aligned} \tag{41}$$

and:

$$\begin{aligned} \hat{V}(\hat{\pi}_{Gwr}) &= \frac{N^2}{nM_0^2}\sum_{i=1}^{n}\frac{1}{n-1}\left(M_i\hat{\pi}_{iG} - \overline{M}\hat{\pi}_{Gwr}\right)^2 \\ &+ \frac{N}{nM_0^2}\sum_{i=1}^{N}\frac{M_i^2}{m_i-1}\left[\frac{\left\{(1-T_i)^{k_{i1}+1}T_i^{k_{i2}+1}(1-P)^{k_{i2}+1}P_i^{k_{i1}+1}\right\}^2}{\varphi_{i2}^2}\hat{\sigma}_{iZ}^2\right], \end{aligned} \tag{42}$$

respectively, where $\overline{M} = M_0/N$.

4. Efficiency Comparisons

4.1. PPSWR Sampling versus Equal Probability Two-Stage Sampling in Yennum et al.'s Model

If we assume $N - 1 \doteq N$, then the difference between the variance of equal probability two-stage sampling, (19), and the variance of PPS with replacement sampling, (6), is given by:

$$\begin{aligned} V(\hat{\pi}_{wr}) - V(\hat{\pi}_{ppswr}) &= \frac{1}{nM_0\overline{M}}\Bigg[\sum_{i=1}^{N} (M_i - \overline{M})^2 \pi_i^2 + \overline{M}\Bigg\{\sum_{i=1}^{N} (M_i - \overline{M})(\pi_i^2 - \pi^2)\Bigg\} \\ &+ \sum_{i=1}^{N} (M_i - \overline{M})^2 \frac{1}{m_i}\left(\pi_i(1-\pi_i) + \frac{\pi_i}{\psi_i^2}A_i + \frac{1}{\psi_i^2}B\right) \\ &+ \overline{M}\Bigg\{\sum_{i=1}^{N} (M_i - \overline{M})\frac{1}{m_i}\left(\pi_i(1-\pi_i) + \frac{\pi_i}{\psi_i^2}A_i + \frac{1}{\psi_i^2}B\right)\Bigg\}\Bigg]. \end{aligned} \tag{43}$$

In (43), we can see that $V(\hat{\pi}_{wr}) = V(\hat{\pi}_{ppswr})$ under the condition $M_i = \overline{M} = M_0/N$; i.e., if the cluster sizes are equal, the selection probabilities of the PPS with replacement sampling are all N^{-1} and equal to those of equal probability two-stage replacement sampling.

If the size of a cluster, M_i is significantly different, then $\sum_{i=1}^{N} (M_i - \overline{M})^2 \pi_i^2$, the first term on the right side of (43), has large values, and the second term, $\sum_{i=1}^{N} (M_i - \overline{M})^2 (\pi_i^2 - \pi^2)$, has relatively small values. Hence, the estimation by PPS with replacement sampling is more efficient than that by equal probability two-stage replacement sampling.

We used the relative efficiency (RE) to compare the efficiency of the two sampling methods—PPS with replacement sampling and equal probability two-stage replacement sampling:

$$RE_1 = \frac{V(\hat{\pi}_{wr})}{V(\hat{\pi}_{ppswr})} \times 100(\%).$$

Values of RE_1 over 100% indicate that the estimator obtained by the PPS with the replacement sampling method was more efficient than the estimator obtained by the equal probability two-stage replacement sampling.

In calculating REs, we set the parameters as follows:

$$\begin{aligned} &M_0 = 10,000, M_1 = 1,000, M_2 = 2,000, M_3 = 3,000, M_4 = 4,000 \\ &m_0 = 1,000, m_1 = 100, m_2 = 200, m_3 = 300, m_4 = 400, \\ &p_1 = 0.235, p_2 = 0.441, p_3 = 0.609, p_4 = 0.715. \end{aligned}$$

From Table 1, when the selection probability W for the first-stage randomization device increased from 0.1 to 0.9 by 0.2 and the second stage randomization devices T increased from 0.6 to 0.8 by 0.1 and P from 0.65 to 0.90 by 0.05, REs increase under the fixed proportion of a sensitive attribute (particularly when the selection probability of the second randomization device T increased), and the RE increased according to the conditions of P and π_i.

Table 1. The relative efficiencies (REs) of a sensitive estimator between the probability proportional to size (PPS) sampling with replacement and the equal probability two-stage sampling with replacement in Yennum et al.'s model to change π_i and W.

π_i	T \ P	W: 0.1			0.3			0.5			0.7			0.9		
		0.6	**0.7**	**0.8**	**0.6**	**0.7**	**0.8**	**0.6**	**0.7**	**0.8**	**0.6**	**0.7**	**0.8**	**0.6**	**0.7**	**0.8**
0.1	0.65	56.59	95.07	123.5	48.08	61.06	91.71	52.73	46.18	55.97	63.18	52.59	39.63	75.51	71.08	61.88
	0.7	54.42	89.17	120	50	54.61	81.69	58.65	48.28	48.17	71.01	60.85	45.08	83.88	80.2	72.61
	0.75	52.93	81.67	114.8	53.61	51.26	70.48	64.76	53.56	45.34	77.67	69.08	54.13	90.33	87.46	81.68
	0.8	52.61	72.72	106.4	58.15	51.63	59.9	70.32	60.37	48.28	83.01	76.38	64.29	95.2	93.08	88.97
	0.85	53.84	63.67	93.17	62.84	55.3	53.48	75.02	67.37	55.79	87.17	82.48	73.91	98.87	97.41	94.69
	0.9	56.57	57.65	74.27	67.15	61.05	54.47	78.79	73.8	65.52	90.32	87.41	82.23	101.6	100.7	99.15
0.2	0.65	82.74	134.4	153.1	50.22	92.87	130.8	60.29	48.44	86.05	90.64	61.69	34.18	117.4	108.9	89.02
	0.7	75.82	129.7	151.4	51.78	76.68	121.8	77.73	48.28	65.5	108.7	84.97	43.35	130.7	125.1	111.8
	0.75	68.56	122.7	148.7	60.97	61.93	108.4	94.15	62.33	48.62	121.2	104.6	67.95	139	135.4	127
	0.8	62.59	111.6	144.1	74.34	56.05	88.85	107.2	82.13	48.28	129.7	118.9	94	144.4	142.1	137
	0.85	61.04	94.57	135	87.93	64.32	66.55	116.7	100.4	68.61	135.4	128.9	114.4	148	146.5	143.6
	0.9	66.6	74.23	114.3	99.4	82.02	61.52	123.5	114.4	95.72	139.3	135.8	128.5	150.4	149.6	148.1
0.3	0.65	106.8	152.4	164.9	54.34	117.7	149.2	70.41	53.88	109.9	119	74.13	31.19	148.1	139.5	115.9
	0.7	98.24	149.4	163.9	53.64	99.37	142.7	100.6	48.28	85.26	139.9	111.1	42.92	159	154.4	141.5
	0.75	87.63	144.7	162.4	70.27	76.57	132.3	124	73.92	55.9	151.7	135.4	85.58	164.8	162.2	155.2
	0.8	76.17	136.6	159.9	94.79	61.53	114	139.2	107.4	48.28	158.6	149.6	122.7	168.2	166.6	162.8
	0.85	69.9	121.6	154.7	116.5	75.47	83.46	148.7	131.9	85.14	162.8	158	145.2	170.2	169.4	167.4
	0.9	78.83	94.88	140.8	131.9	107.7	69.61	154.7	146.9	126.4	165.4	163.1	157.6	171.6	171.1	170.2
0.4	0.65	124.6	162.1	171.2	59.97	134.4	159.3	82.29	61.29	126.3	141.7	88.14	30.06	166.3	159.3	136.7
	0.7	116.5	160	170.5	55.58	117.6	154.6	122.6	48.28	102.2	160.3	133.3	43.6	173.7	170.4	159.7
	0.75	105.4	156.9	169.6	81.25	91.92	146.7	147.4	87.35	65.18	169.2	156.3	103.3	177.3	175.5	170.3
	0.8	91.07	151.5	168	116.6	67.99	131.8	160.8	130.8	48.28	173.8	167.6	144.2	179.1	178.2	175.5
	0.85	80.35	140.6	164.9	142.3	88.79	100.6	168.2	155.2	103.2	176.4	173.5	164	180.2	179.7	178.4
	0.9	93.87	116	156.4	157	133.7	78.99	172.3	167.2	150.6	178	176.6	173.2	180.9	180.6	180.1

On the other hand, RE increased when the first-stage selection probability W was less than 0.5, and the values of T, P, and π_i (from 0.1 to 0.4) decreased, but the RE decreased when the value of W was greater than 0.5 under a fixed value for T, P, and π_i.

Furthermore, the greater the true population proportion of a sensitive attribute π_i, the higher the overall efficiency of Yennum et al.'s model, as shown by the values of the bottom cells in Table 1. This result agrees with the typical sampling survey methodology as the true population proportion of a sensitive attribute π_i increases.

4.2. PPSWR Sampling versus Equal Probability Two-Stage Sampling in Yennum et al.'s Generalized Model

If we assume $N - 1 \doteq N$, then the difference between the variance of equal probability two-stage sampling scheme (41) and the variance of the PPS with replacement sampling scheme (28) is given by:

$$\begin{aligned} V(\hat{\pi}_{Gwr}) - V(\hat{\pi}_{Gppswr}) = & \frac{1}{nM_0\overline{M}}\left[\sum_{i=1}^{N}(M_i - \overline{M})^2\pi_i^2 + \overline{M}\left\{\sum_{i=1}^{N}(M_i - \overline{M})(\pi_i^2 - \pi^2)\right\}\right. \\ & + \sum_{i=1}^{N}(M_i - \overline{M})^2\frac{1}{m_i}\left[\frac{\left\{(1-T_i)^{k_{i1}+1}T_i^{k_{i2}+1}(1-P)^{k_{i2}+1}P_i^{k_{i1}+1}\right\}^2}{\varphi_{i2}^2}\sigma_{iZ}^2\right] \\ & \left. + \overline{M}\left\{\sum_{i=1}^{N}(M_i - \overline{M})\frac{1}{m_i}\left(\frac{\left\{(1-T_i)^{k_{i1}+1}T_i^{k_{i2}+1}(1-P)^{k_{i2}+1}P_i^{k_{i1}+1}\right\}^2}{\varphi_{i2}^2}\sigma_{iZ}^2\right)\right\}\right]. \end{aligned} \tag{44}$$

In (44), we can see that $V(\hat{\pi}_{Gwr}) = V(\hat{\pi}_{Gppswr})$ under the condition $M_i = \overline{M} = M_0/N$, i.e., if the cluster sizes are equal, the selection probabilities of the PPS with replacement sampling are all N^{-1} and equal to those of the equal probability two-stage replacement sampling.

If cluster sizes, M_i, were significantly different, then $\sum_{i=1}^{N} (M_i - \overline{M})^2 \pi_i^2$, the first term of the right-hand side in (44), had large values, and the second term, $\sum_{i=1}^{N} (M_i - \overline{M})^2 (\pi_i^2 - \pi^2)$, had relatively small values. Hence, the estimation by PPS with replacement sampling is more efficient than that by equal probability two-stage replacement sampling.

We used the relative efficiency (RE) to compare the efficiency of the two sampling designs (PPS with replacement sampling and equal probability two-stage replacement sampling):

$$RE_2 = \frac{V(\hat{\pi}_{Gwr})}{V(\hat{\pi}_{Gppswr})} \times 100(\%)$$

Values of RE_2 over 100% indicate that the estimator obtained by PPS with the replacement sampling method was more efficient than the estimator obtained by equal probability two-stage replacement sampling.

Table 2 shows the results of the REs obtained by increasing the true population proportion π_i from 0.1 to 0.4 by 0.1. The selection probabilities of the randomized response model (W, T and P) are shown in Section 4.1.

Table 2. The REs for a sensitive estimator between the PPS with replacement sampling and equal probability two-stage sampling with replacement in Yennum et al.'s generalized model for changing π_i and W.

π_i		**W**														
	T	**0.1**			**0.3**			**0.5**			**0.7**			**0.9**		
	P	**0.6**	**0.7**	**0.8**	**0.6**	**0.7**	**0.8**	**0.6**	**0.7**	**0.8**	**0.6**	**0.7**	**0.8**	**0.6**	**0.7**	**0.8**
0.1	0.65	171.5	172.8	172.8	163	166.9	167.9	145.4	155.4	159.1	104.1	126.5	138.8	48.17	52.15	68.27
	0.7	167.5	169.7	170.1	152.7	160.2	163.1	119.1	139.1	148.9	57.85	83.09	111	78.98	76.63	58.68
	0.75	162.2	165.9	167	135.6	150.2	156.7	79	111	133.1	54.18	48.22	69.55	116.3	124.4	117.6
	0.8	154.2	160.9	163.3	106.2	133.1	147.2	48.74	68.09	105.6	88.1	76.9	50.14	136.5	148.1	152.8
	0.85	139.8	153.1	158.5	64.63	99.62	130.1	64.15	50.65	60.85	116.4	120.1	102.3	147	159.4	168.3
	0.9	108.4	136	150.3	49.94	51.5	89.01	96.84	95.38	69.51	133	144.7	150.1	152.7	165.1	175.3
0.2	0.65	180.9	181.1	181	177	178.1	178.3	168.4	172	173.1	140.7	154.2	160	48.39	59.25	87.17
	0.7	178.9	179.5	179.5	172.1	174.6	175.5	152.4	162.4	166.8	75.03	112.8	137.8	120.5	111	71.17
	0.75	176.4	177.5	177.8	163.1	169.2	171.8	113.3	142.1	156.1	67.28	48.47	88.43	159.6	161.5	149
	0.8	172.6	174.9	175.8	143.4	159.1	166.1	50.3	91.74	133.4	132.9	110.7	52.83	171.6	175.4	173.9
	0.85	165.8	171	173.1	91.83	133.9	154.7	94.45	55.74	74.87	160	159	135.4	176.4	180.4	182.1
	0.9	147.3	162.3	168.5	55.09	58.5	118.4	145.2	139.7	94.09	170.3	174.6	173.7	178.6	182.6	185.2
0.3	0.65	184	184	183.9	181.6	182	182	176.2	178	178.5	157.3	165.5	168.9	48.68	66.76	101.4
	0.7	182.8	183	182.9	178.5	179.7	180.2	165.6	171.4	174	91.45	130.8	151.3	145.6	132.9	81.96
	0.75	181.2	181.7	181.8	172.9	176.1	177.6	134.5	156.5	166.1	81.89	48.76	102.6	174.4	173.9	161.9
	0.8	178.9	180	180.4	159.7	169.3	173.6	52.38	109.5	147.9	155.9	132.1	55.54	181.2	182.4	180.1
	0.85	174.8	177.5	178.6	114.3	150.9	165.5	120.4	61.6	86.84	174.9	172.4	151.2	183.6	185.3	185.4
	0.9	163.6	172.1	175.6	62.66	66.66	135.6	166.4	160.1	112.1	180.7	182.3	180.3	184.7	186.4	187.3
0.4	0.65	185.5	185.5	185.4	183.9	184.1	184	180	181	181.3	166.1	171.6	173.8	49.04	74.03	112.4
	0.7	184.7	184.8	184.7	181.7	182.4	182.6	172.3	176.1	177.9	105.3	142.5	159.4	159.6	146.4	91.15
	0.75	183.5	183.8	183.9	177.7	179.7	180.7	147.7	164.5	171.7	95.48	49.07	113.7	180.5	179.2	168.8
	0.8	181.9	182.6	182.9	168.2	174.7	177.7	54.79	122.6	156.8	167.6	145.4	58.24	184.8	185.2	182.8
	0.85	179.2	180.8	181.6	130.6	160.5	171.4	138.7	67.61	97.01	181	178.3	160.1	186.3	187	186.7
	0.9	171.7	177.1	179.3	71.76	75	146.8	176	170.1	125	184.7	185.3	183.2	186.9	187.8	188.1

In calculating the REs, we set the parameters as follows:

$$\begin{aligned}
&M_0 = 10,000, M_1 = 1,000, M_2 = 2,000, M_3 = 3,000, M_4 = 4,000\\
&m_0 = 1,000, m_1 = 100, m_2 = 200, m_3 = 300, m_4 = 400,\\
&p_1 = 0.235, p_2 = 0.441, p_3 = 0.609, p_4 = 0.715,\\
&k_1 = 2, k_2 = 1.
\end{aligned}$$

From the results of Table 2, the efficiencies vary according to changes in the probabilities of selection during the first stage W and the second stage T and P in the randomization device, but when the first-stage selection probability W is fixed, and the second-stage selection probabilities T and P increase, then the relative efficiency of the PPS sampling is better than that of the equal probability two-stage sampling in Yennum et al.'s model.

5. Conclusions

We extended Yennum et al.'s model, in which geometric distribution is used as a randomization device for a population consisting of different-sized clusters, and clusters are selected by PPS sampling. Estimators for the true population proportion of a sensitive attribute, their variances, and their variance estimators are derived under PPS sampling and equal probability two-stage sampling.

We also applied these sampling designs to the case of Yennum et al.'s generalized model. Numerical studies were carried out to compare the efficiencies of the proposed methods in each case of Yennum et al.'s model and Yennum et al.'s generalized model in cases with a replacement.

Although the experiments were assumed to use a replacement, we expected similar results for a case without replacement, as per typical sampling theory.

From the numerical study, we found that the efficiency of the two-stage sampling for probability proportional to size depends on the given parameter values, but the efficiency of Yennum et al.'s generalized model is preferred for most combinations of parameters over around 80%.

Author Contributions: Conceptualization, G.-S.L.; methodology, C.-K.S.; writing—original draft preparation, K.-H.H.; writing—review and editing, C.-K.S.; project administration and funding acquisition, G.-S.L.

Acknowledgments: We would like to thank the anonymous reviewers for their very careful reading and valuable comments/suggestions.

References

1. Warner, S.L. Randomized response: A survey technique for eliminating evasive answer bias. *J. Am. Stat. Assoc.* **1965**, *60*, 63–69. [CrossRef] [PubMed]
2. Cochran, W.G. *Sampling Techniques*, 3rd ed.; John Wiley and Sons: New York, NY, USA, 1977.
3. Fox, J.A.; Tracy, P.E. *Randomized Response: A Method for Sensitive Survey*; Sage Publications: Newbury Park, CA, USA, 1986.
4. Kuk, A.Y.C. Asking sensitive questions indirectly. *Biometrika* **1990**, *77*, 436–438. [CrossRef]
5. Chaudhuri, A.; Mukerjee, R. *Randomized Response: Theory and Techniques*; Marcel Dekker Inc.: New York, NY, USA, 1988.
6. Ryu, J.B.; Hong, K.H.; Lee, G.S. *Randomized Response Model*; Freedom Academy: Seoul, Korean, 1993.
7. Lee, G.S.; Hong, K.H. Randomized response model by two-stage cluster sampling. *Korean Commun. Stat.* **1998**, *5*, 99–105.
8. Lee, G.S. A Study on the Randomized Response Technique by PPS Sampling. *Korean J. Appl. Stat.* **2006**, *19*, 69–80.

9. Yennum, N.Y.; Sedory, S.A.; Singh, S. Improved strategy to collect sensitive data by using geometric distribution as a randomization device. *Commun. Stat. Theory Methods* **2019**, *48*, 5777–5795. [CrossRef]
10. Hussain, Z.; Shabbir, J.; Pervez, Z.; Shah, S.F.; Khan, M. Generalized geometric distribution of order k: A flexible choice to randomize the response. *Commun. Stat. Simul. Comput.* **2017**, *46*, 4708–4721. [CrossRef]

2

A Functional Data Analysis Approach for the Detection of Air Pollution Episodes and Outliers

Javier Martínez Torres [1,*], Jorge Pastor Pérez [2], Joaquín Sancho Val [3], Aonghus McNabola [4], Miguel Martínez Comesaña [5] and John Gallagher [4]

1 Department of Applied Mathematics I. Telecommunications Engineering School, University of Vigo, 36310 Vigo (Pontevedra), Spain
2 Centro de Evaluación, Formación y Calidad de Aragón, 50018 Zaragoza, Spain; jjpastor@aragon.es
3 Centro Universitario de la Defensa. Academia General Militar, 50090 Zaragoza, Spain; jsanchov@unizar.es
4 Department of Civil, Structural and Environmental Engineering, Trinity College Dublin, University of Dublin, Dublin D02 PN40, Ireland; amcnabol@tcd.ie (A.M.); J.Gallagher@tcd.ie (J.G.)
5 Escuela de Ingeniería Industrial, University of Vigo, 36310 Vigo (Pontevedra), Spain; migmartinez@uvigo.es
* Correspondence: javmartinez@uvigo.es

Abstract: Ground level concentrations of nitrogen oxide (NOx) can act as an indicator of air quality in the urban environment. In cities with relatively good air quality, and where NOx concentrations rarely exceed legal limits, adverse health effects on the population may still occur. Therefore, detecting small deviations in air quality and deriving methods of controlling air pollution are challenging. This study presents different data analytical methods which can be used to monitor and effectively evaluate policies or measures to reduce nitrogen oxide (NOx) emissions through the detection of pollution episodes and the removal of outliers. This method helps to identify the sources of pollution more effectively, and enhances the value of monitoring data and exceedances of limit values. It will detect outliers, changes and trend deviations in NO_2 concentrations at ground level, and consists of four main steps: classical statistical description techniques, statistical process control techniques, functional analysis and a functional control process. To demonstrate the effectiveness of the outlier detection methodology proposed, it was applied to a complete one-year NO_2 dataset for a sub-urban site in Dublin, Ireland in 2013. The findings demonstrate how the functional data approach improves the classical techniques for detecting outliers, and in addition, how this new methodology can facilitate a more thorough approach to defining effect air pollution control measures.

Keywords: air pollution; functional data analysis; non-normal data; statistical process control; outlier

---

1. Introduction

Nowadays, most cities have an increasing environmental problem related to air pollution [1–4]. This specific pollution is a continuing threat to human health and welfare, with a range of different sources generating different pollutants which have distinct health effects on urban populations [5–7]. Detailed air quality monitoring data for pollutants, such as carbon monoxide (CO), nitrogen oxides (NO and NO_2), sulphur dioxide (SO_2), ozone (O_3) and particulate matter (PM_{10} and $PM_{2.5}$), are becoming more important because of the health problems said pollutants can cause in living beings [6]. The measurements of pollutants provide real-time data to inform the public and provide a mechanism of alerting local residents of a possible hazard. In particular, pollutant sources from traffic emissions, such as NOx, which represents a combination of nitrogen oxide (NO) and nitrogen dioxide (NO_2), are typically emitted at ground level from vehicles and are associated with health-related problems [8].

Despite a reduction in emissions from the transport sector, an increasing trend in NO_2 concentrations has been observed in a number of different European countries; for example, the United Kingdom and Ireland [9,10]. Therefore, meeting the standards and air quality guidelines by European and national environmental agencies for pollutants such as NO_2 is becoming more challenging [9], as exceedances of pollutant concentrations can lead to short-term, chronic human health problems [11].

On the other hand, it is understandable that occasional values in polluted air samples behave as outliers in an urban environmental database. They can be classified as local outliers [12,13] or global outliers. Unlike global outliers, local outliers can be detected by comparison with near neighbours. For the purpose of air pollution studies in urban areas, global outliers that deviate from the guide values indicate that there may be a significant source of pollution. Observations which are not excessively high but are different from neighbouring values may also contain information on unusual processes such as pollution. Outliers may merely be noisy observations, or alternatively, they may indicate atypical behaviour in the system. These abnormal values are very important and may lead to useful information or significant discoveries, but also contribute to the selection of the most suitable mitigation techniques or measures [14].

Different techniques of functional data analysis (FDA) have been used in vectorial problems. This new methodology appeared due to the inefficiency of the classical data mining techniques treating vector data [15]. FDA is applicable in a multitude of fields, such as environmental [16–19] and medical research [20], and is applicable for sensors [21,22] and industrial methods [23,24]. The functional model is based on two ideas that make it unique: it takes into account the time correlation structure of the data and leads to a global view of the problem through curves analysis instead individual observations. This analysis is focused on the comparison of the curves using the functional depths, a variable that measures the centrality of a given curve within a set of curves [25]. Functional depth has already been used in several environmental problems [26,27].

The aspiration of this research is to create a model to detect air pollution episodes and identify outliers in gaseous emissions, and to validate this method using real world data from a suburban air quality monitoring site in Dublin, Ireland. Although many methods are known to identify outliers (from the classical Grubbs test [28] to a test proposed in 2019 by [29]), they are all based on the vector approach. This study was carried out, on one hand, with conventional methods, and on the other hand, with a functional approach; a comparative study between the two methodologies is presented. Each method will be presented and the findings will outline the most effective method for detecting outliers in air pollution monitoring data to enhance its capacity for informing new measures to improve local air quality.

2. Methods

2.1. *Case Study—A Sub-Urban Air Quality Monitoring Station in Dublin, Ireland*

Ireland has a range of air quality monitoring stations across the country, which are part of the national ambient air quality monitoring programme (AAMP). The data collated from these monitoring sites are used to inform on air quality at the local and national levels, and are being used for forecast modelling. The Blanchardstown sub-urban site is one of the 17 national sites, managed by the Environmental Protection Agency (EPA), which monitors NO_2 and is classified as a suburban monitoring site. It is located to the west of Dublin city centre in Ireland [30]. The Blanchardstown air quality monitoring station was selected, as it provided continuous, high-resolution NO_2 data emissions over a 1-year period. Its location is adjacent to the major arterial carriageway around Dublin city centre, and as a monitoring location, is therefore affected by traffic emissions. In this manner, NO_2 hourly data was collected throughout 2013, with 96% of data capture and availability from [31], and the information needed about weather conditions in Dublin in 2013 was obtained on [32].

EU legislation for NO_2 limit values (2008 CAFE Directive and S.I. number 180 of 2011) align with the World Health Organisation (WHO) guidelines, with 1-hour and 1-year limit values of 200 $\mu g/m^3$

and 40 μg/m^3 respectively. In circumstances in which the hourly value is exceeded on three consecutive hours, short-term action plans must be implemented by local authorities to mitigate against continued pollution events (limiting traffic flows, restricting construction work, industrial processes, etc.). Despite there being no daily NO_2 limit value in the EU and WHO, some countries have set an average daily limit of 100 μg/m^3 (range from 80–150 μg/m^3) [31].

The sources and trends of NO_2 emissions over the last 20 years has seen a recent increasing trend due to growth in the transport sector and a recovery since the 2008 economic downturn. As such, Ireland's air quality in relation to NO_2 is considered to be deteriorating, as measurement data suggests it may reach limit values and the national emissions ceiling in the coming years.

2.2. Analysis Methods

A range of systems are available to analyse environmental data, such as air quality measurements. These systems can be used to detect uncharacteristic data points by taking into account trends, variations between neighbouring network stations and expected values with respect to the sampling location. An example of these expert systems of data and environmental parameter validation would be the trends analysis throughout R-programming (openair) [33].

With classical analysis, the data are only analysed statically. The proposed methodology includes using a large amount of existing data to extract conclusions. Today, the amount of data that has been stored in environmental databases requires automated analysis techniques. Actually, the analysis methodology presented here is oriented to knowledge discovery in databses (KDD) [34], which provides a complete process for extracting information and also provides a clear methodology for the preparation of data and interpreting the results obtained. KDD involves an iterative and interactive process of searching for models, patterns and parameters that are useful for detection, classification and/or prediction in order to generate knowledge and help in decision making.

2.2.1. Classical Analysis

The classical monitoring strategy for air quality uses individual time series, descriptive statistics, box plots, autocorrelation analysis, etc., to determine if any of the values fall outside of the limits and to analyse trends [35,36]. In general, classical statistical analysis seeks to describe the distribution of a measurable variable (descriptive statistics) and to determine the consistency of a sample drawn from an initial population (inferential statistics). In addition, classical analysis is based on repetition; one must measure properties of objects and try to predict the frequency of occurrence of results when the measuring operation is repeated at random or stochastically.

This type of analysis determines the empirical frequency distribution that yields the absolute or relative frequency of the occurrence of the different possible results of the repeated measurement of a property of an object (discrete case). Instead, if the case is an infinitely repeated and arbitrarily precise measurement and every outcome is diffferent, the relative frequency of a single outcome would not be very instructive; the distribution function is used, which, for every numerical value x of the measured variable, yields the absolute (or relative) frequency of the occurrence of all values smaller than x [37].

2.2.2. Statistical Process Control

By applying statistical process control (SPC) methods to the monitoring of a system, it is possible to detect outliers. This study is concentrated on significantly high and low measurements, even in situations where the values do not exceed the established limit. These methodologies can be used to study individual observations, using individual or average charts.

The dataset should be partitioned into rational subgroups, minimising the probability of large differences between subgroups [38]. The formation of rational subgroups is important, because variation within subgroups can be clustered and the presence of special causes of variability can be easily detected. However, sometimes it is not practical to use rational subgroups; for example, when repeated measurements in the process differ only by laboratory or analytical error. Even when automated inspection is used, because every unit manufactured is analysed.

The rational subgroups represent the way of collecting the data. Usually, they should be gathered so that each of them shows only the inherent variation that is natural for the process (*common cause variation*). Because they contribute to identifying any other source of variation (*special cause variation*) that may badly affect the process, the subgroups should avoid special-cause variation where possible. Moreover, the limits on a control chart, which mark the boundary to identify if a process is too volatile, are calculated on the basis of the variability within each subgroup. For this, only subgroups that reproduce the common cause variation in a process should be selected.

Once the data are correctly structured, a normality test has to be done. If the hypothesis of normality is rejected, there are two possibilities: use modified classical techniques to non-normal distributions [39], or transform the data to normalise the dataset [40]. The second option applies a Box-Cox transformation [41], which smoothes the data structure. The most widely used and known transformation is the Box-Cox transformation, defined as follows:

$$X_j^{(\lambda)} = \begin{cases} \frac{X_j^{\lambda}-1}{\lambda}, & \text{if } \lambda \neq 0 \\ log(X_j), & \text{if } \lambda = 0 \end{cases}$$

where λ maximises the profile likelihood function of the data X_j.

A classical analysis process can be divided in two main stages: the learning stage, when a test of normality is performed and atypical measurements are deleted from the data; and the control stage, when the trends are analysed to encounter *out-of-control* situations. At the first one, it is when the centreline (CL) and control limits are defined. Specifically, the CL is defined with the control sample and represents the objective value. In addition, the warning limits are set at a distance of $\pm 2\sigma$ from it, and control limits at $\pm 3\sigma$, with σ being the standard deviation of the process [42].

Shewhart control charts have been the most widely used due to their good performance in detecting large changes in a process. However, because these charts use the most recent samples, they do not efficiently detect small or progressive changes in a process. In this regard, complementary rules are needed; multiple authors have defined different rules to detect specific deviations [43,44] and to complement the initial control rules [45]. The use of these supplementary rules makes Shewhart's control charts more sensitive and leads to an improvement in one's capacity to detect non-random patterns.

A widely used way to quantify the potential of a control chart with supplementary rules is through the average run length (ARL). The ARL, in control charts, is the average number of points that should be analysed before showing an alert warning that the process is not under control. When this occurs, the efficient thing to do would be to detect it as soon as possible. On the contrary, when the process is statistically stable, it would be appropriate to have few false alarms. This term is directly related to a Type I error (also known as α) and a Type II error (also known as β), which also describe the sensitivity of the method, and it is highly related to the number of false alarms. For that reason, it must be contemplated that if the capacity of this methodology to detect out-of-control situations is high, there will be a lot of false alarms [43].

2.2.3. Functional Data Analysis

The functional data analysis works with the observations that come from a continuous random process that is evaluated at discrete points [46]. Starting from vector samples, the dataset will be transformed into a functional sample. The first step is to construct the most appropriate curves from the initial points that come from the discrete values measured in the study. This process, known as smoothing, converts the vector values into continuous functions over time. This structure of data is essential in the air pollution context because it is taking into account all the values in the day as a set. In this way, a day in which NO_2 values are obtained with a lot of variability but which has an average similar to the other days, is not detected from any vectorial approach. Functional analysis would identify these types of days as candidates for outliers. Furthermore, in several similar studies in which they also tried to detect outliers with data from certain gases (see [23,47]), the superiority of functional approaches was demonstrated.

In a situation where the initial observations $x(t_j)$ are contained in a set of n_p points, $t_j \in \mathbb{R}$ represents the time steps and n_p is the number of observations $(j = 1, 2, \ldots, n_p)$. They can be watched as the individual values of the function $x(t) \in \mathbb{X} \subset F$, F being a functional space. The estimation of $x(t)$ takes into account a functional space $F = span\{\phi_1, \ldots, \phi_{n_b}\}$, where ϕ_k is the set of basis functions $(k = 1, 2, \ldots, n_b)$ and n_b is the number of basis functions required to build a functional sample. There are several types of basis in statistics, but the most used one is the Fourier basis [26,48]. Moreover, with periodic data such as we have in this study, Fourier bases are the most appropriate [49]. Smoothing consists of finding a solution to the regularisation problem [46],

$$\min_{\mathbb{X} \in F} \sum_{j=1}^{n_p} \{z_j - x(t_j)\}^2 + \lambda \Gamma(x) \tag{1}$$

with $z_j = x(t_j) + \epsilon_j$ being the result of observing x at the point t_j; ϵ_j the random noise with zero mean, Γ being a penalising operator focused on obtaining the simplest possible solutions; and λ being a parameter that defines the level of the regularisation. The initial step is the following expansion

$$x(t) = \sum_{k=1}^{n_b} c_k \phi_k(t) \tag{2}$$

where $\{c_k\}_{k=1}^{n_b}$ are coefficients that multiply the selected basis functions. The smoothing problem can be written as follows:

$$\min_c \{(z - \Phi c)^T (z - \Phi c) + \lambda c^T R c\} \tag{3}$$

with a vector of observations $z = (z_1, \ldots, z_{n_p})^T$; a vector of coefficients of the expansion $c = (c_1, \ldots, c_{n_b})^T$; a (n_p, n_b)-matrix Φ whose elements are $\Phi_{j_k} = \phi_k(t_j)$; and a (n_b, n_b)-matrix $\mathbf{R}$ whose elements are:

$$R_{kl} = \langle D^2\phi_k, D^2\phi_l \rangle_{L_2(T)} = \int_T D^2\phi_k(t) D^2\phi_l(t) dt \tag{4}$$

Finally, the problem is solved [46] according to:

$$c = (\Phi'\Phi + \lambda R)^{-1} \Phi' z \tag{5}$$

As soon as the data are in functional form, they can be analysed to identify pollution episodes and detect outliers. The functional data allow us to identify whether different periods of time such as days, weeks or months are above the mean function and how much they are deviating. Moreover, it permits the elimination of outliers which are not real; they are due to system fails. The depth concept provides a way for ordering a set of data, contained in a euclidian space, according their proximity to the centre of the sample.

The concept of depth appeared in multivariate analyses, and was created to measure the centrality of a point among a cloud of them [50,51]. Over the years, this concept began to be introduced into functional data analysis [52]. In this field, depth represents the centrality of a certain curve $\mathbf{x}_i$ and the mean curve is the centre of the sample. Two depth measurements very common in the context of functional data are: Fraiman-Muniz depth (FMD) [25] and H-modal depth (HMD) [52].

Therefore, the identification of outliers with a functional approach is possible with the calculation of depths. In this case, elements that have different behavioural patterns than the rest will be taken into account. The concept of depth makes it possible to work with observations, defined in a given time interval, in the form of curves, instead of having to summarise the observations of the curve into a single value, such as the mean. This method of outlier detection is based on depth measures and centrality: an element that is far from the centre of the sample will have a low depth. Thus, the curves with the least depth are the functional outliers.

Firstly, the cumulative empirical distribution function $F_{n,t}(x_i(t))$ of the values of the curves $\{x_i(t)\}_{i=1}^{n}$ in a certain time $t \in [a,b]$ it is contemplated. It can be defined as

$$F_{n,t}(x_i(t)) = \frac{1}{n}\sum_{k=1}^{n} I(x_k(t) \leq x_i(t)) \tag{6}$$

with $I(\cdot)$ being a indicator function. Subsequently, the FMD for any curve x_i within a set of curves $x_1, \ldots, x_n$ is calculated as

$$FMD_n(x_i(t)) = \int_a^b D_n(x_i(t))dt \tag{7}$$

where $D_n(x_i(t))$ is the depth of the curve $x_i(t)$, $\forall t \in [a,b]$, expressed as

$$D_n(x_i(t)) = 1 - \left|\frac{1}{2} - F_{n,t}(x_i(t))\right| \tag{8}$$

On the other hand, for HMD the functional mode is the element or curve most densely surrounded by the other curves in the dataset. HMD is defined as

$$HMD_n(x_i, h) = \sum_{k=1}^{n} K\left(\frac{\|X_i - X_k\|}{h}\right) \tag{9}$$

with a kernel function $K : R^+ \to R^+$, a bandwidth parameter h and $\|\cdot\|$ as the norm in a functional space. Among all the norms, in the most cases, it is used the norm L_2, expressed as

$$\|x_i(t) - x_j(t)\|_2 = \left(\int_a^b (x_i(t) - x_j(t))^2 dt\right)^{1/2} \tag{10}$$

In addition, also exist several options for the kernel functions $K(\cdot)$. A widely used one is the truncated Gaussian kernel, expressed as

$$K(t) = \frac{2}{\sqrt{2\pi}} \exp\left(-\frac{t^2}{2}\right), \quad t > 0 \tag{11}$$

In this paper, the depth chosen to identified outliers is the HMD. The bandwidth h is the value that leaves, below it, 15% of the data coming from the distribution of $\{\|x_i(t) - x_t(t)\|_2,\ i,j = 1, \ldots, n\}$ [15], and the cut-off C is selected, specifically, to have a 1% Type I error [50], according to

$$Pr(HMD_n(x_i(t)) < C) = 0.01, \quad i = 1, \ldots, n \tag{12}$$

The cut-off C has to be estimated because the distribution of the functional depth is not known. There are several ways to carry out this estimation; however, the bootstrapping method is the most appropriate for the purpose of this research [52,53]. The steps to follow are:

1. Extract, with replacement, a new sample of the original.
2. Estimate the study parameter through the statistic of this new sample.
3. Repeat the above steps a large number of times. This repetition is also known as Monte Carlo simulation; it relies on repetition to extract information from the data (see examples in [54,55]).
4. Obtain the empirical distribution of the statistic.

2.2.4. Functional Strengths

The functional data analysis (FDA) has many strong points, but the ones that allow the FDA to have a better performance in contamination analysis are the following: [18,19,24]:

- It is not necessary to know anything in advance about the distribution of the data.
- The analysis of the time sets as a unit. The sample analysed is structure in complete time units like days or years. Individually distributed values are not taken into account.
- Analysis of homogeneity. Outliers are defined differently; data which do not exceed the limit but which constantly have small deviations should be classified as outliers.
- Trend analysis. With these techniques, besides detect outliers, also it is possible to analyse situations where there are no outliers but small deviations from the normal data behaviour are observed.
- Complete analysis of the time spectrum. In classic studies, generally, the analyses are based on specific values measured in a determined set of points. The FDA, on the other hand, made it possible to work with the entire time spectrum of a continuous mode.

3. Results and Discussion

The results of the three methodologies presented in this document are presented below. All figures were obtained with the R-programming [56] and Python [57] software.

3.1. Classical Analysis

To perform the classical monitoring strategy on air quality the individual time series, descriptive statistics, box plots and autocorrelation analysis were calculated to determine if any of the values fell outside of the limits, and to analyse trends. The descriptive statistical parameters of the dataset are shown in Table 1:

Table 1. Summary descriptive statistics of hourly NO_2 concentrations from Blanchardstown air quality monitoring station in Dublin, Ireland. The statistical quartiles (Q1, Q2, Q3) and the interquartile range (IQR) are also displayed. Take into account that 0 $\mu g/m^3$ represents a missing or wrong value.

max	153.48 $\mu g/m^3$	**Q1**	9.56 $\mu g/m^3$
min	0 $\mu g/m^3$	**Q2**	20.61 $\mu g/m^3$
mean	28.47 $\mu g/m^3$	**Q3**	42.33 $\mu g/m^3$
mode	42.33 $\mu g/m^3$	**IQR**	32.77 $\mu g/m^3$
std	23.91 $\mu g/m^3$	**var**	571.79$\mu g/m^3$
n	8760		

The descriptive statistical parameters in Table 1 show that the limit values are not exceeded. The next step of classical data analysis is present a time series of the hourly data in 2013 (Figure 1), ranging from the maximum value 153.48 $\mu g/m^3$ to the minimum value 0 $\mu g/m^3$. From here it is possible to say that in any moment, the hourly upper limit (200 $\mu g/m^3$) is not exceed and that the data have a high variability.

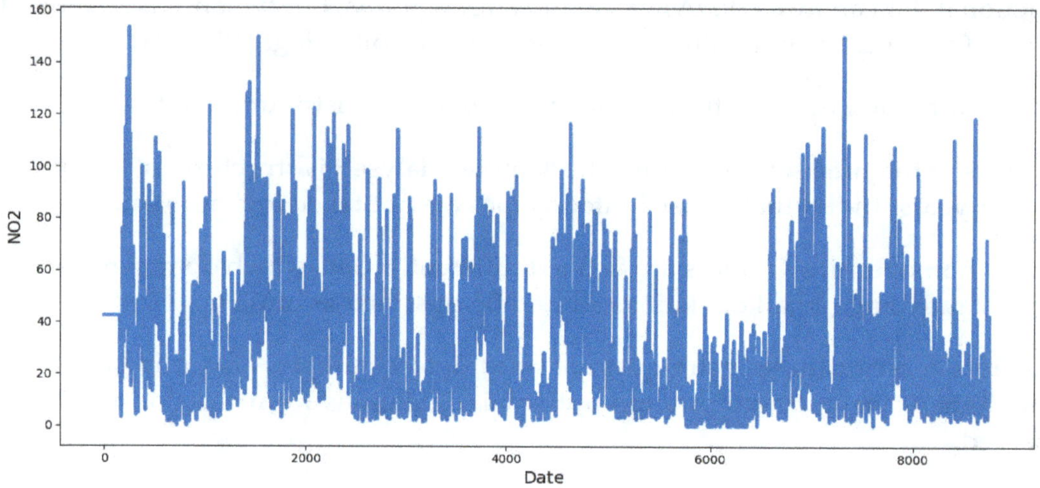

Figure 1. Individual time series of hourly NO_2 concentrations from Blanchardstown air quality monitoring station in Dublin, Ireland. Software: Python [57].

Figure 2 presents a boxplot which graphically characterises the data groups of the NO_2 concentration by quartiles. The diagram graphically displays the values of the first quartile (9.56 $\mu g/m^3$); third quartile (42.33 $\mu g/m^3$); the interquartile range (32.77 $\mu g/m^3$); and some, in red, that are considered atypical.

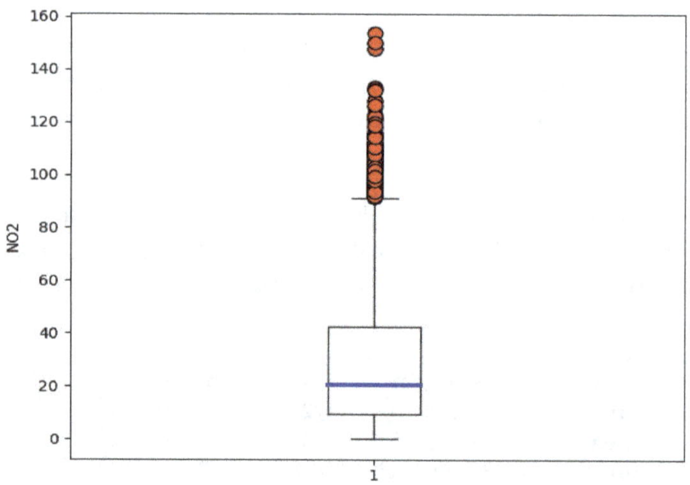

Figure 2. Box-plot of hourly NO_2 concentrations from Blanchardstown air quality monitoring station. The central and blue line represents the median, and the end of the whiskers are the quartiles (25% for the lower part and 75% for the upper part). The red dots represent the outliers. Software: Python [57].

Figure 3 presents the frequency of hourly concentrations of NO_2, which, as can be seen, are biased by 0 values. Another weakness of this analysis is that, when data are poorly collected or no data are available, only two options remain: either delete these observations (data are lost) or replace them with 0 values.

Figure 4 shows the normal probability plot of the data, again affected by 0 values. A Kolmogorov-Smirnov test and Anderson Darling test were applied to compare NO_2 concentrations to a standard normal distribution [54]. The null hypothesis is that the values have a standard normal distribution. The alternative hypothesis is that the values do not have that distribution. The results obtained for both tests were p-values very close to 0, so, with a 5% significance level, statistical evidence of the non-normality of the data has been found. The test statistic is: $\max(F(x) - G(x))$, where F(x) is the empirical cumulative distribution function and G(x) is the standard normal cumulative distribution function.

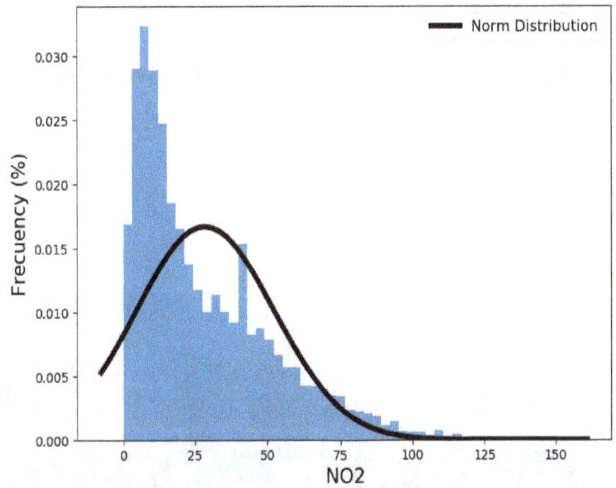

Figure 3. Frequency of hourly NO_2 concentrations from Blanchardstown air quality monitoring station. Comparison of data distribution with normal. Software: Python [57].

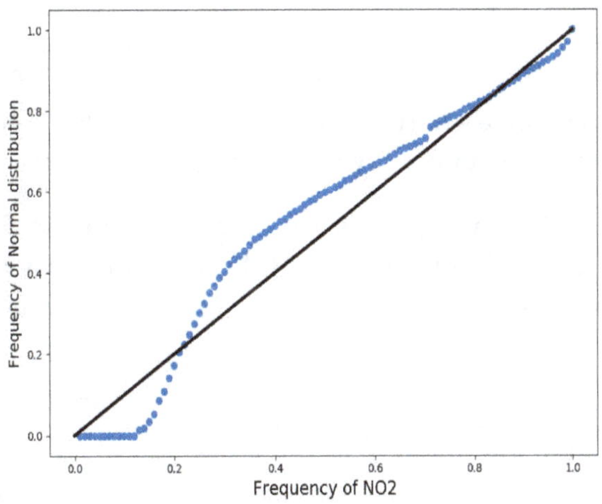

Figure 4. Normal probability plot of hourly NO_2 (QQ-plot) concentrations from Blanchardstown air quality monitoring station. Software: Python [57].

Other tests have been performed to check whether the data approaches any type of distribution: normal, generalised extreme value or Weibull and Rayleigh, but none have been acceptable with a null hypothesis at 5% significance. From the classic analysis of the data it must be concluded that there are no data that are outside the limit values. This classical method is limited to a time series analysis with regard to the assessment of trends (Figure 1), and although it allows for the identification of the main parameters within the data and how the data are distributed, is an incomplete method because it provides us with information that is too simple and does not take into account the correlation between hourly observations.

3.2. Statistical Process Control

3.2.1. Control I-MR Charts with Individual Mean

To analyse the data using the SPC method, an individual-moving range chart (IMR chart) of hourly NO_2 concentrations was made. With the examination of the results shown in Figure 5, it can be observed that the number of false alarms, i.e., outliers, is significant. This problem is attributable to:

- The non-normality of the data, which is clear from the analysis shown in Figure 4.
- The effect of autocorrelation in time series data (Figure 6).
- The existence of greater variability with data of different rational subgroups than within the data inside each analised subgroup.

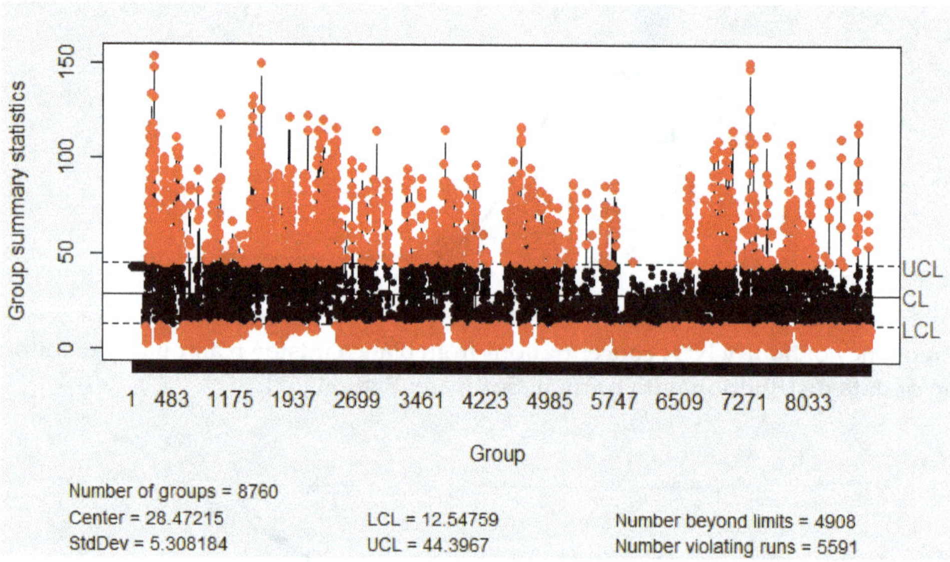

Figure 5. Individual-moving range chart (IMR chart) X/R with mobile range of hourly NO_2 concentrations from Blanchardstown air quality monitoring station. Software: R-programming [56].

By performing an autocorrelation analysis, it can be observed from Figure 6 that the data are very autocorrelated. This is very common in environmental data and shows that the autocorrelation has 24-hour cycles and decreases with time.

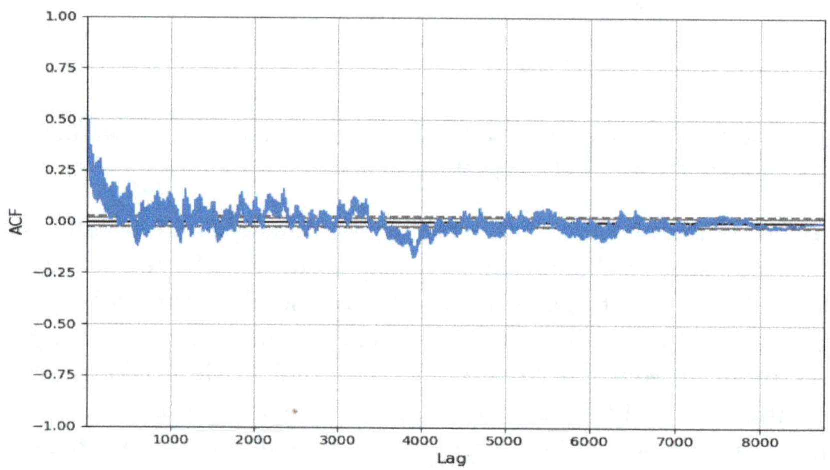

Figure 6. Sample autocorrelation function of hourly values for NO_2 concentrations from Blanchardstown air quality monitoring station. Software: Python [57].

In Figure 6 the correlation of all data for the year is shown, while Figure 7 only shows the data of the first 86 h in order to see in more detail, the 24 h cycles. Due to the non-normality of the data and the data's autocorrelation, the control chart has a large number of false alarms. Therefore, SPC is not a very suitable method with which to detect outliers for NO_2 concentration.

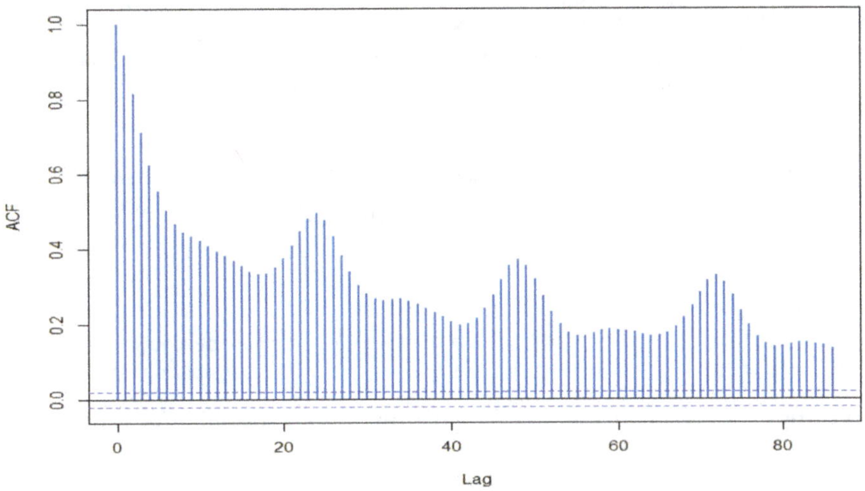

Figure 7. Sample autocorrelation function of hourly values for NO_2 concentrations from Blanchardstown air quality monitoring station over the initial 86 h period. Software: R-programming [56].

3.2.2. Control Charts with Daily Rational Subgroups

The study of datasets choosing days as the rational subgroup of the X/s chart (every day is summed up by one point), is not under control due to the non-normality and the autocorrelation (see Figures 8 and 9). Although the chart is not under control, and there is much variability, none of the 365 days exceed the limit value (100 μg/m^3).

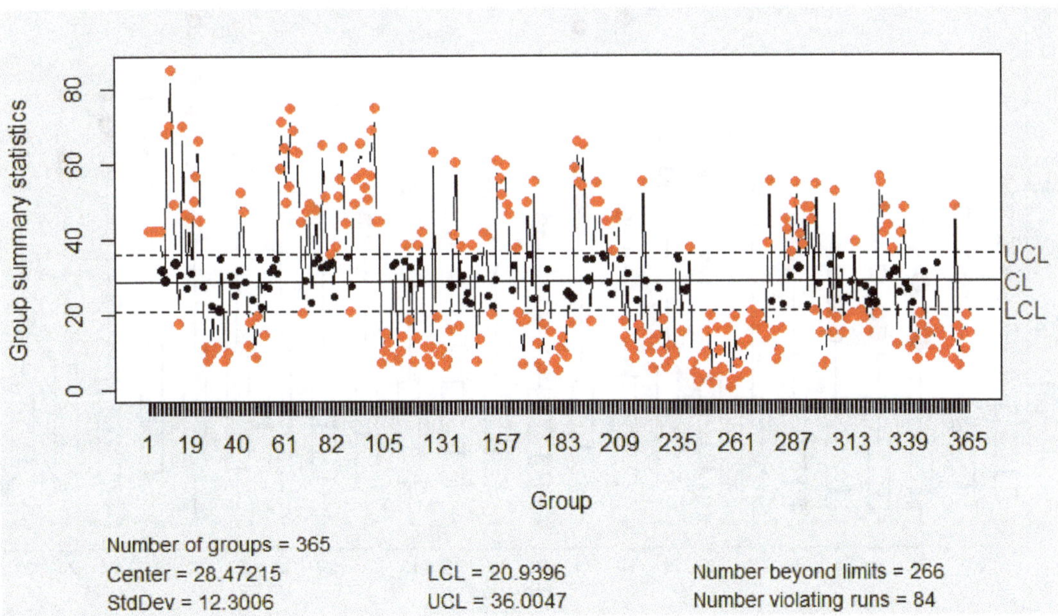

Figure 8. Xbar-chart of hourly NO_2 concentrations with the daily rational subgroup of data. Software: R-programming [56].

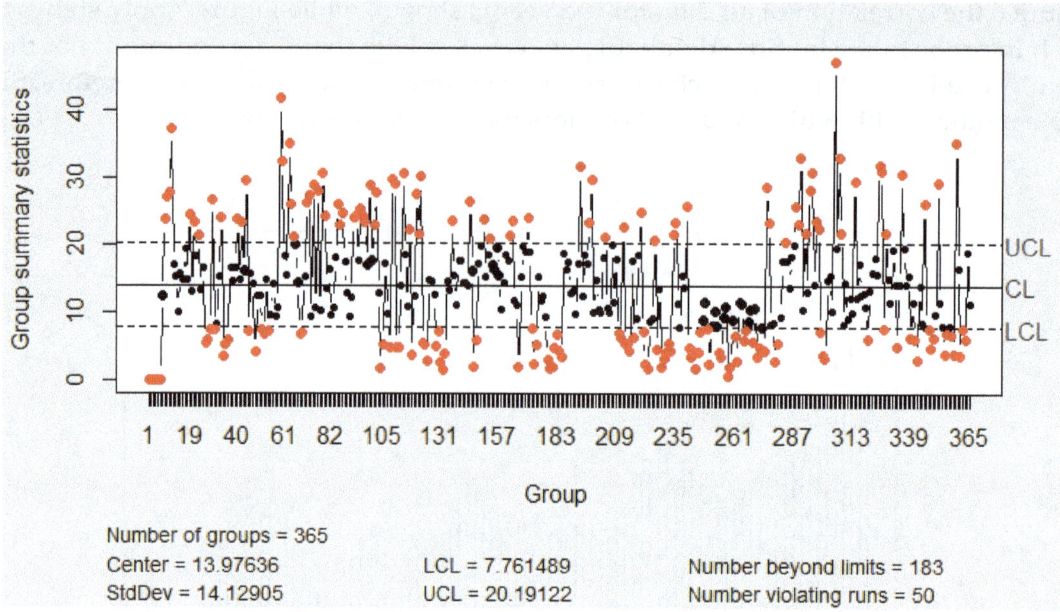

Figure 9. S-chart of hourly NO_2 concentrations with the daily rational subgroup of data. Software: R-programming [56].

3.2.3. Trend Analysis

Using the hourly NO_2 concentration data, a trend analysis was undertaken to examine the diurnal patterns and identify outliers. Figure 10 shows the box plots of hourly emissions over 365 days and represents the mean, the confidence interval of this mean, quartiles and abnormal values in red that. This way it is possible to studying individually the distribution of the NO_2 emissions in each hour of the day.

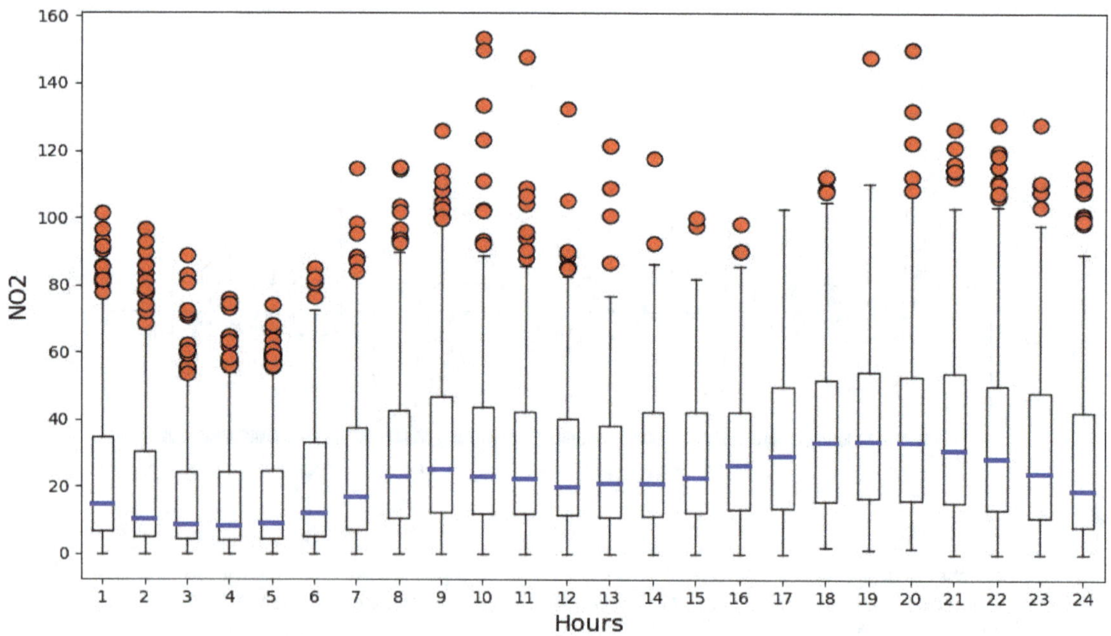

Figure 10. Hourly box plot of NO_2 concentrations from Blanchardstown air quality monitoring station. Software: Python [57].

In Figure 11, where only the mean values have been represented, two maximum values were analysed, at 9 a.m. and 7 p.m., which correspond to heavy traffic hours in this area of Dublin. These two hours, as can be seen in Figure 10, also have high variability (wide boxes), and around them there are numerous outliers.

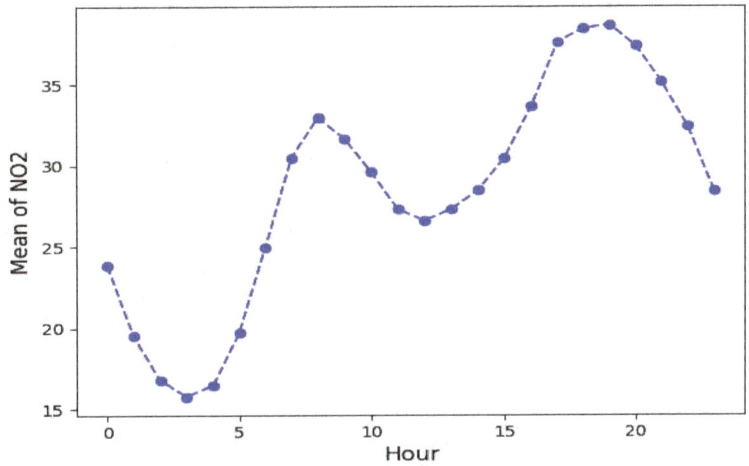

Figure 11. Mean hourly NO_2 concentrations from Blanchardstown air quality monitoring station. Software: Python [57].

It can also be seen in these figures that the absolute minimum is 4 a.m., which also corresponds to the hour with the least variability in the NO_2 concentration. If a daily analysis approach is considered throughout the year (Figure 12), it is clear that Wednesday, Thursday and Friday are the days with the highest NO_2 concentrations.

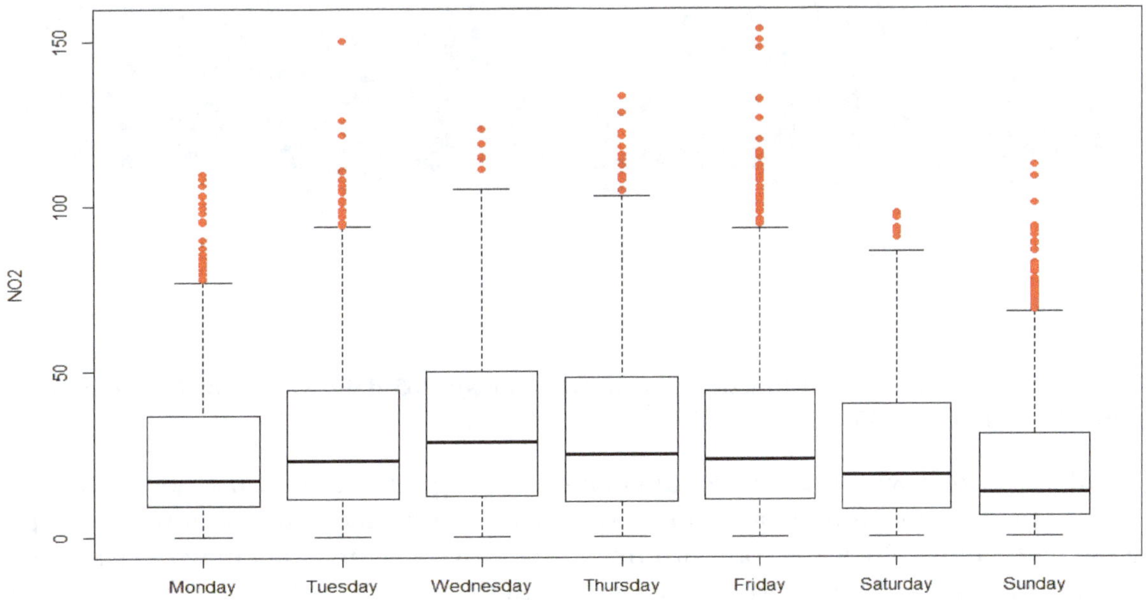

Figure 12. Daily analysis of NO_2 concentrations at Blanchardstown air quality monitoring station. Software: R-programming [56].

Blanchardstown contains the largest shopping centre in Ireland and the most major motorway nearby; because of this, there is a lot of traffic when rush hour and evening shopping are combined in this area. On the other hand, there are lower concentrations during the weekend, although not as low as expected. This is because the largest shopping centre in the country is very busy at weekends, especially during Christmas period.

In this case, a SPC gives us more information than the previous analysis, such as the differences in the NO_2 levels between days or hours. However, it still does not take into account the complete daily behaviour of NO_2 emissions from correlated hourly measurements.

3.3. Functional Analysis of NO_2 in Dublin

The subsequent step in the functional methodology is to compare the results between the classical analysis and the SPC. In the functional methodology, the first thing to do is to build a sample of curves based on the discrete values measured every hour. Figure 13 shows the 365 functions generated with 24 hourly data. Once the data are transformed into functional data, i.e., daily curves of 24 values, each of them takes into account the correlation between the hourly NO_2 values and can be analysed for outliers.

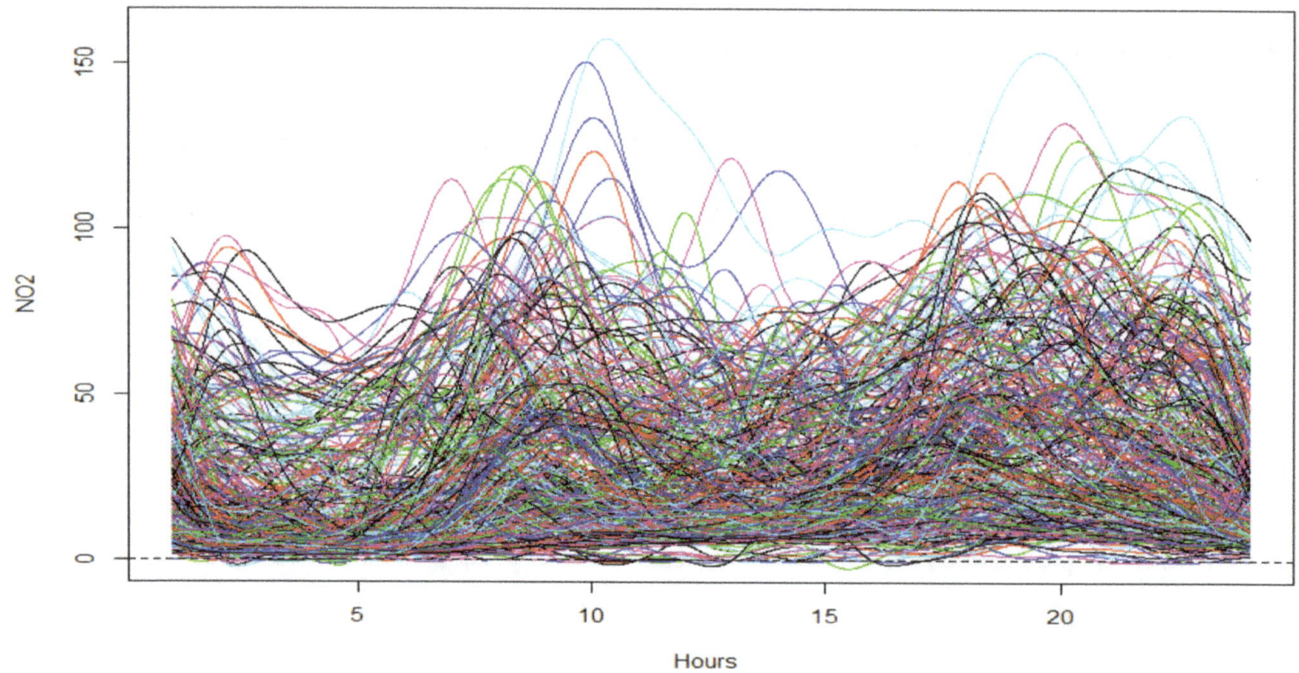

Figure 13. Data represented in functional form (functions): 365 daily curves of NO_2 emissions. Software: R-programming [56].

The results obtained with the functional analysis, taking into account the depths, allow us to identify days with abnormal functional values, even if, discreetly, they are not outliers. Despite not exceeding the daily limit values, the concentration of NO_2 over a whole day may have an abnormal behaviour. For this reason the vectorial analysis, like SPC, does not get to detect these days. In a different way, the functional approach detects any deviation from normal daily behaviour in the emissions of NO_2, without relying on any distribution restrictions. This is shown in Figure 14 where the functional outliers found in this case study are presented.

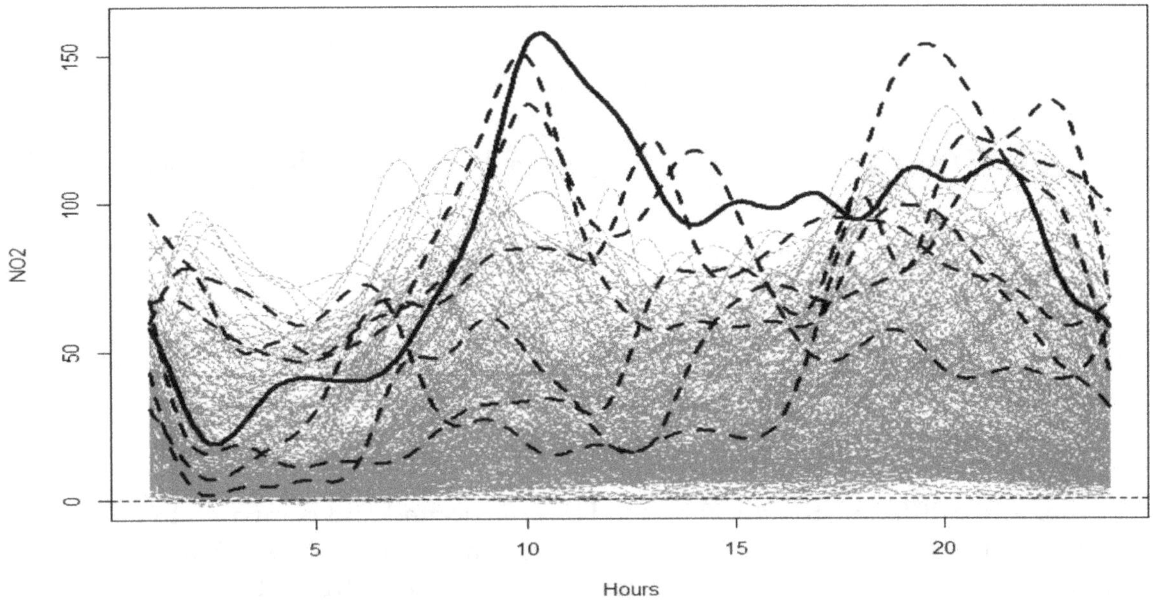

Figure 14. NO_2 functional data; and in the dark grey and dotted line, the functional outliers detected. Highlighted is the 11th day which is also an outlier. Software: R-programming [56].

The data analysed were not discretely outside the limits values; however, functionally excessive variations were observed on specific days. It can be deduced that there were no NO_2 pollution problems in 2013 because hourly, daily and annual NO_2 concentration limits were not exceeded. But it is also important to analyse whether there are hours or days with anomalous trends of NO_2 concentrations, although they remain within legal limits. The NO_x (NO, NO_2) is an pollutant not coming from a natural source ([58]), and for this reason also, its involvement in the mechanisms of depleting the ozone layer are very unfortunate. The FDA methodology has proven to be very effective in detecting days with trends that are not the same as the rest of the data. It is important not only to analyse whether the contamination is within the allowed limits, but also to find days that are different than expected.

For example, on the 11th day, detected by the functional approach and highlighted in the Figure 14, with the results obtained through the SPC, as can be seen in Figure 8, this day has a higher mean, but it is within the limit values. Neither using the classical analysis (Figure 1) with individual time series nor the one with the corresponding statistical parameters, it could be considered an outlier. Figure 14 shows a strength of the functional approach by detecting this curve as an atypical day; it will be possible to study the reasons that lead the NO_2 to behave this way on this particular day. In fact, there are several studies that demonstrate the greater power of the functional approach for detecting outliers than other methodologies (see [17,59]). There are also studies that, specifically, show that the depth measure used here (*h-modal*) is the one that achieves the lowest error rates [50].

To find a reason that explains the anomalous behaviour on those days, it would be necessary to have greater traceability of the most important sources of NO_2 emissions. It would be necessary to have data relating to the weather conditions, traffic movements, industry sources affecting the study area, etc. For example, incorporating weather conditions, i.e., temperatures, sunshine hours and precipitation, could improve the assessment of outliers. The detection of outliers and air pollution episodes can help to separate the causes of normal and specific variability, and is a first step towards the effective design and implementation of mitigation measures. Although the only reason for these outliers is possibly not the weather conditions, as can be seen in Table 2, those days were colder than usual; had very little precipitation; and in general, had fewer sunshine hours than the average for the month.

Table 2. Sample of other environmental characteristics from Dublin (temperature, sunshine and precipitation) which may impact hourly NO_2 concentrations measured at Blanchardstown air quality monitoring station. The averages shown represent the monthly average of each variable.

Nº	Date	Day	Lowest Tª	Average Lowest Tª	Sunshine Hours	Average Sunshine Hours	Precipitations
10	10/01	Thursday	−2.5 °C	2.6 °C	0 h	1.2 h	2 mm
11	11/01	Friday	−0.5 °C	2.6 °C	1 h	1.2 h	0.5 mm
59	28/02	Thursday	1 °C	2 °C	0 h	2.6 h	0 mm
60	01/03	Friday	−3.1 °C	1.2 °C	3 h	2.1 h	0 mm
64	5/03	Tuesday	−3,8 °C	1.2 °C	5 h	2.1 h	0 mm
78	19/03	Tuesday	0 °C	1.2 °C	0 h	2.1 h	5.6 mm
122	02/05	Thursday	−0.4 °C	6.3 °C	5 h	6.3 h	0 mm
193	12/07	Friday	8.7 °C	11.9 °C	14 h	7.7 h	0 mm
305	01/11	Friday	2 °C	4.2 °C	4.20 h	2.7 h	0 mm

4. Conclusions

In this paper, three different analytical methods were adopted and compared to determine their effectiveness to identify pollution episodes and outliers. The data used come from a sub-urban air quality monitoring site in Dublin, Ireland, and cover the whole year 2013 with hourly measurements. Firstly, a classical vectorial approach was applied by analysing the data through time series, boxplots and frequency plots. Secondly, a statistical process control was adopted to study the data grouped by days or hours, and with different control charts (Xbar-chart, S-chart). Finally, to identify air pollution episodes and outliers, a functional data analysis approach was adopted to analyse the daily patterns of NO_2 concentrations.

To effectively support the mitigation of air pollution and provide true air quality conditions, a new approach and set of tools are required to effectively assess local air pollution. With this in mind, the classical or vectorial approach is too simplistic, despite providing significant information for decision making. It has several weaknesses related to the time correlation structure of the data, and the inability to detect real outliers, days with behaviours far from the standard, just because they do not exceed limit values. As such, more advanced and modern techniques can provide deeper insights to support controlling air pollution episodes. Statistical process control presents similar problems. Even though it manages to take into account the correlation between data, the lack of normality causes a significant increase in false alarms for days that are within the legal limits; this method labels them as outliers. This is because it works with discrete observations and it is not able to extract continuous information from the data.

Alternately, the functional data analysis method, adapted to study pollution episodes, as shown in this paper, presents important improvements as a method that can be extrapolated to any city in the world. In short, it is not restricted to certain characteristics of the data distribution; takes into account complete time units; works with the entire time spectrum of a continuous mode; and the detection of outliers is more effective, which is very important in pollution issues. On the contrary, working from a functional point of view can also have its drawbacks, such as the need for large amounts of data (it is not always possible to get them), or other data constraints, such as continuity, positivity and monotonicity. But as explained, the fact that all the hourly values of a day are not outliers does not mean that that the day is not an outlier. The outlier search with the FDA allows one to detect days that, for some reason, have different sets of measurements of NO_2 than the rest (usually higher or with strange peaks). Being able to study the reasons that make these daily functions behave differently from others makes it possible to prevent them; to try to reduce the variability in emissions; and in short, to improve air pollution control. With a classical method there would be no possibility of relating certain events to lost hours that are labelled as outliers.

The FDA's approach is presented in this document as a methodology for more effective assessment of air pollution, which is hazardous to the health of living beings, and to inform effective mitigation measures in the future.

Author Contributions: The tasks of this paper were divided as follows: on the one hand, J.P.P., J.S.V. and M.M.C. focused on the development of the mathematical methodology; on the other hand, J.M.T., A.M. and J.G. focused on the environmental part of the work. All authors have read and agreed to the published version of the manuscript.

Acknowledgments: This paper was supported by the Spanish Government (Industry and Competitiveness Ministry) under the project RTI2018-096296-B-C21.

References

1. García Nieto, P. Parametric study of selective removal of atmospheric aerosol by coagulation, condensation and gravitational settling. *Int. J. Environ. Health Res.* **2001**, *11*, 151–162. [CrossRef] [PubMed]
2. Akkoyunku, A.; Ertürk, F. Evaluation of air pollution trends in Istanbul. *Int. J. Environ. Health Res.* **2003**, *18*, 388–398.
3. Karaca, F.; Alagha, O.; Ertürk, F. Statistical characterization of atmospheric PM10 and PM2.5 concentrations at a non-impacted suburban site of Istanbul, Turkey. *Chemosphere* **2005**, *59*, 183–190. [CrossRef] [PubMed]
4. García Nieto, P. Study of the evolution of aerosol emissions from coal-fired power plants due to coagulation, condensation, and gravitational settling and health impact. *J. Environ. Manag.* **2006**, *79*, 372–382. [CrossRef] [PubMed]
5. RCPCH. *Every Breath We Take: The Lifelong Impact of Air Pollution*; Royal College of Physicians: London, UK, 2016. Available online: https://www.rcplondon.ac.uk/projects/outputs/every-breath-we-take-lifelong-impact-air-pollution (accessed on 7 February 2020).
6. WHO. *Review of Evidence on Health Aspects of Air Pollution—REVIHAAP Project*; World Health Organization: Geneva, Switzerland, 2013. Available online: http://www.euro.who.int/__data/assets/pdf_file/0004/193108/REVIHAAP-Final-technical-report.pdf (accessed on 7 February 2020).
7. Kumar, P.; Druckman, A.; Gallagher, J.; Gatersleben, B.; Allison, S.; Eisenman, T.S.; Hoang, U.; Hama, S.; Tiwari, A.; Sharma, A.; et al. The nexus between air pollution, green infrastructure and human health. *Environ. Int.* **2019**, *133*, 105181. [CrossRef]
8. EPA. United States Environmental Protection Agency. 2019. Available online: https://www.epa.gov/ (accessed on 19 December 2019).
9. AQEG. *Trends in Primary Nitrogen Dioxide in the UK*. Air Quality Expert Group. 2007. Available online: https://uk-air.defra.gov.uk/assets/documents/reports/aqeg/primary-no-trends.pdf (accessed on 7 February 2020).
10. EPA. *Ireland's Transboundary Gas Emissions*; Environmental Protection Agency: Washington, DC, USA, 2018. Available online: http://www.epa.ie/pubs/reports/air/airemissions/Irelands%20Air%20Pollutant%20Emissions%202016.pdf (accessed on 7 February 2020).
11. Costa, S.; Ferreira, J.; Silveira, C.; Costa, C.; Lopes, D.; Relvas, H.; Borrego, C.; Roebeling, P.; Miranda, A.I.; Teixeira, J.P. Integrating Health on Air Quality Assessment—Review Report on Health Risks of Two Major European Outdoor Air Pollutants: PM and NO_2. *J. Toxicol. Environ. Heal. Part B* **2014**, *17*, 307–340. [CrossRef]
12. Cooper, C.; Alley, F. *Air Pollut. Control*; Waveland Press: New York, NY, USA, 2002.
13. Lutgens, F.; Tarbuck, E. *The Atmosphere: An Introduction to Meteorology*; Prentice Hall: New York, NY, USA, 2001.
14. Jeanjean, A.; Gallagher, J.; Monks, P.; Leigh, R. Ranking current and prospective NO_2 pollution mitigation strategies: An environmental and economic modelling investigation in Oxford Street, London. *Environ. Pollut.* **2017**, *225*, 587–597. [CrossRef]
15. Cuevas, A.; Fraiman, R. A plug-in approach to support estimation. *Ann. Stat.* **1997**, *25*, 2300–2312. [CrossRef]
16. Matías, J.; Ordóñez, C.; Taboada, J.; Rivas, T. Functional support vector machines and generalized linear models for glacier geomorphology analysis. *Int. J. Comput. Math.* **2009**, *86*, 275–285. [CrossRef]

17. Martínez, J.; Garcia Nieto, P.; Alejano, L.; Reyes, A. Detection of outliers in gas emissions from urban areas using functional data analysis. *J. Hazard. Mater.* **2011**, *186*, 144–149.
18. Martínez, J.; Saavedra, A.; García Nieto, P.; Piñeiro, J.; Iglesias, C.; Taboada, J.; Sancho, J.; Pastor, J. Air quality parameters outliers detection using functional data analysis in the Langreo urban area (Northern Spain). *Appl. Math. Comput.* **2014**, *241*, 1–10. [CrossRef]
19. Sancho, J.; Iglesias, C.; Piñeiro, J.; Martínez, J.; Pastor, J.; Araújo, M.; Taboada, J. Study of water quality in a spanish river based on statistical process control and functional data analysis. *Math. Geosci.* **2016**, *48*, 163–186. [CrossRef]
20. Dombeck, D.; Graziano, M.; Tank, D. Functional clustering of neurons in motor cortex determined by cellular resolution imaging in awake behaving mice. *J. Neurosci.* **2009**, *29*, 13751–13760. [CrossRef] [PubMed]
21. Wu, D.; Huang, S.; Xin, J. Dynamic compensation for an infrared thermometer sensor using least-squares support vector regression (LSSVR) based functional link artificial neural networks (FLANN). *Meas. Sci. Technol.* **2008**, *19*, 105202.1–105202.6. [CrossRef]
22. Ordoñez, C.; Martínez, J.; Cos Juez, J.; Sánchez Lasheras, F. Comparison of GPS observations made in a forestry setting using functional data analysis. *Int. J. Comput. Math.* **2011**, *89*, 402–408. [CrossRef]
23. Ordóñez, C.; Martínez, J.; Saavedra, A.; Mourelle, A. Intercomparison Exercise for Gases Emitted by a Cement Industry in Spain: A Functional Data Approach. *J. Air Waste Manag. Assoc. (1995)* **2011**, *61*, 135–141. [CrossRef] [PubMed]
24. Sancho, J.; Pastor, J.; Martínez, J.; García, M. Evaluation of Harmonic Variability in Electrical Power Systems through Statistical Control of Quality and Functional Data Analysis. *Procedia Eng.* **2013**, *63*, 295–302. [CrossRef]
25. Fraiman, R.; Muniz, R. Trimmed means for functional data. *Test* **2001**, *10*, 419–440. [CrossRef]
26. Piñeiro, J.; Martínez, J.; García Nieto, P.; Alonso, J.; Díaz, C.; Taboada, J. Analysis and detection of outliers in water quality parameters from different automated monitoring stations in the Miño river basin (NW Spain). *Ecol. Eng.* **2013**, *60*, 60–66.
27. Sancho, J.; Martínez, J.; Pastor, J.; Taboada, J.; Piñeiro, J.; García Nieto, P. New methodology to determine air quality in urban areas based on runs rules for functional data. *Atmos. Environ.* **2014**, *83*, 185–192. [CrossRef]
28. Grubbs, F.E. Procedures for Detecting Outlying Observations in Samples. *Technometrics* **1969**, *11*, 1–21. [CrossRef]
29. Jäntschi, L. A test detecting the outliers for continuous distributions based on the cumulative distribution function of the data being tested. *Symmetry* **2019**, *11*, 835. [CrossRef]
30. EPA. *Air Quality in Ireland 2018*; Environmental Protection Agency: Washington, DC, USA, 2019. Available online: http://www.epa.ie/pubs/reports/air/quality/Air%20Quality%20In%20Ireland%202018.pdf (accessed on 7 February 2020).
31. EPA. *Air Quality in Ireland 2013: Key Indicators of Ambient Air Quality*; Environmental Protection Agency: Washington, DC, USA, 2014. Available online: https://www.epa.ie/pubs/reports/air/quality/Air%20Quality%20Report%202013.pdf (accessed on 7 February 2020).
32. Romer, U. Weather Online(Ireland). 2013. Available online: https://www.weatheronline.co.uk/weather/maps/current?TYP=tmin&KEY=IE&LANG=en&ART=tabelle&JJ=xxxx&SORT=2&INT=24 (accessed on 19 December 2019).
33. Carslaw, D.; Ropkins, K. Openair—An R package for air quality data analysis. *Environ. Model. Softw.* **2012**, *27–28*, 52–61. [CrossRef]
34. Piatesky-Shapiro, G.; Frawley, W. *Knowledge Discovery in Databases*; MIT Press: Cambridge, MA, USA, 1991.
35. Takeuchi, J.; Yamanishi, K. A unifying framework for detecting outliers and change points from time series. *IEEE Trans. Knowl. Data Eng.* **2006**, *18*, 482–492. [CrossRef]
36. Sim, C.; Gan, F.; Chang, T. Outlier Labeling With Boxplot Procedures. *J. Am. Stat. Assoc.* **2005**, *100*, 642–652. [CrossRef]
37. Montgomery, D. *Design and Analysis of Experiments*; John Wiley & Sons, Inc.: Hoboken, NJ, USA, 2013; Chapter 1–2, pp. 1–65.
38. Shewhart, W. *Economic Control of Quality of Manufactured Product*; Van Nostrand Company: New York, NY, USA, 1931.
39. Chen, Y.K. An evolutionary economic-statistical design for VSIXcontrol charts under non-normality. *J. Adv. Manuf. Technol.* **2003**, 22, 602–610. [CrossRef]

40. Freeman, J.; Modarres, R. Inverse Box-Cox: the power-normal distribution. *Stat. Probab. Lett.* **2006**, *76*, 764–772. [CrossRef]
41. Box, G.; Cox, D. An analysis of transformations. *J. R. Stat. Soc. Ser. B (Stat. Methodol.)* **1964**, *26*, 211–252. [CrossRef]
42. Grant, E.; Leavenworth, R. *Statistical Quality Control*; McGraw-Hill: New York, NY, USA, 1998.
43. Champ, C.; Woodall, W. Exact results for Shewhart control charts with supplementary runs rules. *Technometrics* **1987**, *29*, 393–399. [CrossRef]
44. Zhang, M.; Lin, W.; Klein, S.; Bacmeister, J.; Bony, S.; Cederwall, R.; Del Genio, A.; Hack, J.; Loeb, N.; Lohmann, U.; et al. Comparing clouds and their seasonal variations in 10 atmospheric general circulation models with satellite measurements. *J. Geophys. Res.* **2005**, *110*. [CrossRef]
45. Electric, W. *Statistical Quality Control Handbook*; AT&T Technologics: Indianapolis, Indiana, 1956.
46. Ramsay, J.; Silverman, B. *Functional Data Analysis*; Springer: New York, NY, USA, 2005.
47. Sánchez-Lasheras, F.; Ordóñez, C.; Garcia Nieto, P.J.; García-Gonzalo, E. Detection of outliers in pollutant emissions from the Soto de Ribera coal-fired plant using Functional Data Analysis: A case study in northern Spain. *Proceedings* **2018**, *2*, 1473. [CrossRef]
48. Muñiz, C.D.; García Nieto, P.J.; Alonso Fernández, J.R.; Torres, J.V.; Taboada, J. Detection of outliers in water quality monitoring samples using functional data analysis in San Esteban estuary (Northern Spain). *Sci. Total. Environ.* **2012**, *439*, 54–61. [CrossRef] [PubMed]
49. Kamada, M.; Toraich, K.; Mori, R. Periodic spline orthonormal bases. *J. Approx. Theory* **1988**, *55*, 27–34. [CrossRef]
50. Febrero, M.; Galeano, P.; González-Manteiga, W. Outlier detection in functional data by depth measures, with application to identify abnormal NOx levels. *Environmetrics* **2008**, *19*, 331–345. [CrossRef]
51. Zuo, Y.; Serfling, R. General notions of statistical depth function. *Ann. Stat.* **2000**, *28*, 461–482. [CrossRef]
52. Cuevas, A.; Febrero, M.; Fraiman, R. On the use of the bootstrap for estimating functions with functional data. *Comput. Stat. Data Anal.* **2006**, *51*, 1063–1074. [CrossRef]
53. Cuevas, A. A partial overview of the theory of statistics with functional data. *J. Stat. Plan. Inference* **2014**, *147*, 1–23. [CrossRef]
54. Jäntschi, L.; Bolboacă, S.D. Computation of Probability Associated with Anderson-Darling Statistic. *Mathematics* **2018**, *6*, 88. [CrossRef]
55. Jäntschi, L.; Bolboacă, S.D. Rarefaction on natural compound extracts diversity among genus. *J. Comput. Sci.* **2014**, *5*, 363–367. [CrossRef]
56. R Core Team. *R: A Language and Environment for Statistical Computing*; R Foundation for Statistical Computing: Vienna, Austria, 2014.
57. Van Rossum, G.; Drake, F.L., Jr. *Python Reference Manual*; Centrum voor Wiskunde en Informatica: Amsterdam, The Netherlands, 1995.
58. Cosma, C.; Suciu, I.; Jäntschi, L.; Bolboaca, S. Ion-Molecule Reactions and Chemical Composition of Emanated from Herculane Spa Geothermal Sources. *Int. J. Mol. Sci.* **2008**, *9*, 1024–1033. [CrossRef]
59. Yu, G.; Zou, C.; Wang, Z. Outlier Detection in Functional Observations With Applications to Profile Monitoring. *Technometrics* **2012**, *54*, 308–318. [CrossRef]

3

Skew-Reflected-Gompertz Information Quantifiers with Application to Sea Surface Temperature Records

Javier E. Contreras-Reyes [1,*], Mohsen Maleki [2] and Daniel Devia Cortés [3]

1 Departamento de Estadística, Facultad de Ciencias, Universidad del Bío-Bío, Concepción 4081112, Chile
2 Department of Statistics, College of Sciences, Shiraz University, Shiraz 71946 85115, Iran; m.maleki.stat@gmail.com
3 Departamento de Evaluación de Pesquerías, Instituto de Fomento Pesquero, Valparaíso 2361827, Chile; ddeviac@gmail.com
* Correspondence: jcontreras@ubiobio.cl or jecontrr@uc.cl.

Abstract: The Skew-Reflected-Gompertz (SRG) distribution, introduced by Hosseinzadeh et al. (J. Comput. Appl. Math. (2019) 349, 132–141), produces two-piece asymmetric behavior of the Gompertz (GZ) distribution, which extends the positive to a whole dominion by an extra parameter. The SRG distribution also permits a better fit than its well-known classical competitors, namely the skew-normal and epsilon-skew-normal distributions, for data with a high presence of skewness. In this paper, we study information quantifiers such as Shannon and Rényi entropies, and Kullback–Leibler divergence in terms of exact expressions of GZ information measures. We find the asymptotic test useful to compare two SRG-distributed samples. Finally, as a real-world data example, we apply these results to South Pacific sea surface temperature records.

Keywords: Skew-Reflected-Gompertz distribution; Gompertz distribution; entropy; Kullback–Leibler divergence; sea surface temperature

---

1. Introduction

The Skew-Reflected-Gompertz (SRG) distribution was recently introduced by [1] and corresponds to an extension of the Gompertz distribution [2], named after Benjamin Gompertz (1779–1865). It extends the positive dominion $\mathbb{R}_+$ to the whole of $\mathbb{R}$ by an extra parameter, ε, $-1 < \varepsilon < 1$, and produces two-piece asymmetric behavior of Gompertz (GZ) density. The SRG distribution has as particular cases the Reflected-GZ and GZ distributions, when $\varepsilon \to 1$ and $\varepsilon \to -1$, respectively. The SRG distribution family can also represent a suitable competitor against the skew-normal (SN, [3]) and epsilon-skew-normal (ESN, [4]) distributions as a way to fit asymmetrical datasets. Indeed, refs. [5,6] dealt with the frequentist and Bayesian inferences of ESN distribution. Contributions by [1] provided probability density function (pdf), cumulative distribution function (cdf), quantile function, moment-generating function (MGF), stochastic representation, the Expectation-Maximization (EM) algorithm for SRG parameter estimates and the Fisher information matrix (FIM).

Moreover, several recent investigations confirmed the usefulness of entropic quantifiers in the study of asymmetric distributions [3,7,8] and their applications to topics such as thermal wake [9], marine fish biology [3,8], sea surface temperature (SST), relative humidity measured in the Atlantic Ocean [10], and more. We build on the study of [3], which developed hypothesis testing for normality, i.e., if the shape parameter is close to zero. They considered the Kullback–Leibler (KL) divergence in terms of moments and cumulants of the modified SN distribution. Posteriorly, we consider a real-world data set of the anchovy condition factor for testing the shape parameter to decide if a food deficit produced by environmental conditions such as El Niño exists [11].

This work arose from a motivation to tackle the problem of determining the adequate pdf of SST [9,10]. Indeed, probabilistic modelling of SST is key for accurate predictions [9]. Therefore, we propose that the SRG model based on two-piece distributions could be more suitable for interpreting annual bimodal and asymmetric SST data. We also considered the existent results of Shannon and Rényi entropies, and KL divergence for GZ distributions for developed entropic quantifiers for SRG distributions. Posteriorly, we considered SST along the South Pacific and Chilean coasts from 2012 to 2014 to illustrate our results. Specifically, we introduced hypothesis testing developed by [12] for the SRG distribution, which is useful to compare two data sets with bimodal and asymmetric behavior such as SST.

2. The Skew-Reflected-Gompertz Distribution

The Gompertz (GZ, [2]) distribution is a continuous probability distribution with the following pdf

$$f(x|\sigma,\eta) = \frac{\eta}{\sigma}e^{\frac{x}{\sigma}}e^{-\eta(e^{\frac{x}{\sigma}}-1)}, \quad x \geq 0, \tag{1}$$

where $\sigma > 0$ and $\eta > 0$ are the scale and shape parameters, respectively, and are denoted by $X \sim GZ(\sigma,\eta)$. The mean and variance of X are

$$\begin{aligned} E(X) &= \sigma e^{\eta} Ei(-\eta), \\ Var(X) &= \sigma^2 e^{\eta}\tau, \end{aligned} \tag{2}$$

respectively; where $Ei(z) = \int_{-z}^{\infty} \frac{e^{-u}}{u}du$, $\tau = -2\eta F(-\eta) + \gamma^2 + \frac{\pi^2}{6} + 2\gamma\log\eta + (\log\eta)^2 - e^{\eta}[Ei(-\eta)]^2$, $\gamma = 0.5772156649$ is the Euler constant and

$$F(z) = \sum_{k=0}^{+\infty} \frac{z^k}{k!(k+1)^3}.$$

The SRG distribution is an extension of the GZ proposed by [1]. If Y follows, the SRG distribution is denoted by $Y \sim \text{SRG}(\mu,\sigma,\eta,\varepsilon)$ and has pdf

$$g(y|\mu,\sigma,\eta,\varepsilon) = \begin{cases} \frac{1}{2}f\left(\frac{\mu-y}{1+\varepsilon}\middle|\sigma,\eta\right), & y \leq \mu, \\ \frac{1}{2}f\left(\frac{y-\mu}{1-\varepsilon}\middle|\sigma,\eta\right), & y > \mu, \end{cases} \tag{3}$$

where $\mu \in \mathbb{R}$ is the location parameter and $\varepsilon \in (-1,1)$ is the slant parameter. Note that SRG is the GZ distribution when $\mu = 0$ and $\varepsilon \to -1$, GZ distribution with negative support when $\varepsilon \to 1$, and Reflected-GZ distribution when $\varepsilon = 0$. Also, the Reflected-GZ distribution corresponds to a particular case of a more general class of two-piece asymmetric distributions proposed by [13,14]. The mean, variance and MGF of Y are

$$\begin{aligned} E(Y) &= \mu - 2\varepsilon\sigma e^{\eta} Ei(-\eta), \\ Var(Y) &= \sigma^2\{\tau e^{\eta} + 2(1-\varepsilon^2)e^{2\eta}[Ei(-\eta)]^2\}, \\ M_Y(t) &= \frac{1}{2}\eta e^{\eta+\mu t}[(1-\varepsilon)\digamma_{-\sigma t}(\eta) + (1+\varepsilon)\digamma_{\sigma t}(\eta)], \end{aligned} \tag{4}$$

respectively; where $\digamma_s(z) = \int_1^{\infty} v^{s+1}e^{-vz}dv$. Jafari et al. [15] provide the MGF of X using expansion series. However, (4) is considered a clearer expression that depends only on integral $\digamma_s(z)$. See Section 4.1 for some details of the MLE EM-based algorithm related to SRG parameters.

According to [1], the SRG distribution can be re-parametrized in terms of GZ and Reflected-GZ distributions as

$$g(y|\mu,\sigma_+,\sigma_-,\eta) = p_1 f(\mu-y|\sigma_+,\eta)I_{(-\infty,\mu]}(y) + p_2 f(y-\mu|\sigma_-,\eta)I_{(\mu,+\infty)}(y), \tag{5}$$

where $\sigma_{\pm} = \sigma(1 \pm \varepsilon)$, $p_1 + p_2 = 1$, and $p_1 = \sigma_+/(\sigma_+ + \sigma_-) = (1+\varepsilon)/2$. Let $\mathbf{Y} = (Y_1, \ldots, Y_n)^\top$ be an i.i.d sample from the SRG distribution with parameters $(\mu, \sigma_{\pm}, \eta)$ and latent vectors $\mathbf{Z} = (\mathbf{Z}_1, \ldots, \mathbf{Z}_n)$, thus (5) can be equivalently represented as $(-1)^j(Y_i - \mu)|Z_{ij} = 1 \sim GZ(\sigma_{\pm}, \eta)$, $i = 1, \ldots, n$, $j = 1, 2$, where $\mathbf{Z}_i = (Z_{i1}, Z_{i2})^\top \sim Mult(1, p_1, p_2)$ is a multinomial vector, $P(Z_{i1} = z_{i1}, Z_{i2} = z_{i2}) = p_1^{z_{i1}} p_2^{z_{i2}}$, $z_{ij} = \{0, 1\}$, and $z_{i1} + z_{i2} = 1$. Given that $P(Z_{i1} = 1) = P(Z_{i1} = 1, Z_{ik} = 0; \forall j \neq k)$, the complete log-likelihood function is

$$\begin{aligned} \ell(\mu, \sigma_+, \sigma_-, \eta|\mathbf{Y}, \mathbf{Z}) &= -n\log(2\sigma) + n(\eta + \log\eta) \\ &\quad + \sum_{i=1}^{n}\left[z_{i1}\left(\frac{\mu - y_i}{\sigma_+} - \eta e^{\frac{\mu - y_i}{\sigma_+}}\right) + z_{i2}\left(\frac{y_i - \mu}{\sigma_-} - \eta e^{\frac{y_i - \mu}{\sigma_-}}\right)\right]. \end{aligned} \tag{6}$$

Conditional expectations of latent variables $\mathbf{Z}_i$ are given by

$$\widehat{z}_{i1} = E[Z_{i1}|\widehat{\mu}, \widehat{\sigma}_+, \widehat{\sigma}_-, y_i] = \widehat{p}_1 \frac{f(\widehat{\mu} - y_i|\widehat{\sigma}_+, \widehat{\eta})}{g(y_i|\widehat{\mu}, \widehat{\sigma}_+, \widehat{\sigma}_-, \widehat{\eta})} I_{(-\infty, \widehat{\mu}]}(y_i), \tag{7}$$

$$\widehat{z}_{i2} = 1 - \widehat{z}_{i1}, \quad i = 1, \ldots, n. \tag{8}$$

The E- and M-steps on the $(k+1)$th iteration of the EM algorithm are
E-step. From (6)–(8), we have

$$\begin{aligned} Q(\mu, \sigma_+, \sigma_-, \eta|\mu^{(k)}, \sigma_+^{(k)}, \sigma_-^{(k)}, \eta^{(k)}) &= E[\ell(\mu, \sigma_+, \sigma_-, \eta|\mathbf{Y}, \mathbf{Z})|\mu^{(k)}, \sigma_+^{(k)}, \sigma_-^{(k)}, \eta^{(k)}] \\ &= -n\log(2\sigma) + n(\eta + \log\eta) \\ &\quad + \sum_{i=1}^{n}\left[\widehat{z}_{i1}^{(k)}\left(\frac{\mu - y_i}{\sigma_+} - \eta e^{\frac{\mu - y_i}{\sigma_+}}\right) + \widehat{z}_{i2}^{(k)}\left(\frac{y_i - \mu}{\sigma_-} - \eta e^{\frac{y_i - \mu}{\sigma_-}}\right)\right]. \end{aligned}$$

and
M-step. Update $\sigma_{\pm}$, by solving the following equation

$$\sum_{i=1}^{n} \widehat{z}_{ij}^{(k)}\left(\eta^{(k)}\frac{|y_i - \mu^{(k)}|}{\sigma_{\pm}^2} e^{\frac{|y_i - \mu^{(k)}|}{\sigma_{\pm}}} - \frac{|y_i - \mu^{(k)}|}{\sigma_{\pm}^2}\right) = \frac{n}{2\sigma}.$$

Update μ by solving the following equation

$$\widehat{\mu}^{(k+1)} = \mathrm{argmax}_{\mu} \sum_{i=1}^{n}\left\{\widehat{z}_{i1}^{(k)}\left(\frac{\mu - y_i}{\widehat{\sigma}_+^{(k+1)}} - \eta e^{\frac{\mu - y_i}{\widehat{\sigma}_+^{(k+1)}}}\right) + \widehat{z}_{i2}^{(k)}\left(\frac{\mu - y_i}{\widehat{\sigma}_-^{(k+1)}} - \eta e^{\frac{\mu - y_i}{\widehat{\sigma}_-^{(k+1)}}}\right)\right\}.$$

Update η by

$$\widehat{\eta} = n\left(\sum_{i=1}^{n}\left\{\widehat{z}_{i1}^{(k)} e^{\frac{\mu - y_i}{\widehat{\sigma}_+^{(k+1)}}} + \widehat{z}_{i2}^{(k)} e^{\frac{\mu - y_i}{\widehat{\sigma}_-^{(k+1)}}}\right\}\right)^{-1}.$$

The EM-algorithm must be iterated until the sufficient convergence rule is satisfied:

$$\|(\widehat{\mu}^{(k+1)}, \widehat{\sigma}_+^{(k+1)}, \widehat{\sigma}_-^{(k+1)}, \widehat{\eta}^{(k+1)}) - (\widehat{\mu}^{(k)}, \widehat{\sigma}_+^{(k)}, \widehat{\sigma}_-^{(k)}, \widehat{\eta}^{(k)})\| < \tau,$$

for a tolerance τ close to zero. The FIM for standard deviations of MLEs $(\widehat{\mu}, \widehat{\sigma}, \widehat{\eta}, \widehat{\varepsilon})$ and additional details of the EM-algorithm are described in [1].

3. Entropic Quantifiers

In the next section, we present the main results of entropic quantifiers for SRG distribution.

3.1. Shannon Entropy

The Shannon entropy (SE), introduced by [16] in the context of univariate continuous distributions, quantifies the information contained in a random variable X with pdf $f(x)$ through the expression

$$H(X) = -\int_{-\infty}^{+\infty} f(x) \log f(x) dx. \tag{9}$$

The SE concept is attributed to the uncertainty of the information presented in X [17]. Propositions 1 and 2 present the SE for GZ and SRG distributions, respectively.

Proposition 1. *[15]. The SE of $X \sim GZ(\sigma, \eta)$ is*

$$H(X) = \log\left\{\frac{B(1,1)}{\eta}\right\} - \sigma\eta - \frac{E(X)}{\sigma} + \sigma\eta M_X(\sigma^{-1}),$$

where $B(\cdot,\cdot)$ is the usual Beta function and $E(X)$ is given in (2).

Substituting $\mu = 0$ and $\varepsilon = -1$ into (4) (i.e., reducing SRG to its special case GZ), we obtain $M_X(\sigma^{-1}) = \eta e^{\eta} F_{-1}(\eta) = 1$. Therefore, $H(X)$ in Proposition 1 is reduced to

$$H(X) = -\log \eta - e^{\eta} E_i(-\eta), \tag{10}$$

i.e., the SE of the GZ random variable only depends on shape parameter η.

Proposition 2. *The SE of $Y \sim SRG(\mu, \sigma, \eta, \varepsilon)$ is*

$$H(Y) = \frac{1+\varepsilon}{2}\left\{H(X_{+\varepsilon}) - \log\left(\frac{1+\varepsilon}{2}\right)\right\} + \frac{1-\varepsilon}{2}\left\{H(X_{-\varepsilon}) - \log\left(\frac{1-\varepsilon}{2}\right)\right\},$$

where $X_{\pm\varepsilon} \sim GZ(\sigma(1 \pm \varepsilon), \eta)$ and $H(X_{\pm\varepsilon})$ are obtained using Proposition 1.

Proof. From (3) and (9), we obtained

$$\begin{aligned}
H(Y) &= -\int_{-\infty}^{+\infty} g(y|\mu,\sigma,\eta,\varepsilon) \log g(y|\mu,\sigma,\eta,\varepsilon) dy \\
&= -\frac{1}{2}\int_0^{+\infty} f\left(\frac{x}{1+\varepsilon}\middle|\sigma,\eta\right) \log\left\{\frac{1}{2} f\left(\frac{x}{1+\varepsilon}\middle|\sigma,\eta\right)\right\} dx \\
&\quad -\frac{1}{2}\int_0^{+\infty} f\left(\frac{x}{1-\varepsilon}\middle|\sigma,\eta\right) \log\left\{\frac{1}{2} f\left(\frac{x}{1-\varepsilon}\middle|\sigma,\eta\right)\right\} dx \\
&= -\frac{1}{2}\int_0^{+\infty} (1+\varepsilon) f(x|\sigma(1+\varepsilon),\eta) \log\left\{\frac{1+\varepsilon}{2} f(x|\sigma(1+\varepsilon),\eta)\right\} dx \\
&\quad -\frac{1}{2}\int_0^{+\infty} (1-\varepsilon) f(x|\sigma(1-\varepsilon),\eta) \log\left\{\frac{1-\varepsilon}{2} f(x|\sigma(1-\varepsilon),\eta)\right\} dx,
\end{aligned}$$

which concludes the proof. □

From (10), given that $H(X_{\pm\varepsilon})$ only depends on shape parameter η, we obtain $H(X_{\pm\varepsilon}) = H(X)$, and $H(Y)$ only depends on η and ε parameters. Therefore,

$$H(Y) = -\log \eta - e^{\eta} E_i(-\eta) - \frac{1+\varepsilon}{2} \log\left(\frac{1+\varepsilon}{2}\right) - \frac{1-\varepsilon}{2} \log\left(\frac{1-\varepsilon}{2}\right). \tag{11}$$

Figure 1 illustrates SE behavior for random variable Y. We observed that SE increases when η decreases. For each η, SE is maximized and minimized at $\varepsilon = 0$ (Reflected-GZ) and $\varepsilon \to -1$

(Truncated-GZ and GZ), respectively. More details appear in [3,8] for the SE expressions of other asymmetric distributions.

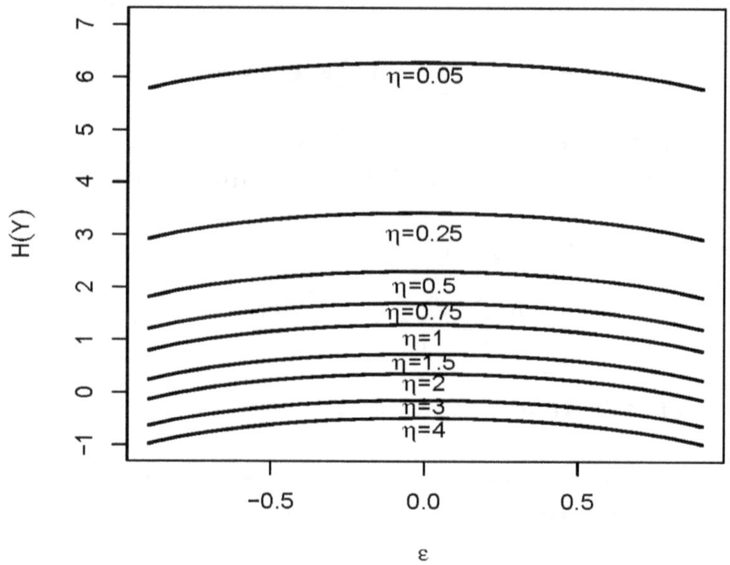

Figure 1. Shannon entropy of Skew-Reflected-Gompertz (SRG) distributions for $\varepsilon \in (-1, 1)$ and several values of η.

3.2. *Rényi Entropy*

The αth-order Rényi entropy (RE), introduced by [18] in the context of univariate continuous distributions, extends the concept of SE information contained in a random variable X with pdf $f(x)$ through a level α, $\alpha \in \mathbb{N}$, $\alpha > 0$, and the expression

$$R_\alpha(X) = \frac{1}{1-\alpha} \log \int_{-\infty}^{+\infty} [f(x)]^\alpha dx. \tag{12}$$

RE information can be negative and is ordered with respect to α, i.e., $R_{\alpha_1}(X) \geq R_{\alpha_2}(X)$ for any $\alpha_1 < \alpha_2$ (see, e.g., [7] and other properties of RE). From (12), the SE is obtained by the limit of $H(X) = \lim_{\alpha \to 1} R_\alpha(X)$ by applying l'Hôpital's rule to $R_\alpha(X)$ with respect to α (see e.g., [7]). The RE of the GZ and SRG distributions is presented in Propositions 3 and 4, respectively.

Proposition 3. *[15,19]. The RE of $X \sim GZ(\sigma, \eta)$ with $\alpha > 1$, $\alpha \in \mathbb{N}$, is*

$$R_\alpha(X) = -\frac{\log \alpha}{1-\alpha} + \log \frac{\eta}{\sigma} + \frac{1}{1-\alpha} \log \left\{ \sum_{j=0}^{\alpha-1} \binom{\alpha-1}{j} \frac{\Gamma(j+1)}{(\alpha\eta)^j} \right\},$$

where $\Gamma(u) = \int_0^\infty t^{u-1} e^{-t} dt$ is the gamma function.

Proposition 4. *The RE of $Y \sim SRG(\eta, \varepsilon)$ with $\alpha > 1$, $\alpha \in \mathbb{N}$, is*

$$R_\alpha(Y) = \frac{1}{1-\alpha} \log \left\{ \left(\frac{1+\varepsilon}{2} \right)^\alpha e^{(1-\alpha) R_\alpha(X_{+\varepsilon})} + \left(\frac{1-\varepsilon}{2} \right)^\alpha e^{(1-\alpha) R_\alpha(X_{-\varepsilon})} \right\},$$

where $X_{\pm\varepsilon} \sim GZ(\sigma(1 \pm \varepsilon), \eta)$ and $R_\alpha(X_{\pm\varepsilon})$ are obtained using Proposition 3.

Proof. From (3) and (12), we obtained

$$\begin{aligned} R_\alpha(Y) &= \frac{1}{1-\alpha}\log\int_{-\infty}^{+\infty}[g(y|\mu,\sigma,\eta,\varepsilon)]^\alpha dy, \\ &= \frac{1}{1-\alpha}\log\left\{\int_0^{+\infty}\left[\frac{1}{2}f\left(\frac{x}{1+\varepsilon}\Big|\sigma,\eta\right)\right]^\alpha dx+\int_0^{+\infty}\left[\frac{1}{2}f\left(\frac{x}{1-\varepsilon}\Big|\sigma,\eta\right)\right]^\alpha dx\right\}, \\ &= \frac{1}{1-\alpha}\log\left\{\left(\frac{1+\varepsilon}{2}\right)^\alpha\int_0^{+\infty}[f(x|\sigma(1+\varepsilon),\eta)]^\alpha dx+\left(\frac{1-\varepsilon}{2}\right)^\alpha\int_0^{+\infty}[f(x|\sigma(1-\varepsilon),\eta)]^\alpha dx\right\}, \end{aligned}$$

which concludes the proof. □

Figure 2a illustrates the behavior of RE for random variable Y when $\alpha = 2$ (quadratic RE). As in the SE case, we also observed that RE increases when η decreases and reaches maximum and minimum at $\varepsilon = 0$ (Reflected-GZ) and $\varepsilon \to -1$ (Truncated-GZ and GZ), respectively. When $\alpha = 5$ (or $\alpha > 2$) (see Figure 2b), RE decays faster than in the quadratic RE case as $\varepsilon \to -1$. More details appear in [7] for the RE expressions of other asymmetric distributions.

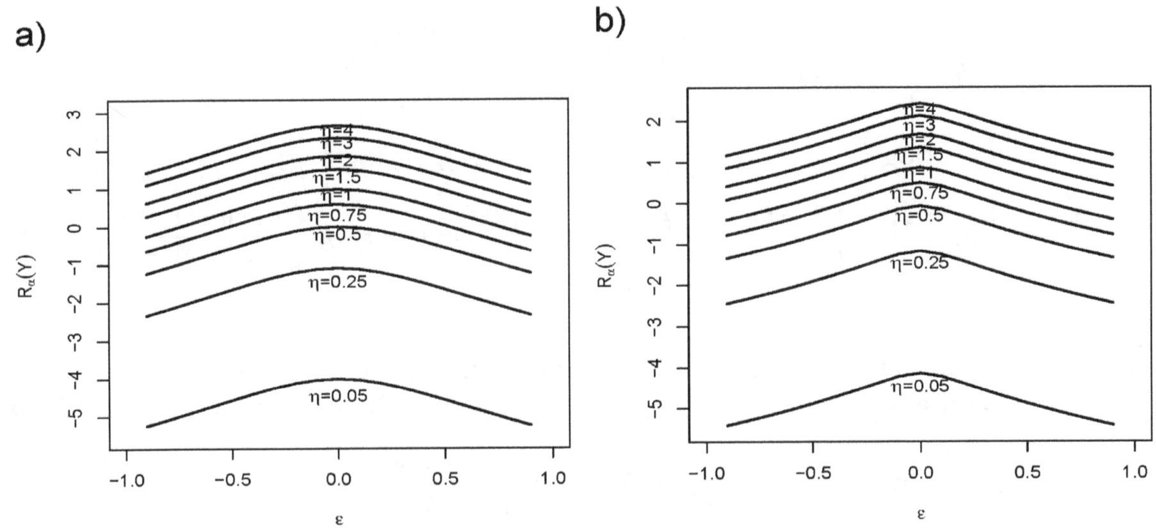

Figure 2. Rényi entropy of SRG distributions for $\sigma = 1$, $-1 < \varepsilon < 1$, several values of η and (**a**) $\alpha = 2$ and (**b**) $\alpha = 5$ values.

3.3. Kullback–Leibler Divergence

The Kullback–Leibler (KL) divergence introduced by [20] in the context of univariate continuous distributions, extends the concept of SE between two random variables X_1 and X_2 with pdfs $f_1(x_1)$ and $f_2(x_2)$, respectively, through the expression

$$K(X_1,X_2)=\int_{-\infty}^{+\infty}f_1(x)\log\left\{\frac{f_1(x)}{f_2(x)}\right\}dx. \tag{13}$$

The KL divergence measures the disparity between the pdfs of X_1 and X_2, and is non-negative, non-symmetric and zero only if $X_1 = X_2$ in distribution. Also, the KL divergence does not satisfy the triangular inequality (see, e.g., [8,17] for other properties of KL and other divergences). The KL divergence for two GZ and two SRG distributions are presented in Propositions 5 and 6.

Proposition 5. *[21]. The KL divergence between $X_1 \sim GZ(\sigma_1,\eta_1)$ and $X_2 \sim GZ(\sigma_2,\eta_2)$ is*

$$K(X_1,X_2)=\log\left\{\frac{e^{\eta_1}\sigma_2\eta_1}{e^{\eta_2}\sigma_1\eta_2}\right\}+e^{\eta_1}\left[\left(\frac{\sigma_1}{\sigma_2}-1\right)E_i(-\eta_1)+\frac{\eta_2}{\eta_1^{\sigma_1/\sigma_2}}\Gamma\left(\frac{\sigma_1}{\sigma_2}-1,\eta_1\right)\right]-(\eta_1+1),$$

where $\Gamma(u,v) = \int_v^{\infty} t^{u-1}e^{-t}dt$ *is the upper incomplete gamma function.*

Proposition 6. *The KL divergence between* $Y_1 \sim SRG(0,\sigma_1,\eta_1,\varepsilon_1)$ *and* $Y_2 \sim SRG(0,\sigma_2,\eta_2,\varepsilon_2)$ *is*

$$K(Y_1,Y_2) = \frac{1+\varepsilon_1}{2}\left[\log\left\{\frac{1+\varepsilon_1}{1+\varepsilon_2}\right\} + K(X_{+\varepsilon_1},X_{+\varepsilon_2})\right] + \frac{1-\varepsilon_1}{2}\left[\log\left\{\frac{1-\varepsilon_1}{1-\varepsilon_2}\right\} + K(X_{-\varepsilon_1},X_{-\varepsilon_2})\right],$$

where $X_{\pm\varepsilon_i} \sim GZ(\sigma_i(1\pm\varepsilon_i),\eta_i)$, $i = 1,2$, *and* $K(X_{\pm\varepsilon_1},X_{\pm\varepsilon_2})$ *are obtained using Proposition 5.*

Proof. From (3) and (13), we obtained

$$\begin{aligned}
K(Y_1,Y_2) &= \int_{-\infty}^{+\infty} g(x|0,\sigma_1,\eta_1,\varepsilon_1)\log\left\{\frac{g(x|0,\sigma_1,\eta_1,\varepsilon_1)}{g(x|0,\sigma_2,\eta_2,\varepsilon_2)}\right\}dx,\\
&= \frac{1}{2}\int_0^{+\infty} f\left(\frac{x}{1+\varepsilon_1}\Big|\sigma_1,\eta_1\right)\log\left\{\frac{f\left(\frac{x}{1+\varepsilon_1}\Big|\sigma_1,\eta_1\right)}{f\left(\frac{x}{1+\varepsilon_2}\Big|\sigma_2,\eta_2\right)}\right\}dx\\
&\quad+\frac{1}{2}\int_0^{+\infty} f\left(\frac{x}{1-\varepsilon_1}\Big|\sigma_1,\eta_1\right)\log\left\{\frac{f\left(\frac{x}{1-\varepsilon_1}\Big|\sigma_1,\eta_1\right)}{f\left(\frac{x}{1-\varepsilon_2}\Big|\sigma_2,\eta_2\right)}\right\}dx,\\
&= \frac{1+\varepsilon_1}{2}\left[\log\left\{\frac{1+\varepsilon_1}{1+\varepsilon_2}\right\} + \int_0^{+\infty} f\left(x|\sigma_1(1+\varepsilon_1),\eta_1\right)\log\left\{\frac{f\left(x|\sigma_1(1+\varepsilon_1),\eta_1\right)}{f\left(x|\sigma_2(1+\varepsilon_2),\eta_2\right)}\right\}dx\right]\\
&\quad+\frac{1-\varepsilon_1}{2}\left[\log\left\{\frac{1-\varepsilon_1}{1-\varepsilon_2}\right\} + \int_0^{+\infty} f\left(x|\sigma_1(1-\varepsilon_1),\eta_1\right)\log\left\{\frac{f\left(x|\sigma_1(1-\varepsilon_1),\eta_1\right)}{f\left(x|\sigma_2(1-\varepsilon_2),\eta_2\right)}\right\}dx\right],
\end{aligned}$$

which concludes the proof. □

More details appear in [3,8] for the KL divergence expressions of other asymmetric distributions. Using Proposition 6, the asymptotic KL divergence between $Y \sim \text{SRG}(0,\sigma,\eta,\varepsilon)$ and $X \sim \text{GZ}(\sigma,\eta)$ is

$$K(Y,X) \approx \frac{1+\varepsilon}{2}\left[\lim_{\varepsilon_2\to-1}\log\left(\frac{1+\varepsilon}{1+\varepsilon_2}\right) + K(X_{+\varepsilon},X)\right] + \frac{1-\varepsilon}{2}\left[\log\left(\frac{1-\varepsilon}{2}\right) + K(X_{-\varepsilon},X)\right],$$

as $\varepsilon_2 \to -1$. However, we see that $\log\left(\frac{1+\varepsilon}{1+\varepsilon_2}\right) = +\infty$ as $\varepsilon_2 \to -1$ and $K(Y,X)$ is not finite. However, from Proposition 6 the asymptotic KL divergence between Y_1 and Y_2 is

$$K(Y_1,Y_2) \approx K(X,Y) = \log\left(\frac{2}{1-\varepsilon}\right) + K(X,X_{-\varepsilon}), \tag{14}$$

as $\varepsilon_1 \to -1$, where $X_{-\varepsilon} \sim \text{GZ}(\sigma(1-\varepsilon),\eta)$. Therefore, while $K(Y,X)$ is not finite, $K(X,Y)$ is finite and can be used to study the disparity of ε from -1. Thus, hypothesis testing for $H_0 : \varepsilon = -1$ can be addressed. Besides, we further study hypothesis testing for scale and shape parameters between two SRG distributions in Section 3.4. From (14), we also took that $K(Y_1,Y_2) \approx K(X,X_1)$ as $\varepsilon \to -1$, with $X_1 \sim \text{GZ}(2\sigma,\eta)$.

Figure 3 illustrates the KL divergence between two SRG distributions. We observed that for the critical points of $(\varepsilon_1,\varepsilon_2) \to \{(-1,1);\ (1,-1)\}$, the KL divergence reaches the highest values and is close to zero in the other values [panels (a) and (b)]. For large η's [panel (c)], the KL divergence is zero for a concentrated region of the dominion where $\varepsilon_1 = \varepsilon_2$.

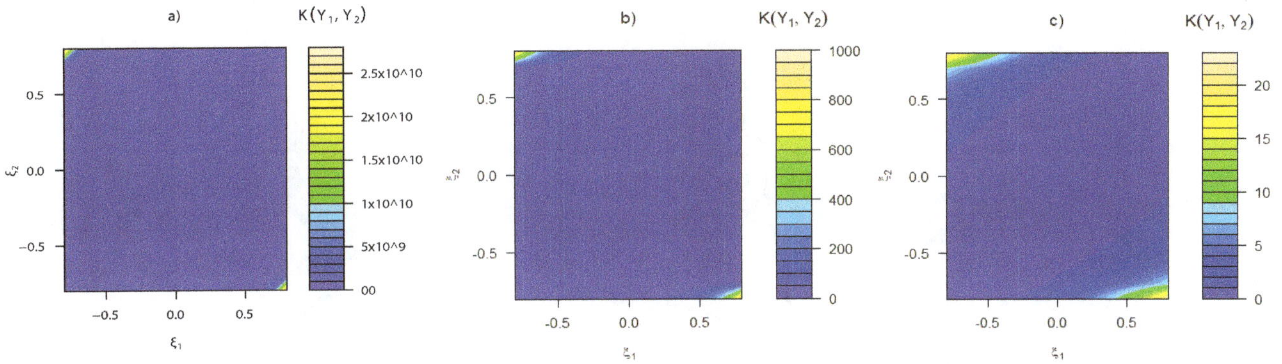

Figure 3. Plots of Kullback–Leibler (KL) divergence between $Y_1 \sim \text{SRG}(0, \sigma_1, \eta_1, \varepsilon_1)$ and $Y_2 \sim \text{SRG}(0, \sigma_2, \eta_2, \varepsilon_2)$ for values $\sigma_1 = \sigma_2 = 1$ and **(a)** $\eta_1 = \eta_2 = 0.25$; **(b)** $\eta_1 = \eta_2 = 3$; and **(c)** $\eta_1 = \eta_2 = 10$.

All information quantifiers and the EM algorithm for SRG distribution were implemented in [22].

3.4. Asymptotic Test

Consider two independent samples of sizes n_1 and n_2 from Y_1 and Y_2, respectively; where $\boldsymbol{\theta}, \boldsymbol{\theta}' \in \Theta \subset \mathbb{R}^p$, and X_1 and X_2 have pdfs $g(y; \boldsymbol{\theta}_1)$ and $g(y; \boldsymbol{\theta}_2)$, respectively; with $\boldsymbol{\theta}_i = (\sigma_i, \eta_i, \varepsilon_i)$, $i = 1, 2$. Suppose partition $\boldsymbol{\theta}_i = (\boldsymbol{\theta}_{i1}, \boldsymbol{\theta}_{i2})$, and assume $\boldsymbol{\theta}_{21} = \boldsymbol{\theta}_{11} \in \Theta_1 \subset \mathbb{R}^r$, so that $\boldsymbol{\theta}_{i2} \in \Theta \cap \Theta_1^c \subset \mathbb{R}^{p-r}$. Let $\widehat{\boldsymbol{\theta}}_i = (\widehat{\boldsymbol{\theta}}_{11}, \widehat{\boldsymbol{\theta}}_{i2})$ be the MLE of $\boldsymbol{\theta}_i = (\boldsymbol{\theta}_{11}, \boldsymbol{\theta}_{i2})$ for $i = 1, 2$, which corresponds to the MLE of the full model parameters $(\boldsymbol{\theta}_1, \boldsymbol{\theta}_2)$ under the null hypothesis $H_0 : \boldsymbol{\theta}_{21} = \boldsymbol{\theta}_{11}$. Thus, part b) of Corollary 1 in [12] establishes that if the null hypothesis $H_0 : \boldsymbol{\theta}_{22} = \boldsymbol{\theta}_{12}$ holds and $\frac{n_1}{n_1+n_2} \underset{n_1, n_2 \to \infty}{\longrightarrow} \lambda$, with $0 < \lambda < 1$, then

$$K_0 = \frac{2n_1 n_2}{n_1 + n_2} K(\widehat{\boldsymbol{\theta}}_1, \widehat{\boldsymbol{\theta}}_2) \underset{n_1, n_2 \to \infty}{\overset{d}{\longrightarrow}} \chi^2_{p-r}, \tag{15}$$

where $r = 3$ is the number of parameters of the SRG distribution (location parameter is not considered for KL divergence). Thus, a test of level α for the above homogeneity null hypothesis consists of rejecting H_0 if $K_0 > \chi^2_{p-r,1-\alpha}$, where $\chi^2_{p-r,\alpha}$ is the αth percentile of the χ^2_{p-r}-distribution.

As [3] stated, the proposed asymptotic test is only valid for regular conditions of the SRG distribution, in particular for a non-singular FIM. Therefore, given that the SRG distributions' FIM is singular at $\varepsilon \to \pm 1$ [1], the SRG model does not serve for testing the null hypothesis using (15) when ε is close to -1 or 1.

4. Application

4.1. Sea Surface Temperature Data

The spatial information and SST data analyzed in this study were recorded by a scientific observer (whose labor concerns biological sampling of fishes, incidental captures of birds, turtles and marine mammals. Biological sampling was complemented with information such as time, longline and hook features, number of buoys, baits, etc.) (SO) in the Chilean longline fleet (industrial and artisanal), which was oriented to capture swordfish (*Xiphias gladius*, [23]) from 2012 to 2014 (obtaining a sampling of 83% in 2012, 55% in 2013, 90% in 2014, and 75% in 2012–2014). The covered area of the study was at 21°31′–36°39′ LS and 71°08′–85°52′ LW (see Figure 4).

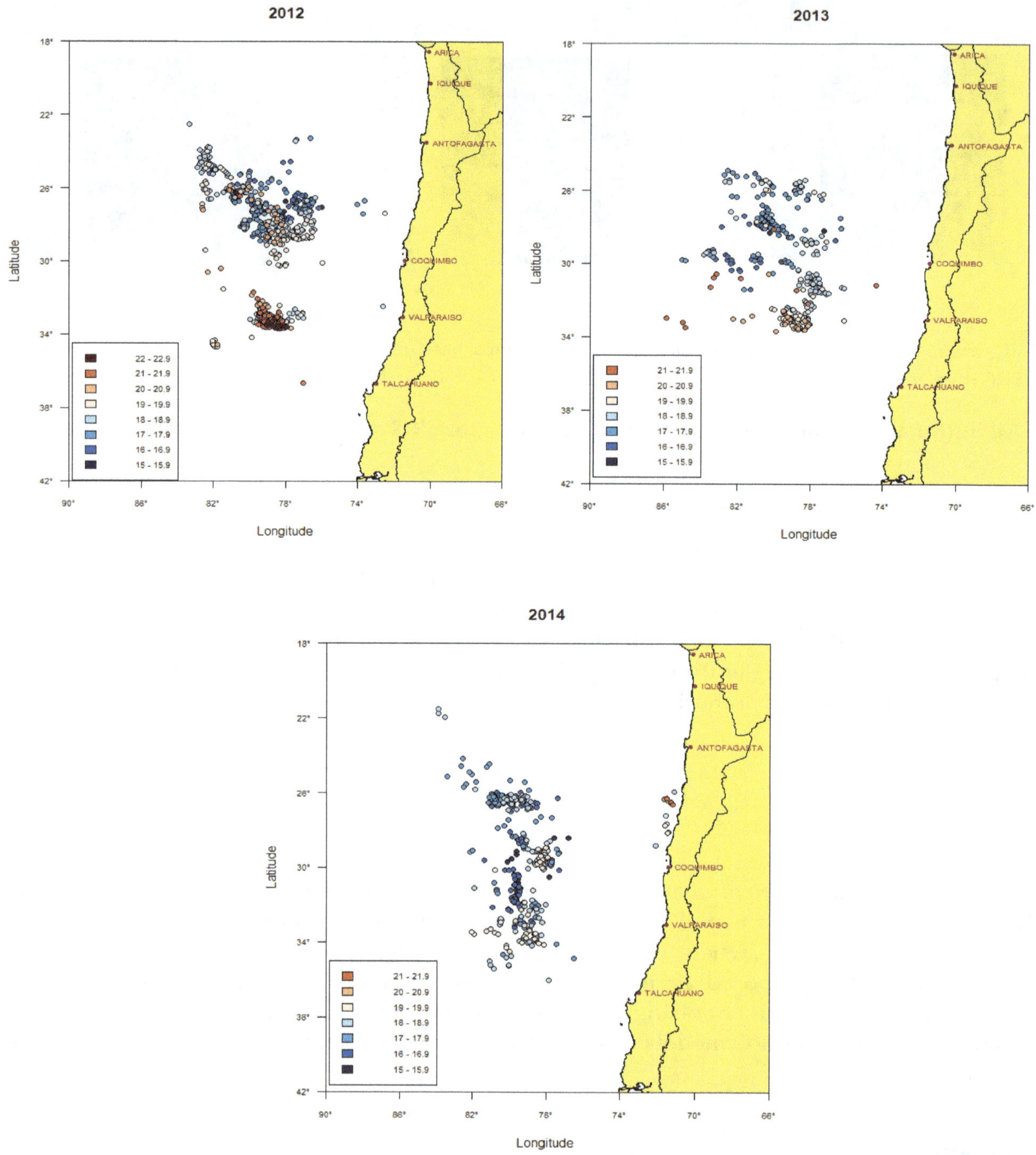

Figure 4. Spatial distribution of Sea Surface Temperature (SST) observations by year (21°31′–36°39′ LS, 71°08′–85°52′ LW).

SST records in swordfish captures are crucial for distributional analysis and fish abundance. Specifically, variations in SST are physical factors that control productivity, growth and migration of species [24]. In addition, SST is strongly correlated with atmospheric pressure at sea level and thus climatic time scales. Therefore, changes in SST overlap with ecosystem changes [25]. However, SST influence on ecosystems is not clear because other physical processes such as superficial warming, horizontal advection of currents, upwelling, etc. [11], modify SST. Therefore, SST anomalies could be symptomatic rather than causal.

4.1.1. SRG Parameter Estimates

Considering the smallest Akaike (AIC) and Schwarz (BIC) information criteria, we observed in Table 1 that SRG performs better than the SN and ESN models (see Appendices A and B, respectively). In addition, Table 1 shows the estimated parameters (based on the EM algorithm presented in Section 2) for SST datasets by year assuming SRG distribution. In 2012, a negative ε estimate corresponds to asymmetry to the right, and in 2013 and 2014 negative ε and η close to zero produce a two-piece distribution to fit "cold" and "warm" temperatures (Figure 5).

Table 1. Parameter estimates and their respective standard deviations (SD) for SST by year based on SRG, epsilon-skew-normal (ESN) and skew-normal (SN) models. For each model, log-likelihood function $\ell(\theta), \theta = (\mu, \sigma, \eta, \varepsilon)$, Akaike's (AIC) and Bayesian (BIC) information criteria, and goodness-of-fit tests (Kolmogorov–Smirnov (K–S), Anderson–Darling (A–D), and Cramer–von Mises, (C–V)) are also reported with respective p-values in parentheses.

Year	Model	Param.	Estim.	(S.D)	$\ell(\theta)$	AIC	BIC	K–S	A–D	C–V
2012 ($n = 774$)	SRG	μ	17.992	0.103	−1401.896	2811.793	2830.399	0.044 (0.095)	2.014 (0.090)	0.214 (0.242)
		σ	2.590	0.067						
		η	1.444	0.027						
		ε	−0.207	0.075						
	ESN	θ	18.000	0.031	−1507.534	3021.069	3035.023	0.118 (<0.01)	26.417 (<0.01)	2.059 (<0.01)
		ϖ	1.657	0.015						
		ϵ	−0.418	0.069						
	SN	ξ	16.777	0.114	−1404.581	2815.161	2829.116	0.041 (0.143)	1.752 (0.126)	0.198 (0.271)
		ω	5.199	0.043						
		λ	2.527	0.311						
2013 ($n = 415$)	SRG	μ	17.935	0.061	−687.420	1382.839	1398.942	0.082 (0.010)	2.632 (0.042)	0.491 (0.041)
		σ	1.112	0.026						
		η	0.432	0.021						
		ε	−0.108	0.029						
	ESN	θ	17.600	0.046	−716.375	1438.750	1450.827	0.089 (<0.01)	7.721 (<0.01)	0.970 (0.002)
		ϖ	1.328	0.026						
		ϵ	−0.376	0.092						
	SN	ξ	16.598	0.200	−691.531	1389.063	1401.140	0.066 (0.054)	2.002 (0.092)	0.328 (0.113)
		ω	3.812	0.054						
		λ	2.421	0.617						
2014 ($n = 439$)	SRG	μ	17.454	0.048	−653.082	1314.164	1330.502	0.092 (<0.01)	2.848 (0.033)	0.533 (0.032)
		σ	0.896	0.020						
		η	0.375	0.020						
		ε	−0.106	0.025						
	ESN	θ	17.200	0.053	−703.748	1413.496	1425.750	0.109 (<0.01)	11.996 (<0.01)	1.529 (<0.01)
		ϖ	0.956	0.035						
		ϵ	−0.384	0.090						
	SN	ξ	16.146	0.098	−666.984	1339.968	1352.222	0.096 (<0.01)	4.055 (<0.01)	0.711 (0.011)
		ω	3.245	0.045						
		λ	3.434	0.618						

To evaluate the goodness-of-fit test, the Kolmogorov–Smirnov (K–S), Anderson–Darling (A–D), and Cramer–von Mises (C–V) tests were considered for all models, commonly used to analyze the goodness-of-fit test of a particular distribution see, e.g., [26]). Considering a 95% confidence level, SRG fits perform well for 2012 and 2013, and on a 90% confidence level, the SRG fit performs well for 2014.

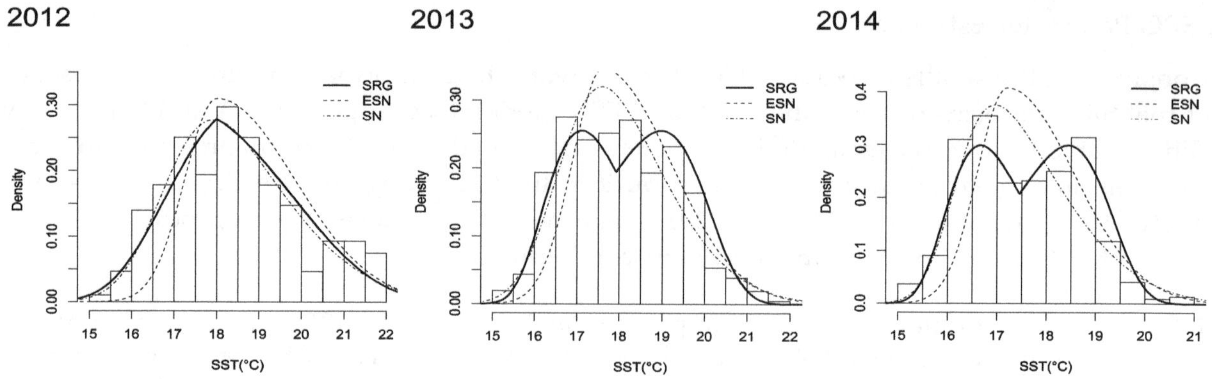

Figure 5. MLE fit of SRG, ESN and SN models for SST data by year.

4.1.2. Information Quantifiers and Asymptotic Test

Parameters estimated from the SRG model and presented in Table 1 are used to perform the quantifiers of Sections 3.1–3.3 for SST in each year and for the asymptotic test of Section 3.4 for comparing SST between two years. The results of these analyses are shown in Table 2. In Table 2, $K_0 = \widehat{K}(Y_1, Y_2)$ represents the KL divergence between the years Y_1 (column) and Y_2 (row).

The first quantifiers (SE and RE) illustrate that the highest information of SST is obtained by SE and increases with the increment of years. For all RE, the highest information of SST is obtained in 2012 and is negative for 2013 and 2014 and similar during that period. Differences in information between SE and RE are produced by the independency of SE with parameter σ, while RE depends on three parameters as in Proposition 4.

In addition, the asymptotic test presented in Table 2 is analogous for all the years in both groups. In fact, the null hypothesis $H_0 : \boldsymbol{\theta}_1 = \boldsymbol{\theta}_2$ is rejected at a 95% confidence level. This rejection is reinforced by high values of statistics K_0, produced by a high sample size of both groups (n_1 and n_2).

Table 2. SRG Shannon, $H(Y)$, and Rényi, $R_\alpha(Y)$, $\alpha = 2, 3, 4$, entropies for SST data. For each year, the KL divergence $K_0 = \widehat{K}(Y_1, Y_2)$, statistic and its respective p-values of Equation (15) are reported. All reported K_0 estimates considered the estimated parameters and sample size n in Table 1.

Year	Quantifier	2012	2013	2014
	$H(Y)$	0.765	0.781	2.754
	$R_2(Y)$	0.384	−0.362	−0.365
	$R_3(Y)$	0.252	−0.417	−0.418
	$R_4(Y)$	0.163	−0.457	−0.457
	K_0	-	0.266	0.911
2012	Statistic	-	143.740	520.41
	p-value	-	<0.01	<0.01
	K_0	0.080	-	0.071
2013	Statistic	43.192	-	30.233
	p-value	<0.01	-	<0.01
	K_0	0.143	0.043	-
2014	Statistic	80.327	18.282	-
	p-value	<0.01	<0.01	-

5. Conclusions

We have presented a methodology to compute the Shannon and the Rényi entropy and the Kullback–Leibler divergence for the family of Skew-Reflected-Gompertz distributions. Our methods consider the information quantifiers previously computed for the Gompertz distribution. Explicit formulas for Shannon and Rényi entropies (in terms of the Gompertz, Shannon and

Rényi entropies, respectively), and the Kullback–Leibler divergence (using incomplete gamma function) facilitate easy computational implementation. Additionally, given the regularity conditions accomplished by the Skew-Reflected-Gompertz distribution, specifically by the Fisher information matrix convergence when ε is in $(-1, 1)$, an asymptotic test for comparing two groups of datasets was developed.

The statistical application to South Pacific sea surface temperature was given. We first carried out SRG goodness-of-fit tests in samples over three years, where we find strong evidence (a 95% confidence level) for 2012, and moderate evidence (a 90% confidence level) for 2013 and 2014. The results show that the proposed methodology serves to compare two sets of samples, Skew-Reflected-Gompertz distributed. The proposed asymptotic test is therefore useful to detect anomalies in sea surface temperature, linked to extreme events influenced by environmental conditions [11,24,25]. We encourage researchers to consider the proposed methodology for further investigations related to environmental datasets [1].

Author Contributions: J.E.C.-R. and M.M. wrote the paper and contributed reagents/analysis/materials tools; J.E.C.-R. and D.D.C. conceived, designed and performed the experiments and analyzed the data. All authors have read and approved the final manuscript.

Acknowledgments: We are grateful to the Instituto de Fomento Pesquero (IFOP) for providing access to the data used in this work. Special thanks to Fernando Espíndola for his helpful insights and discussion on an early version of this paper. The SST datasets and R codes used in this work are available upon request to the corresponding author. The authors thank the editor and two anonymous referees for their helpful comments and suggestions.

Abbreviations

The following abbreviations are used in this manuscript:

A–D	Anderson–Darling
AIC	Akaike's information criterion
BIC	Bayesian information criterion
C–V	Cramer–von Mises
CDF	Cumulative distribution function
EM	Expectation maximization
ESN	Epsilon-skew-normal
FIM	Fisher information matrix
GZ	Gompertz
K–S	Kolmogorov–Smirnov
KL	Kullback–Leibler
MGF	Moment-generating function
MLE	Maximum Likelihood Estimator
PDF	Probability density function
RE	Rényi entropy
SD	Standard deviation
SE	Shannon entropy
SN	Skew-normal
SRG	Skew-Reflected-Gompertz
SST	Sea surface temperature

Appendix A. The Epsilon-Skew-Normal Distribution

The epsilon-skew-normal distribution [4,27] in its location-scale version is denoted as $\text{ESN}(\theta, \varpi, \epsilon)$. It can be derived from a more general class of two-piece asymmetric distributions proposed by [14], by considering the standardized normal kernel $\phi(\cdot)$ (zero mean and variance 1), denoted as $N(0, 1)$, as the density f and the functions $a(\epsilon) = 1 + \epsilon$ and $b(\epsilon) = 1 - \epsilon$. If $Z \sim \text{ESN}(\theta, \varpi, \epsilon)$, thus Z has pdf given by

$$h(z|\theta, \varpi, \epsilon) = \begin{cases} \phi\left(\frac{\theta - z}{\varpi(1+\epsilon)}\right), & z \leq \theta, \\ \phi\left(\frac{z-\theta}{\varpi(1-\epsilon)}\right), & z > \theta, \end{cases} \tag{A1}$$

where $Z = \theta + \varpi X$ for location $\theta \in \mathbb{R}$ and scale $\varpi > 0$ parameters. The mean and variance of Z are

$$\begin{aligned} E(Z) &= \theta - 4\varpi\epsilon/\sqrt{2\pi}, \\ Var(Z) &= \frac{\varpi^2}{\pi}[(3\pi - 8)\epsilon^2 + \pi], \end{aligned}$$

and the MGF of X is given by

$$M_X(t) = (1+\epsilon)e^{\frac{(1+\epsilon)^2 t^2}{2}}\Phi[-(1+\epsilon)t] + (1-\epsilon)e^{\frac{(1-\epsilon)^2 t^2}{2}}\Phi[(1-\epsilon)t],$$

where $\Phi(\cdot)$ is the cdf of standardized Gaussian distribution.

Appendix B. The Skew-Normal Distribution

Let X be a skew-normal (SN, [28]) random variable denoted as $X \sim \text{SN}(\xi, \omega, \lambda)$. The pdf of X is given by

$$f(x; \lambda) = 2\phi(z)\Phi(\lambda z), \tag{A2}$$

with $z = (x - \xi)/\omega$. The SN model with the density (A2) is explained by its stochastic representation

$$X \stackrel{d}{=} \xi + \delta|U_0| + \sqrt{1-\delta^2}U, \tag{A3}$$

where $\delta = \lambda/\sqrt{1+\lambda^2}$, X is represented as a linear combination of Gaussian U and a half-Gaussian $|U_0|$ variable, and $U_0 \sim N(0,1)$ and $U \sim N(0, \omega^2)$ are independent (Theorem 1 of [29]). From (A3), the mean and variance of X are $E(X) = \xi + \sqrt{2/\pi}\delta$ and $Var(X) = \omega^2 - (2/\pi)\delta^2$, respectively.

References

1. Hoseinzadeh, A.; Maleki, M.; Khodadadi, Z.; Contreras-Reyes, J.E. The Skew-Reflected-Gompertz distribution for analyzing symmetric and asymmetric data. *J. Comput. Appl. Math.* **2019**, *349*, 132–141. [CrossRef]
2. Gompertz, B. On the nature of the function expressive of the law of human mortality, and on a new mode of determining the value of life contingencies. *Philos. Trans. R. Soc. Lond.* **1825**, *115*, 513–583. [CrossRef]
3. Arellano-Valle, R.B.; Contreras-Reyes, J.E.; Stehlík, M. Generalized skew-normal negentropy and its application to fish condition factor time series. *Entropy* **2017**, *19*, 528. [CrossRef]
4. Mudholkar, G.S.; Hutson, A.D. The epsilon-skew-normal distribution for analyzing near-normal data. *J. Stat. Plan. Inference* **2000**, *83*, 291–309. [CrossRef]
5. Maleki, M.; Mahmoudi, M.R. Two-Piece Location-Scale Distributions based on Scale Mixtures of Normal family. *Commun. Stat. Theor. Meth.* **2017**, *46*, 12356–12369. [CrossRef]
6. Moravveji, B.; Khodadai, Z.; Maleki, M. A Bayesian Analysis of Two-Piece distributions based on the Scale Mixtures of Normal Family. *Iran. J. Sci. Technol. Trans. A* **2019**, *43*, 991–1001. [CrossRef]
7. Contreras-Reyes, J.E. Rényi entropy and complexity measure for skew-gaussian distributions and related families. *Physica A* **2015**, *433*, 84–91. [CrossRef]
8. Contreras-Reyes, J.E. Analyzing fish condition factor index through skew-gaussian information theory quantifiers. *Fluct. Noise Lett.* **2016**, *15*, 1650013. [CrossRef]

9. Wang, Y.Q.; Derksen, R.W. The confirmation of the α–β model and the maximum entropy formulation in a thermal wake. *Environmetrics* **1998**, *9*, 269–282. [CrossRef]
10. De Queiroz, M.M.; Silva, R.W.; Loschi, R.H. Shannon entropy and Kullback–Leibler divergence in multivariate log fundamental skew-normal and related distributions. *Can. J. Stat.* **2016**, *44*, 219–237. [CrossRef]
11. Di Lorenzo, E.; Combes, V.; Keister, J.E.; Strub, P.T.; Thomas, A.C.; Franks, P.J.; Ohman, M.D.; Furtado, J.C.; Bracco, A.; Bograd, S.J.; et al. Synthesis of Pacific Ocean climate and ecosystem dynamics. *Oceanography* **2013**, *26*, 68–81. [CrossRef]
12. Salicrú, M.; Menéndez, M.L.; Pardo, L.; Morales, D. On the applications of divergence type measures in testing statistical hypothesis. *J. Multivar. Anal.* **1994**, *51*, 372–391. [CrossRef]
13. Maleki, M.; Contreras-Reyes, J.E.; Mahmoudi, M.R. Robust Mixture Modeling Based on Two-Piece Scale Mixtures of Normal Family. *Axioms* **2019**, *8*, 38. [CrossRef]
14. Arellano-Valle, R.B.; Gómez, H.W.; Quintana, F.A. Statistical inference for a general class of asymmetric distributions. *J. Stat. Plan. Inference* **2005**, *128*, 427–443. [CrossRef]
15. Jafari, A.A.; Tahmasebi, S.; Alizadeh, M. The beta-Gompertz distribution. *Rev. Colomb. Estad.* **2014**, *37*, 141–158. [CrossRef]
16. Shannon, C.E. A mathematical theory of communication. *Bell Syst. Tech. J.* **1948**, *27*, 379–423. [CrossRef]
17. Cover, T.M.; Thomas, J.A. *Elements of Information Theory*; Wiley & Son, Inc.: New York, NY, USA, 2006.
18. Rényi, A. *Probability Theory*; Dover Publications: New York, NY, USA, 2012.
19. Abu-Zinadah, H.H.; Aloufi, A.S. Some characterizations of the exponentiated Gompertz distribution. *Int. Math. Forum* **2014**, *9*, 1427–1439. [CrossRef]
20. Kullback, S.; Leibler, R.A. On information and sufficiency. *Ann. Math. Stat.* **1951**, *22*, 79–86. [CrossRef]
21. Bauckhage, C. Characterizations and Kullback–Leibler Divergence of Gompertz Distributions. *arXiv* **2014**, arXiv:1402.3193.
22. R Core Team. *A Language and Environment for Statistical Computing*; R Foundation for Statistical Computing: Vienna, Austria, 2018; ISBN 3-900051-07-0.
23. Barría, P.; González, A.; Cortés, D.D.; Mora, S.; Miranda, H.; Cerna, F.; Cid, L.; Ortega, J.C. *Seguimiento Pesquerías Recursos Altamente Migratorios, 2016. Convenio de Desempeño 2016*; Informe Final, Subsecretaría de Economía y EMT; Instituto de Fomento Pesquero: Valparaíso, Chile, 2017.
24. Alheit, J.; Bernal, P. Effects of physical and biological changes on the biomass yield of the Humboldt Current ecosystem. In *Large Marine Ecosystems—Stress, Mitigation and Sustainability*; American Association for the Advancement of Science: Washington, DC, USA, 1993; pp. 53–68.
25. Oerder, V.; Bento, J.P.; Morales, C.E.; Hormazabal, S.; Pizarro, O. Coastal Upwelling Front Detection off Central Chile (36.5–37°S) and Spatio-Temporal Variability of Frontal Characteristics. *Remote Sens.* **2018**, *10*, 690. [CrossRef]
26. Lenart, A.; Missov, T.I. Goodness-of-fit tests for the Gompertz distribution. *Commun. Stat. Theor. Meth.* **2016**, *45*, 2920–2937. [CrossRef]
27. Bondon, P. Estimation of autoregressive models with epsilon-skew-normal innovations. *J. Multivar. Anal.* **2009**, *100*, 1761–1776. [CrossRef]
28. Azzalini, A. A Class of Distributions which includes the Normal Ones. *Scand. J. Stat.* **1985**, *12*, 171–178.
29. Henze, N. A probabilistic representation of the 'skew-normal' distribution. *Scand. J. Stat.* **1986**, *13*, 271–275.

4

Classical Lagrange Interpolation based on General Nodal Systems at Perturbed Roots of Unity

Elías Berriochoa [1,*,†], Alicia Cachafeiro [1,*,†], Alberto Castejón [1,†] and José Manuel García-Amor [2,†]

[1] Departamento de Matemática Aplicada I, Universidad de Vigo, 36201 Vigo, Pontevedra, Spain; acaste@uvigo.es

[2] Departamento de Matemáticas, Instituto E. S. Valle Inclán, 36001 Pontevedra, Spain; garciaamor@edu.xunta.es

* Correspondence: esnaola@uvigo.es (E.B.); acachafe@uvigo.es (A.C.).

† These authors contributed equally to this work.

Abstract: The aim of this paper is to study the Lagrange interpolation on the unit circle taking only into account the separation properties of the nodal points. The novelty of this paper is that we do not consider nodal systems connected with orthogonal or paraorthogonal polynomials, which is an interesting approach because in practical applications this connection may not exist. A detailed study of the properties satisfied by the nodal system and the corresponding nodal polynomial is presented. We obtain the relevant results of the convergence related to the process for continuous smooth functions as well as the rate of convergence. Analogous results for interpolation on the bounded interval are deduced and finally some numerical examples are presented.

Keywords: lagrange interpolation; unit circle; nodal systems; separation properties; perturbed roots of the unity; convergence

---

1. Introduction

The polynomial interpolation is a classical subject that has been widely studied under different approaches like Lagrange, Hermite, Birkhoff, Pál-type interpolation and some others. Although it is obvious that the subject is important by itself, its numerous numerical applications like numerical integration or numerical derivation are not less important and indeed the polynomial interpolation continues being a subject of current research.

Lagrange interpolation is a very good tool although it is known that for this interpolation scheme and for good nodal systems such as the classical Chebyshev nodes there exists a continuous function on $[-1,1]$ for which the Lagrange interpolation polynomial diverges (see [1]). A similar problem has been posed for arbitrary arrays and it was proved in [2] that for each nodal array in $[-1,1]$, there exists a continuous function such that the Lagrange polynomial interpolation diverges almost everywhere. In any case, recalling the words written by Trefethen in his paper [3] we can say that there is nothing wrong with Lagrange polynomial interpolation. "Yet the truth is, polynomial interpolants in Chebyshev points always converge if f is a little bit smooth". As a consequence, to obtain better results one needs to assume better properties on the function to be interpolated like bounded variation or a condition on its modulus of continuity. Thus one of the most important questions in relation with the interpolation of functions is the choice of the interpolation arrays or nodal systems for which one can expect to obtain pointwise or uniform convergence to the function to be interpolated and another important issue is to determine the class of functions for which some type of convergence can be guaranteed. The nodal systems strongly normal or normal, introduced by Fejér, play an important role in the interpolation theory, although from a practical point of view, the difficulty of testing the definition makes the applications of these systems quite limited. For these systems Grünwald studies in [4] the convergence of the polynomials of Lagrange interpolation for functions satisfying a Lipschitz condition.

Most of the research obtain results on the convergence from the distribution properties of the nodal points. Indeed it was Fejér, who was the first to invert the problem, trying to deduce separation properties of the nodal systems from the interpolation results. The importance of this idea is avaled by the fact that the required interpolation properties are easily verified.

In [5] it is proved that strongly normal distributions on $[-1,1]$ give quasi uniformly nodal systems on the unit circle, that is the length of the arcs between two consecutive nodes has the order of $\frac{1}{n}$. Although the situation more widely studied corresponds to the bounded interval, there are important results in some other situations in the complex plane like the unit circumference, (see [6]). By taking into account that continuous functions on the unit circle can be approximated by Laurent polynomials, the interpolation polynomials on the unit circumference are constructed in this Laurent space. In this field of research, [7] deserves to be highlighted. There, the nodal points are the roots of complex numbers with modulus 1 and in this case it is obtained a result about convergence of the interpolants for continuous functions with a suitable modulus of continuity. Moreover, in the same paper the problem of extending the results to general nodal systems. Indeed, since the roots of complex numbers with modulus 1 can be interpreted like the zeros of the para-orthogonal polynomials with respect to the Lebesgue measure, now the natural extension is to consider the zeros of the para-orthogonal polynomials with respect to other measures.

In [8] we have generalized the results given in [7] for these new nodal systems. First we work with nodal systems characterized by fulfilling some properties of boundedness, which are suggested by those fulfilled by the roots of unimodular complex numbers, obtaining a result of convergence for continuous functions with a suitable modulus of continuity. Next, by taking into account that the zeros of the para-orthogonal polynomials with respect to measures in the Szegő class (see [9]) with analytic extension up to $|z| > 1$ satisfy the properties that we need, we obtain a similar result about convergence for these type of nodal systems.

In [10] we have studied the Lagrange interpolation process for piecewise continuous functions with suitable properties and by using as nodal points the zeros of the para-orthogonal polynomials with respect to analytic weights, which constitutes a novel approach to the Lagrange interpolation theory.

Another extension to more general nodal systems is given in [11] where it has opened a new trend to interpolation at perturbed roots of unity and the functions to be interpolated belong to the disc algebra.

Now, in the present paper we assume a distribution for the nodes that can be obtained through a perturbation of the uniform distribution and, in particular of the roots of the unity, and which is more general than that given in [11]. Thus in the present paper we start from a different point of view because we base it on properties satisfied by the nodal systems and we do not need to consider orthogonality nor para-orthogonality with respect to any measure on the unit circle. The interpolation arrays are described by a separation property and the main goal is to obtain the properties that play a role in the Lagrange process, as well as to present some relevant examples.

The organization of the paper is the following. In Section 2 we introduce the nodal systems that we use throughout all the paper and we prove the main properties that they satisfy in several propositions. Section 3 is devoted to the Lagrange interpolation problem. First we obtain the Lebesgue constant of the process and then we study the convergence of the Lagrange interpolation polynomials related to continuous functions with appropriate modulus of continuity. Secondly we analyze the rate of convergence when we deal with smooth functions, (see [12]) and we also deduce analogous behavior for interpolation on the bounded interval. The last section is devoted to give some numerical examples.

2. Some General Nodal Systems on the Unit Circle

The aim of this paper is to study interpolation problems on the unit circle $\mathbb{T} = \{z \in \mathbb{C} : |z| = 1\}$ by using nodal systems satisfying some suitable properties.

We denote the nodal polynomials by $W_n(z)$ and their zeros by $\{\alpha_{j,n}\}_{j=1}^{n}$, that is, we assume that $W_n(z) = \prod_{j=1}^{n}(z-\alpha_{j,n})$, where $|\alpha_{j,n}| = 1$ for $j = 1, \cdots, n$, and $\alpha_{j,n} \neq \alpha_{k,n}$ for $j \neq k$. For simplicity, in the sequel we omit the subscript n and we write α_j instead of $\alpha_{j,n}$ for $j = 1, \cdots, n$. We denote the length of the shortest arc between any two points of the unit circle, z_1 and z_2, by $\widehat{z_1 - z_2}$, and we use the Landau's notation for complex sequences, denoting by $a_n = \mathcal{O}(b_n)$ if $|\frac{a_n}{b_n}|$ is bounded.

Throughout all the paper we assume that the zeros $\{\alpha_j\}_{j=1}^{n}$ of the nodal polynomials $W_n(z)$ satisfy the following separation property: there exists a positive constant A such that for $n > \frac{A}{\pi}$ the length of the shortest arc between two consecutive nodes α_j and α_{j+1}, satisfies:

$$\widehat{\alpha_j - \alpha_{j+1}} = \frac{2\pi}{n} + \frac{A(j)}{n^2} \text{ with } |A(j)| \leq A \;\; \forall j = 1, \cdots, n, \tag{1}$$

where $\alpha_{n+1} = \alpha_1$, that is, $\widehat{\alpha_j - \alpha_{j+1}} = \frac{2\pi}{n} + \mathcal{O}(\frac{1}{n^2})$.

We use the same $\mathcal{O}$ to denote different sequences. Unless we mention otherwise, the bounds we obtain from (1) will be uniform.

We also consider other nodal polynomials, $\widetilde{W}_n(z)$, well connected with $W_n(z)$. If we denote $\widetilde{W}_n(z) = z^n - \lambda$, with $\lambda = \alpha_1^n$, then $\widetilde{W}_n(z) = \prod_{j=1}^{n}(z-\beta_j)$, where

$$\beta_j = \sqrt[n]{\lambda},\; j = 1, \cdots, n, \text{ and } \alpha_1 = \beta_1.$$

Hence it is clear that the separation property satisfied by the zeros $\{\beta_j\}$ of $\widetilde{W}_n(z)$ is

$$\widehat{\beta_j - \beta_{j+1}} = \frac{2\pi}{n} \;\; \forall j = 1, \cdots, n. \tag{2}$$

In this section we obtain in several propositions the main properties of the nodal polynomials $W_n(z)$. First we recall the following well known relations between arcs and strings that we are going to use throughout the whole paper and which is based on the convex character of the arcsin function: If z_1 and z_2 belong to $\mathbb{T}$ then

$$\frac{2}{\pi}(\widehat{z_1 - z_2}) \leq |z_1 - z_2| \leq (\widehat{z_1 - z_2}). \tag{3}$$

Proposition 1. *If $\{\alpha_j\}_{j=1}^{n}$ and $\{\beta_j\}_{j=1}^{n}$, with $\alpha_1 = \beta_1$, are the nodal points satisfying the separation properties (1) and (2) and we assume they are numbered in the clockwise sense, then*

(i)

$$(j-1)(\frac{2\pi}{n} - \frac{A}{n^2}) \leq \widehat{\alpha_1 - \alpha_j} \leq (j-1)(\frac{2\pi}{n} + \frac{A}{n^2}), \text{ for } j \geq 1.$$

(ii)

$$(j+1)(\frac{2\pi}{n} - \frac{A}{n^2}) \leq \widehat{\alpha_{n-j} - \alpha_1} \leq (j+1)(\frac{2\pi}{n} + \frac{A}{n^2}), \text{ for } j \geq 0.$$

(iii)

$$\widehat{\alpha_j - \beta_j} \leq (j-1)\frac{A}{n^2}, \text{ for } j \geq 1.$$

(iv)

$$\widehat{\alpha_{n-j} - \beta_{n-j}} \leq (j+1)\frac{A}{n^2}, \text{ for } j \geq 0.$$

Proof. (i) By applying (1) we have $\frac{2\pi}{n} - \frac{A}{n^2} \leq \widehat{\alpha_1 - \alpha_2} \leq \frac{2\pi}{n} + \frac{A}{n^2}$ and for $j \geq 3$ it holds $\frac{2\pi}{n} - \frac{A}{n^2} \leq \widehat{\alpha_{j-1} - \alpha_j} \leq \frac{2\pi}{n} + \frac{A}{n^2}$. Then if we sum, it is straightforward (i).

(ii) Proceeding in the same way we get $\frac{2\pi}{n} - \frac{A}{n^2} \leq \widehat{\alpha_n - \alpha_1} \leq \frac{2\pi}{n} + \frac{A}{n^2}$ and for $j \geq 0$ it holds $\frac{2\pi}{n} - \frac{A}{n^2} \leq \widehat{\alpha_{n-j-1} - \alpha_{n-j}} \leq \frac{2\pi}{n} + \frac{A}{n^2}$. Hence, by computing the sums we have (ii).

(iii) We know that $\alpha_1 = \beta_1$ and we distinguish two possibilities depending on the position of β_j related to α_j. If

$$\widehat{\alpha_1 - \beta_j} + \widehat{\beta_j - \alpha_j} = \widehat{\alpha_1 - \alpha_j},$$

that is,

$$\widehat{\beta_1 - \beta_j} + \widehat{\beta_j - \alpha_j} = \widehat{\alpha_1 - \alpha_j}$$

and we use that $\widehat{\beta_1 - \beta_j} = (j-1)\frac{2\pi}{n}$, which is a consequence of (2), and we also take into account (i) we get

$$\widehat{\alpha_j - \beta_j} \leq (j-1)\frac{A}{n^2}.$$

The second case corresponds to $\widehat{\alpha_1 - \alpha_j} + \widehat{\alpha_j - \beta_j} = \widehat{\beta_1 - \beta_j}$ and it can be deduced in the same way.

(iv) We proceed like in (iii) distinguishing the following cases $\widehat{\alpha_{n-j} - \alpha_1} = \widehat{\alpha_{n-j} - \beta_{n-j}} + \widehat{\beta_{n-j} - \alpha_1}$ or $\widehat{\beta_{n-j} - \alpha_{n-j}} + \widehat{\alpha_{n-j} - \alpha_1} = \widehat{\beta_{n-j} - \beta_1}$. □

Notice that we can write (iii) and (iv) as follows $\widehat{\alpha_j - \beta_j} = (j-1)\mathcal{O}(\frac{1}{n^2})$ and $\widehat{\alpha_{n-j} - \beta_{n-j}} = (j+1)\mathcal{O}(\frac{1}{n^2})$.

Proposition 2. *Let us assume that the zeros of the nodal polynomials $W_n(z)$ satisfy the separation property (1). Then it holds*

$$|W_n(z)| < 2e^A, \quad \forall z \in \mathbb{T}. \tag{4}$$

Moreover, it also holds

$$\frac{|W_n'(z)|}{n} < 2e^A \text{ and } \frac{|W_n''(z)|}{n^2} < 2e^A, \quad \forall z \in \mathbb{T}.$$

Proof. Since $W_n(\alpha_j) = 0$ for every j, let us take $z \in \mathbb{T}$, such that z is not a nodal point. In order to obtain the result we renumber the nodes in the clockwise sense in such a way that α_1 is the nodal point nearest to z. We distinguish two cases depending on whether the node closest to z is turning in the clockwise sense or in the counter clockwise sense from z. If we assume that the situation is given in Figure 1, that is, α_1 is turning in the counter clockwise sense from z, then we have

$$\widehat{z - \alpha_1} < \frac{\widehat{\alpha_1 - \alpha_2}}{2} \leq \frac{\pi}{n} + \frac{A}{2n^2}.$$

Now we consider the polynomial $\widetilde{W}_n(z) = \prod_{j=1}^{n}(z - \beta_j)$, with $\beta_1 = \alpha_1$ and satisfying (2), introduced at the beginning of the section. Using property (1) we have $\frac{\pi}{n} + \frac{A}{2n^2} < \frac{2\pi}{n} = \widehat{\alpha_1 - \beta_2}$ and then $z - \beta_2 \neq 0$.

Now, for every j it holds $\frac{z-\alpha_j}{z-\beta_j} = 1 + \frac{\beta_j - \alpha_j}{z-\beta_j}$ and therefore, by using Proposition 1,

$$\left|\frac{z-\alpha_j}{z-\beta_j}\right| \le 1 + \frac{\pi}{2}\frac{\widehat{\beta_j - \alpha_j}}{\widehat{z-\beta_j}} \le 1 + \frac{\pi}{2}\frac{(j-1)\frac{A}{n^2}}{\widehat{z-\beta_j}}.$$

Besides, from property (1) we have $\widehat{z-\beta_2} > \frac{\pi}{2n}$ and for $j \ge 3$, $\widehat{z-\beta_j} > \frac{\pi}{2n} + \frac{(j-2)2\pi}{n} > \frac{(j-2)2\pi}{n}$.
Hence $\left|\frac{z-\alpha_j}{z-\beta_j}\right| \le 1 + \frac{A}{n}$.

Proceeding in the same way and taking into account that for $j \ge 0$ it holds that $\widehat{\beta_{n-j} - \alpha_{n-j}} \le (j+1)\frac{A}{n^2}$ and $\widehat{z-\beta_{n-j}} > (j+1)\frac{2\pi}{n}$ we obtain

$$\left|\frac{z-\alpha_{n-j}}{z-\beta_{n-j}}\right| \le 1 + \frac{\pi}{2}\frac{\widehat{\beta_{n-j} - \alpha_{n-j}}}{\widehat{z-\beta_{n-j}}} \le 1 + \frac{\pi}{2}\frac{(j+1)\frac{A}{n^2}}{(j+1)\frac{2\pi}{n}} = 1 + \frac{A}{n}.$$

Therefore we have that

$$\frac{|W_n(z)|}{|\widetilde{W}_n(z)|} = \prod_{j=2}^{n}\left|\frac{z-\alpha_j}{z-\beta_j}\right| < e^A \ \forall z,$$

and since $|\widetilde{W}_n(z)| \le 2$, then we get $|W_n(z)| < 2e^A$.

Notice that if the node closest to z, α_1, is in the clockwise sense from z, we can proceed in a similar way. Indeed $\widehat{z-\alpha_1} < \frac{\widehat{\alpha_1 - \alpha_n}}{2} \le \frac{\pi}{n} + \frac{A}{2n^2}$ and since $\frac{\pi}{n} + \frac{A}{2n^2} < \widehat{\beta_n - \alpha_1} = \frac{2\pi}{n}$ then $z - \beta_n \ne 0$.

The second statement, related to the first and second derivatives of the nodal polynomial, is a consequence of Bernstein's theorem, (see [13]). □

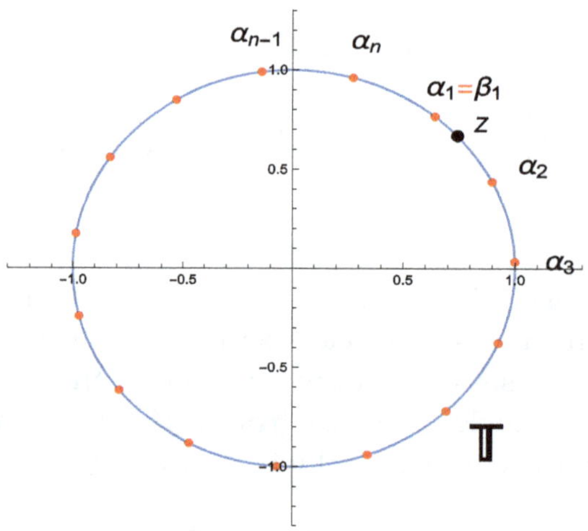

Figure 1. An arbitrary point z and the nodal system.

Proposition 3. *Let us assume that the zeros of the nodal polynomials $W_n(z)$ satisfy the separation property (1). Then there exists a positive constant $C > 0$ such that for n large enough and for every $j = 1, \cdots, n$, it holds that*

$$\frac{|W_n'(\alpha_j)|}{n} > C. \tag{5}$$

Proof. For simplicity we take $j = 1$ and we try to bound from below $\dfrac{|W_n'(\alpha_1)|}{n}$.

Thus we consider the polynomial $\widetilde{W}_n(z)$ satisfying (2), that is, $\widetilde{W}_n(z) = z^n - \alpha_1^n = \prod_{j=1}^{n}(z - \beta_j)$ with $\alpha_1 = \beta_1$.

Since $\widetilde{W}_n'(z) = nz^{n-1}$ then $|\widetilde{W}_n'(\alpha_1)| = n$ and so our aim is to prove that $\dfrac{|W_n'(\alpha_1)|}{|\widetilde{W}_n'(\alpha_1)|} > C$.

Now, by taking into account that

$$\frac{|W_n'(\alpha_1)|}{|\widetilde{W}_n'(\alpha_1)|} = \prod_{j=2}^{n}\left|\frac{\alpha_1 - \alpha_j}{\alpha_1 - \beta_j}\right|,$$

we study the quotients $\left|\dfrac{\alpha_1 - \alpha_j}{\alpha_1 - \beta_j}\right|$ and $\left|\dfrac{\alpha_1 - \alpha_{n-j}}{\alpha_1 - \beta_{n-j}}\right|$.

On the one hand,

$$\left|\frac{\alpha_1 - \alpha_j}{\alpha_1 - \beta_j}\right| = \left|1 + \frac{\beta_j - \alpha_j}{\alpha_1 - \beta_j}\right| \geq 1 - \left|\frac{\beta_j - \alpha_j}{\alpha_1 - \beta_j}\right|,$$

and by applying (3) and Proposition 1, we have for $j \geq 2$,

$$\left|\frac{\beta_j - \alpha_j}{\alpha_1 - \beta_j}\right| \leq \frac{\pi}{2}\frac{\widehat{\beta_j - \alpha_j}}{\widehat{\alpha_1 - \beta_j}} \leq \frac{\frac{\pi}{2}(j-1)\frac{A}{n^2}}{(j-1)\frac{2\pi}{n}} = \frac{A}{4n},$$

and therefore $\left|\dfrac{\alpha_1 - \alpha_j}{\alpha_1 - \beta_j}\right| \geq 1 - \dfrac{A}{4n}$.

On the other hand,

$$\left|\frac{\alpha_1 - \alpha_{n-j}}{\alpha_1 - \beta_{n-j}}\right| = \left|1 + \frac{\beta_{n-j} - \alpha_{n-j}}{\alpha_1 - \beta_{n-j}}\right| \geq 1 - \left|\frac{\beta_{n-j} - \alpha_{n-j}}{\alpha_1 - \beta_{n-j}}\right|$$

and since for $j \geq 0$ we have

$$\left|\frac{\beta_{n-j} - \alpha_{n-j}}{\alpha_1 - \beta_{n-j}}\right| \leq \frac{\pi}{2}\frac{\widehat{\beta_{n-j} - \alpha_{n-j}}}{\widehat{\alpha_1 - \beta_{n-j}}} \leq \frac{\frac{\pi}{2}(j+1)\frac{A}{n^2}}{(j+1)\frac{2\pi}{n}} = \frac{A}{4n},$$

then $\left|\dfrac{\alpha_1 - \alpha_{n-j}}{\alpha_1 - \beta_{n-j}}\right| \geq 1 - \dfrac{A}{4n}$. Hence

$$\frac{|W_n'(\alpha_1)|}{|\widetilde{W}_n'(\alpha_1)|} = \prod_{j=2}^{n}\left|\frac{\alpha_1 - \alpha_j}{\alpha_1 - \beta_j}\right| \geq (1 - \frac{A}{4n})^{n-1},$$

that is, $|W_n'(\alpha_1)| \geq (1 - \frac{A}{4n})^{n-1}n$. Thus, given $\varepsilon > 0$ if $C = e^{-\frac{A}{4}} - \varepsilon > 0$, then for n large enough it holds that $|W_n'(\alpha_1)| > Cn$.

Notice that for another value of j there is no any significant difference. Indeed to obtain that $|W_n'(\alpha_i)| > Cn$, we take the auxiliary polynomial $\widetilde{W}_n(z) = z^n - \alpha_i^n = \prod_{j=1}^{n}(z - \beta_j)$ with $\alpha_i = \beta_i$, we renumber the nodes as in the previous proof and we proceed in a similar way. □

Proposition 4. *Let us assume that the zeros of the nodal polynomials $W_n(z)$ satisfy the separation property (1). Then there exists a positive constant $D > 0$ such that*

$$\frac{|W_n(z)|^2}{n^2}\sum_{j=1}^{n}\frac{1}{|z-\alpha_j|^2} < D, \ \forall z \in \mathbb{T}. \tag{6}$$

Proof. Following the same steps of the proof of Lemma 2 in [14] we have

$$|W_n(z)|^2\sum_{j=1}^{n}\frac{1}{|z-\alpha_j|^2} = |zW_n(z)W_n'(z) + z^2(W_n''(z)W_n(z) - (W_n'(z))^2)|$$

and therefore by using (4) and its consequences in Proposition 2 we get

$$\frac{|W_n(z)|^2}{n^2}\sum_{j=1}^{n}\frac{1}{|z-\alpha_j|^2} \le \frac{|W_n(z)|}{n}\frac{|W_n'(z)|}{n} + \frac{|W_n''(z)|}{n^2}|W_n(z)| + \frac{|W_n'(z)|^2}{n^2} <$$
$$\frac{B}{n}B + B^2 + B^2, \text{ where } B = 2e^A.$$

□

Remark 1. *The nodal systems considered in [15] satisfy condition (1). Indeed they are the para-orthogonal polynomials related to measures in the Baxter class, (see [16]). In that work it is also assumed the additional condition that the sequence $\{(\phi_n^*)'\}$ is uniformly bounded on $\mathbb{T}$, where $\{\phi_n\}$ is the sequence of monic orthogonal polynomials related to the measure and $\{\phi_n^*\}$ is the sequence of the reciprocal polynomials, (see [9]). In that situation studied in [15], properties (4)–6) also hold. Now, in the present paper we start from a different point of view because we base it on properties satisfied by the nodal systems and we do not need to consider orthogonality nor para-orthogonality with respect to any measure.*

3. Lagrange Interpolation. Convergence in Case of Smooth Continuous Functions

To compute the interpolation polynomials, first we recall some well known definitions related to interpolation problems on the unit circle. We work in the space of Laurent polynomials and, in particular, in the subspaces $\Lambda_{p,q}[z] = span\{z^k : p \le k \le q\}$, with p and q integers $p \le q$.

Let us continue denoting by $\{\alpha_j\}_{j=1}^n$ the zeros of the the nodal polynomial $W_n(z)$. If $\{u_j\}_{j=1}^n$ are arbitrary complex numbers, the Laurent polynomial of Lagrange interpolation $\mathcal{L}_{-E[\frac{n}{2}],E[\frac{n-1}{2}]}(z) \in \Lambda_{-E[\frac{n}{2}],E[\frac{n-1}{2}]}[z]$ characterized by satisfying

$$\mathcal{L}_{-E[\frac{n}{2}],E[\frac{n-1}{2}]}(\alpha_j) = u_j, \text{ for } j = 1, \cdots, n,$$

has the following expression

$$\mathcal{L}_{-E[\frac{n}{2}],E[\frac{n-1}{2}]}(z) = \frac{W_n(z)}{z^{E[\frac{n}{2}]}}\sum_{j=1}^{n}\frac{\alpha_j^{E[\frac{n}{2}]}}{W_n'(\alpha_j)(z-\alpha_j)}u_j.$$

If F is a function and $u_j = F(\alpha_j)$ we denote the corresponding Laurent polynomial $\mathcal{L}_{-E[\frac{n}{2}],E[\frac{n-1}{2}]}(F,z)$. If n is odd, since $E[\frac{n-1}{2}] = E[\frac{n}{2}]$, then the interpolation polynomial $\mathcal{L}_{-E[\frac{n}{2}],E[\frac{n}{2}]}(z) \in \Lambda_{-E[\frac{n}{2}],E[\frac{n}{2}]}[z]$ and when n is even, taking into account that $E[\frac{n-1}{2}] = E[\frac{n}{2}] - 1$, then the Laurent polynomial of Lagrange interpolation $\mathcal{L}_{-E[\frac{n}{2}],E[\frac{n}{2}]-1}(z) \in \Lambda_{-E[\frac{n}{2}],E[\frac{n}{2}]-1}[z]$.

Without loss of generality, to fix ideas and to simplify the notation we assume that the number of nodes is even, $2n$, in which case the interpolation polynomial $\mathcal{L}_{n,n-1}$ belongs to the space $\Lambda_{-n,n-1}$ and it can be written in terms of the fundamental polynomials as follows:

$$\mathcal{L}_{-n,n-1}(z) = \frac{W_{2n}(z)}{z^n} \sum_{j=1}^{2n} \frac{\alpha_j^n}{W'_{2n}(\alpha_j)(z-\alpha_j)} u_j. \tag{7}$$

In order to compute the interpolation polynomials for applications and examples it is more convenient to use the barycentric expression, which is given by

$$\mathcal{L}_{-n,n-1}(z) = \frac{\sum_{j=1}^{2n} \frac{w_j}{z-\alpha_j} u_j}{\sum_{j=1}^{2n} \frac{w_j}{z-\alpha_j}}, \tag{8}$$

with $w_j = \frac{\alpha_j^n}{W'_{2n}(\alpha_j)}$, (see [17]).

This last expression has some advantages due to its numerical stability in the sense established in [18]. In this article author claims literally:

The Lagrange representation of the interpolating polynomial can be rewritten in two more computationally attractive forms: a modified Lagrange form and a barycentric form. We give an error analysis of the evaluation of the interpolating polynomial using these two forms. The modified Lagrange formula is shown to be backward stable. The barycentric formula has a less favourable error analysis, but is forward stable for any set of interpolating points with a small Lebesgue constant. Therefore the barycentric formula can be significantly less accurate than the modified Lagrange formula only for a poor choice of interpolating points.

So with a good Lebesgue constant (see next Theorem 1) we have good accuracy, at least as good as the intensively used Lagrange interpolation on the Chebyshev nodal systems.

Following [10] we can obtain the Lebesgue constant, (see [19]), and the convergence of this interpolatory process. Notice that this is a novelty result for our general nodal systems satisfying property (1), although the techniques that we use are the same as in [10].

Theorem 1. *There exists a positive constant $L > 0$ such that for every function F bounded on $\mathbb{T}$ it holds that*

$$|\mathcal{L}_{-n,n-1}(F,z)| \le L \parallel F \parallel_\infty \log n,$$

for every $z \in \mathbb{T}$, where $\parallel \parallel_\infty$ denotes the supremum norm on $\mathbb{T}$.

Proof. Let z be an arbitrary point of $\mathbb{T}$ and assume that z is not a nodal point. Then, if we continue assuming the even case, from (7) we get

$$|\mathcal{L}_{-n,n-1}(z)| \le \sum_{j=1}^{2n} \left| \frac{W_{2n}(z)F(\alpha_j)}{W'_{2n}(\alpha_j)(z-\alpha_j)} \right|,$$

and by our hypothesis about F and by Proposition 3 we have

$$|\mathcal{L}_{-n,n-1}(z)| \le \frac{\parallel F \parallel_\infty}{2nC} \sum_{j=1}^{2n} \left| \frac{W_{2n}(z)}{z-\alpha_j} \right|.$$

If we assume that the nodal points closest to z are α_1 and α_{2n} then by applying (1) we obtain that for $j > 1$ it holds

$$\widehat{z-\alpha_j} > (j-1)\left(\frac{2\pi}{2n} + \mathcal{O}\left(\frac{1}{4n^2}\right)\right).$$

Thus, by using (3) we obtain

$$\frac{1}{|z-\alpha_j|} < \frac{\pi}{2} \frac{2n}{(j-1)} \frac{1}{(2\pi + \mathcal{O}(\frac{1}{2n}))} = \frac{nE}{j-1},$$

for some positive constant E.

Proceeding in the same way we get $z-\widehat{\alpha_{2n-j}} > j(\frac{2\pi}{2n}+\mathcal{O}(\frac{1}{4n^2}))$ and therefore $\frac{1}{|z-\alpha_{2n-j}|} < \frac{nE}{j}$.

We also obtain

$$\left|\frac{W_{2n}(z)}{z-\alpha_1}\right|, \left|\frac{W_{2n}(z)}{z-\alpha_{2n}}\right| < 2nK,$$

for some positive constant K.

Indeed $|W_{2n}(z)| = |W_{2n}(z)-W_{2n}(\alpha_1)| = |W_{2n}(e^{i\theta})-W_{2n}(e^{i\theta_1})| \leq 2\max_{z\in\mathbb{T}}|W'_{2n}(z)||\theta-\theta_1| \leq \max_{z\in\mathbb{T}}|W'_{2n}(z)|\pi|z-\alpha_1| \leq 2nK|z-\alpha_1|$.

Hence

$$|\mathcal{L}_{-n,n-1}(z)| \leq \frac{\| F \|_\infty}{2nC}\left(\left|\frac{W_{2n}(z)}{z-\alpha_1}\right| + \sum_{j=2}^{n}\left|\frac{W_{2n}(z)}{z-\alpha_j}\right| + \sum_{j=1}^{n-1}\left|\frac{W_{2n}(z)}{z-\alpha_{2n-j}}\right| + \left|\frac{W_{2n}(z)}{z-\alpha_{2n}}\right|\right) \leq$$

$$\frac{\| F \|_\infty}{2nC}\left(4nK + 2\sum_{j=2}^{n}\frac{2e^A nE}{j-1}\right) = \frac{2\| F \|_\infty}{C}\left(K + \sum_{j=2}^{n}\frac{e^A E}{j-1}\right) \leq 2\| F \|_\infty P(1+H_{n-1,1}),$$

with $H_{n-1,1}$ the harmonic number equal to $\sum_{j=1}^{n-1}\frac{1}{j}$ and P a positive constant. □

Remark 2. *When the values of $F(\alpha_i)$ are affected by any type of error, which we can suppose is bounded by some $\epsilon > 0$, then the previous result ensures us, taking into account the linearity of the interpolation process, that the final result is affected by an error bounded by $L\log(n)\,\epsilon$, that is, it is at least so good as the intensively used Lagrange interpolation on the Chebyshev nodal systems.*

For applying the interpolation it could be very useful the following results concerning the convergence and the rate of convergence for smooth continuous functions (see [10,12]).

Theorem 2. *(i) Let $F(z)$ be a function defined on $\mathbb{T}$. If F is continuous with modulus of continuity $\omega(F,\delta) = o(|\log\delta|^{-1})$, then $\mathcal{L}_{-n,n-1}(F,z)$ converges uniformly to F on $\mathbb{T}$.*

(ii) Let $F(z)$ be a function defined on $\mathbb{T}$. If $F(z) = \sum_{-\infty}^{\infty} A_k z^k$ with $|A_k| \leq K\frac{1}{|k|^c}$ for $k \neq 0$, with $c > 1$ then $\mathcal{L}_{-n,n-1}(F,z)$ uniformly converges to F on $\mathbb{T}$ and the rate of convergence is $\mathcal{O}\left(\frac{\log n}{n^{c-1}}\right)$.

(iii) If $F(z)$ is an analytic function in an open annulus containing $\mathbb{T}$, then $\mathcal{L}_{-n,n-1}(F,z)$ uniformly converges to F on $\mathbb{T}$. Besides, the rate of convergence is geometric.

Proof. The results are consequence of the preceding Theorem 1 and they are also based on the properties satisfied by our nodal systems. Thus one can obtain these results following the same steps as in the proof of Theorems 3 and 4 in [10], where one can see the details. □

The Case of the Bounded Interval

We recall that the Lagrange interpolation polynomial $\ell_{n-1}(x)$ related to a nodal system $\{x_j\}_{j=1}^n$ in $[-1,1]$ and satisfying the conditions $\{v_j\}_{j=1}^n$ is given by

$$\ell_{n-1}(x) = \sum_{j=1}^{n}\frac{w_n(x)}{w'_n(x_j)(x-x_j)}v_j,$$

where $w_n(x) = \prod_{j=1}^{n}(x-x_j)$. When $v_j = f(x_j)$ for a function f, we write $\ell_{n-1}(f,x)$.

In this subsection we consider the nodal polynomial $w_n(x) = \prod_{j=1}^{n}(x - x_j)$ with $\{x_j\}_{j=1}^{n} \subset [-1,1]$ and numbered in the following way: $1 \geq x_1 > x_2 > \cdots > x_{n-1} > x_n \geq -1$.

We also assume that the nodes satisfy the following separation property:

There exists a positive constant A such that for $n > \dfrac{A}{\pi}$

(i) If $x_1 = 1$ and $x_n = -1$ then $\arccos x_{j+1} - \arccos x_j = \dfrac{\pi}{n} + \dfrac{a(j)}{n^2}$, with $|a(j)| \leq A$, $\forall j = 1, \cdots, n-1$.

(ii) If $x_1 < 1$ and $x_n = -1$ then $\arccos x_{j+1} - \arccos x_j = \dfrac{\pi}{n} + \dfrac{a(j)}{n^2}$, with $|a(j)| \leq A$, $\forall j = 1, \cdots, n-1$, and $2\arccos x_1 = \dfrac{\pi}{n} + \dfrac{a(0)}{n^2}$, with $|a(0)| \leq A$.

(iii) If $x_1 = 1$ and $x_n > -1$ then $\arccos x_{j+1} - \arccos x_j = \dfrac{\pi}{n} + \dfrac{a(j)}{n^2}$, with $|a(j)| \leq A$, $\forall j = 1, \cdots, n-1$, and $2(\pi - \arccos x_n) = \dfrac{\pi}{n} + \dfrac{a(n)}{n^2}$, with $|a(n)| \leq A$.

(iv) If $x_1 < 1$ and $x_n > -1$ then $\arccos x_{j+1} - \arccos x_j = \dfrac{\pi}{n} + \dfrac{a(j)}{n^2}$, with $|a(j)| \leq A$, $\forall j = 1, \cdots, n-1$, and $2\arccos x_1 = \dfrac{\pi}{n} + \dfrac{a(0)}{n^2}$, with $|a(0)| \leq A$, and $2(\pi - \arccos x_n) = \dfrac{\pi}{n} + \dfrac{a(n)}{n^2}$, with $|a(n)| \leq A$.

Under the above assumptions we obtain the following results about the convergence and the rate of convergence for the interpolation polynomials with these nodal systems.

Theorem 3. *If f is a continuous function on $[-1,1]$ and $\omega(f,\delta) = o(|\log\delta|^{-1})$, then the interpolation polynomial $\ell_{n-1}(f,x)$ fulfilling the interpolation conditions*

$$\ell_{n-1}(f, x_j) = f(x_j) \text{ for } j = 1, \cdots, n \tag{9}$$

converges uniformly to f on $[-1,1]$.

Proof. Let us define a continuous function F on $\mathbb{T}$ by $F(z) = F(\bar{z}) = f(x)$ with $x = \dfrac{z + \frac{1}{z}}{2}$. Then it is clear that its modulus of continuity satisfies

$$\omega(F,\delta) = \sup_{z_1,z_2\in\mathbb{T}, |z_1 - z_2|<\delta} |F(z_1) - F(z_2)| \leq \sup_{x_1,x_2\in[-1,1], |x_1-x_2|<\delta} |f(x_1) - f(x_2)| = \omega(f,\delta).$$

To fix ideas we assume that $x_1 \neq 1$ and $x_n \neq -1$, that is, case (iv). By applying Szegő transformation $w_n(\dfrac{z+\frac{1}{z}}{2}) = \dfrac{1}{2^n z^n}\prod_{j=1}^{n}(z - \alpha_j)\prod_{j=1}^{n}(z - \overline{\alpha_j})$, where $\dfrac{\alpha_j + \overline{\alpha_j}}{2} = x_j$, that is, $\alpha_j = e^{i\theta_j}$ with $\theta_j = \arccos x_j$. Hence we consider the nodal polynomial $W_{2n}(z) = 2^n z^n w_n(\dfrac{z+\frac{1}{z}}{2}) = \prod_{j=1}^{n}(z - \alpha_j)\prod_{j=1}^{n}(z - \overline{\alpha_j})$.

Now our nodal system is constituted by $\{\alpha_j\}_{j=1}^{n} \cup \{\overline{\alpha_j}\}_{j=1}^{n}$ and the arguments are $\{\theta_j\}_{j=1}^{n} \cup \{-\theta_j\}_{j=1}^{n}$. If we renumber the arguments in such a way that $-\theta_n = \theta_{n+1}, \cdots, -\theta_1 = \theta_{2n}$, then it holds that

$$\theta_{j+1} - \theta_j = \frac{2\pi}{2n} + \frac{A(j)}{n^2},$$

with $|A(j)| \leq A$ for $j = 1, \cdots, 2n$ and $\theta_{2n+1} = \theta_1$.

Let $\mathcal{L}_{-n,n-1}(F,z)$ be the Lagrange interpolation polynomial satisfying the conditions

$$\mathcal{L}_{-n,n-1}(F,\alpha_j) = \mathcal{L}_{-n,n-1}(F,\overline{\alpha_j}) = f(x_j), \quad \text{for } j = 1,\cdots,n.$$

Since F satisfies the hypothesis of Theorem 2 (i) then $\mathcal{L}_{-n,n-1}(F,z)$ converges uniformly to F on $\mathbb{T}$.

Analizying the expression of $\mathcal{L}_{-n,n-1}(F,z)$ and by taking into account that $W_{2n}(z)$ as well as $W'_{2n}(z)$ have real coefficients we get that $\mathcal{L}_{-n,n-1}(F,z)$ has real coefficients. Since it is clear that $\mathcal{L}_{-n,n-1}(F,\frac{1}{z}) \in \Lambda_{-(n-1),n}$ satisfies the same interpolation conditions, now we consider the algebraic polynomial $\frac{1}{2}\left(\mathcal{L}_{-n,n-1}(F,z) + \mathcal{L}_{-n,n-1}(F,\frac{1}{z})\right)$, which has real coefficients and satisfies the interpolation conditions (9). Since the polynomial satisfying (9) is uniquely determined, then $\frac{1}{2}\left(\mathcal{L}_{-n,n-1}(F,z) + \mathcal{L}_{-n,n-1}(F,\frac{1}{z})\right) = \ell_{n-1}(f,x)$ and it converges uniformly to f on $[-1,1]$.

When $x_1 = 1$ (case (iii)), or $x_n = -1$ (case (ii)), or $x_1 = 1$ and $x_n = -1$ (case (i)), one proceeds in a similar way and the auxiliary nodal polynomials are given by $W_{2n-1}(z) = \frac{2^n z^n}{z-1} w_n(\frac{z+\frac{1}{z}}{2}) = (z-1)\prod_{j=2}^{n}(z-\alpha_j)\prod_{j=2}^{n}(z-\overline{\alpha_j})$ or $W_{2n-1}(z) = \frac{2^n z^n}{z+1} w_n(\frac{z+\frac{1}{z}}{2}) = (z+1)\prod_{j=1}^{n-1}(z-\alpha_j)\prod_{j=1}^{n-1}(z-\overline{\alpha_j})$ or $W_{2n-2}(z) = \frac{2^n z^n}{(z-1)(z+1)} w_n(\frac{z+\frac{1}{z}}{2}) = (z-1)(z+1)\prod_{j=2}^{n-1}(z-\alpha_j)\prod_{j=2}^{n-1}(z-\overline{\alpha_j})$, respectively. Notice that in the four cases the nodal polynomials have real coefficients. □

Notice that some well known results related to Lagrange interpolation with Chebyshev and Chebyshev extended nodes are particular cases of the above theorem (see [9,20]).

Notice that from the proof of the above theorem and by applying Theorem 1 we can obtain an analogous bound, the Lebesgue constant, for this interpolatory process, that is, there exists a positive constant M such that

$$|\ell_{n-1}(f,x)| \leq M \parallel f \parallel_\infty \log n. \tag{10}$$

In order to obtain information concerning the rate of convergence, first we recall the following result about the expansion of an analytic function in a Jacobi series (see [9,21]). For a more actual version see [12].

Theorem 4. *Let $f(x)$ be analytic on the closed segment $[-1,1]$. The expansion of f in a Jacobi series, $f(x) \sim \sum_{n=0}^{\infty} a_n P_n^{(\alpha,\beta)}(x)$, is convergent in the interior of the greatest ellipse with foci at ± 1, in which f is regular. The expansion is divergent in the exterior of this ellipse and the sum R of the semi-axes of the ellipse of convergence is $R = \liminf \frac{1}{\sqrt[n]{|a_n|}}$.*

Thus, in our conditions we have the following results, which are in concordance with Theorem 2.

Theorem 5. *(i) If f is a function defined on $[-1,1]$ by $f(x) = \sum_{k=0}^{\infty} a_k T_k(x)$, where $T_k(x)$ is the Chebyshev polynomial of degree k and with $|a_k| \leq \frac{K}{k^s}$, with $k \neq 0$, $K > 0$ and $s > 1$, then the Lagrange interpolation polynomial $\ell_{n-1}(f,x)$ converges to f with rate of convergence $\mathcal{O}\left(\frac{\log n}{n^{s-1}}\right)$.*

(ii) If f is analytic on the closed segment $[-1,1]$, then the Lagrange interpolation polynomial $\ell_{n-1}(f,x)$ converges to f with rate of convergence geometric.

Proof. (i) If we decompose f like $f(x) = f_{1,n-1}(x) + f_{2,n-1}(x) = \sum_{k=0}^{n-1} a_k T_k(x) + \sum_{k=n}^{\infty} a_k T_k(x)$, we have that $\ell_{n-1}(f_{1,n-1}, x) = f_{1,n-1}(x)$ and $|f_{2,n-1}(x)| \leq \sum_{k=n}^{\infty} |a_k| \leq K \sum_{k=n}^{\infty} \frac{1}{k^s} < \frac{K}{(s-1)(n-1)^{s-1}}$, where the last inequality follows from the integral criterion. Thus, by applying the Lebesgue constant obtained before in (10), we get $|\ell_{n-1}(f_{2,n-1}, x)| \leq M \frac{K}{(s-1)(n-1)^{s-1}} \log n$. Hence

$$|f(x) - \ell_{n-1}(f, x)| = |f_{1,n-1}(x) - \ell_{n-1}(f_{1,n-1}, x) + f_{2,n-1}(x) - \ell_{n-1}(f_{2,n-1}, x)| =$$
$$|f_{2,n-1}(x) - \ell_{n-1}(f_{2,n-1}, x)| \leq \frac{K}{(s-1)(n-1)^{s-1}}(1 + M \log n) \leq T \frac{\log n}{n^{s-1}},$$

for some $T > 0$.

(ii) Since f is analytic on $[-1, 1]$, then it can be analytically continued to a neighborhood of $[-1, 1]$ in the complex plane. Hence the expansion in Chebyshev series $\sum_{n=0}^{\infty} a_n T_n(x)$ converges to f in the interior of the greatest ellipse with foci at ± 1, known as Bernstein ellipse E_R and the sum R of the semi-axes of the ellipse of convergence is $R = \liminf \frac{1}{\sqrt[n]{|a_n|}}$. Then it holds that $|a_n| \leq P r^n$, for some $0 < r < 1$ and $P > 0$.

Proceeding in the same way as before we have $f(x) = f_{1,n-1}(x) + f_{2,n-1}(x) = \sum_{k=0}^{n-1} a_k T_k(x) + \sum_{k=n}^{\infty} a_k T_k(x)$, $\ell_{n-1}(f_{1,n-1}, x) = f_{1,n-1}(x)$ and $|f_{2,n-1}(x)| \leq \sum_{k=n}^{\infty} |a_k| \leq P \frac{r^n}{1-r}$. Thus, by applying the Lebesgue constant we get $|\ell_{n-1}(f_{2,n-1}, x)| \leq MP \frac{r^n}{1-r} \log n$ and therefore

$$|f(x) - \ell_{n-1}(f, x)| = |f_{2,n-1}(x) - \ell_{n-1}(f_{2,n-1}, x)| \leq Q r^n \log n \leq Q r_1^n,$$

for some $Q > 0$ and $0 < r < r_1 < 1$. □

4. Numerical Examples

We have carried out different numerical experiments to visualize the main contributions of this article. The first examples correspond to the three cases of Theorem 2 and in all of them we work in the following way:

1. We construct the nodal systems in a quite random way. We consider four arcs or sections in the unit circumference $\mathbb{T}$. The first one begins in $\alpha_1 = 1$ and its $\frac{n}{4}$ nodes are constructed in counter clockwise sense separated by an angular length $\frac{2\pi}{n} + \epsilon$, where the ϵ are random errors determined by using the uniform distribution in $[\frac{A}{n^2} 2\pi, \frac{2A}{n^2} 2\pi]$. The fourth section begins in $\alpha_1 = 1$ and its $\frac{n}{4}$ nodes are constructed in clockwise sense with arcs of angular length $\frac{2\pi}{n} + \epsilon$, where the ϵ are random errors determined by using the uniform distribution in $[\frac{A}{n^2} 2\pi, \frac{2A}{n^2} 2\pi]$. The second section begins after the first one and its $\frac{n}{4}$ nodes are constructed in counter clockwise sense with arcs of angular length $\frac{2\pi}{n} + \epsilon$, where the ϵ are random errors determined by using the uniform distribution in $[-\frac{2A}{n^2} 2\pi, -\frac{A}{n^2} 2\pi]$. Finally, in the third section the arcs between the nodes are all equal.
   Obviously we use different values of n and we must remark that we obtain always the same results, really we must say similar results because due to our random choice we never have the same nodal system.

2. We consider a test function $F(z)$, that we detail in each example, and we always plot $F(z)$ in black.
3. We consider the Lagrange interpolation polynomial $\mathcal{L}_{-n,n-1}(F,)$, which interpolates the test function at the nodal system. We always plot $\Re(\mathcal{L}_{-n,n-1}(F,))$ in red.

These examples are devoted to visualize the items (i), (ii) and (iii) respectively of Theorem 2.

Example 1. *In this example we work with* $F(z) = 1 + 20\left(\frac{z+z^{-1}}{2}\right)\sin\left(\frac{2}{z+z^{-1}}\right)$ *for* $z \in \mathbb{T}$, *which satisfies the hypotheses of Theorem 2 (i). We take* $n = 1000$, $A = 2$ *and we use (8) to obtain* $\mathcal{L}_{-n,n-1}(F,)$.

We represent the function $F(e^{i\theta})$ *which takes real values and, as we have said, the real part of the interpolation polynomial. Notice that due to its variability, F is a quite difficult function to interpolate. Indeed, it is easy to check that* $F(e^{i\theta})$ *is not differentiable at* $\frac{\pi}{2}$.

We present in Figure 2 two graphics. On the left we have a general panoramic of the interpolation along $\mathbb{T}$ *and we have added the interpolation points in green. We must point up that the interpolatory process is successful where the function has no variability. However, we have an unsuccessful situation where the function has great variability.*

In the graphic on the right we have a detailed situation between 1.2 and 2, that is near $\frac{\pi}{2}$, *which can help us to understand the problem. According to the theory presented, we must increase the number of nodes to obtain better results in this region.*

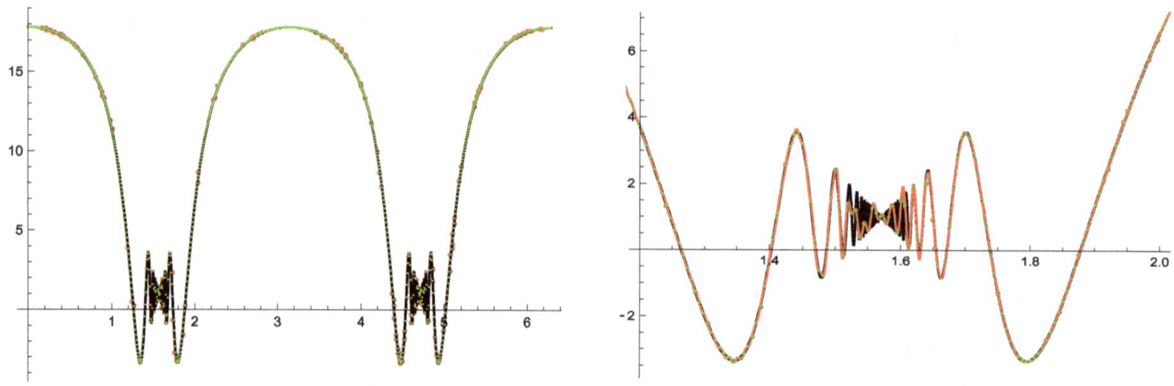

Figure 2. $F(z)$ and $\Re(\mathcal{L}_{-n,n-1}(F,z))$ with F(z)= $1 + 20(\frac{z+z^{-1}}{2})\sin(\frac{2}{z+z^{-1}})$, $z = e^{i\theta}$, $\theta \in [0, 2\pi]$, $\theta \in [1.2, 2]$ and $n = 1000$.

Example 2. *Now we consider the function defined on* $\mathbb{T}$ *by* $F(z) = \sum_{k=1}^{\infty} \frac{1}{k^6}(z^k + z^{-k})$, *which satisifies the hypotheses of Theorem 2. In the next Figure 3 we plot on the left* $F(e^{i\theta})$ *and* $\Re(\mathcal{L}_{-n,n-1}(F, e^{i\theta}))$ *for* $\theta \in [0, 2\pi]$ *and* $n = 60$. *Notice that they are indistinguishable. On the right we plot the errors given by* $\Re(\mathcal{L}_{-n,n-1}(F, e^{i\theta})) - F(e^{i\theta})$ *with* $\theta \in [0, 2\pi]$. *We point out that the errors are less or equal than* 2×10^{-8}.

In the next example we also construct an alternative interpolation polynomial based on the equispaced nodal system on $\mathbb{T}$, but using the values of the function on our nodal system. We do this because a natural criticism to our method could be that with errors as $\mathcal{O}(1/n^2)$ we can be so close to the equispaced nodal system to accept this approximation. We denote by $\mathcal{A}_{-n,n-1}(F,)$ this alternative interpolation polynomial.

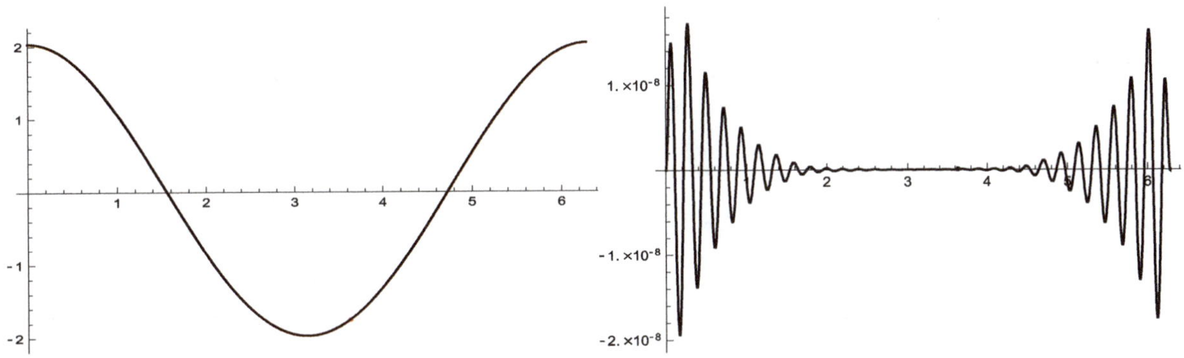

Figure 3. $F(z)$ and $\Re(\mathcal{L}_{-n,n-1}(F,z))$ and $\Re(\mathcal{L}_{-n,n-1}(F,z)) - F(z)$ with $F(z) = \sum_{k=1}^{\infty} \frac{1}{k^6}(z^k + z^{-k})$, $z = e^{i\theta}$, $\theta \in [0, 2\pi]$ and $n = 60$.

Example 3. *In this example we take* $F(z) = e^z$, $n = 24$, $A = 2$ *and we use (8) to obtain the interpolation polynomials* $\mathcal{L}_{-n,n-1}(F,)$ *and* $\mathcal{A}_{-n,n-1}(F,)$. *Taking into account that F is analytic we know that F and* $\mathcal{L}_{-n,n-1}(F,)$ *must be close. In Figure 4 we plot* $\Re(F)$ *in black,* $\Re(\mathcal{L}_{-n,n-1}(F,))$ *in red and* $\Re(\mathcal{A}_{-n,n-1}(F,))$ *in green for* $z = e^{i\theta}$ *with* $\theta \in [0, 2\pi]$. *On the left hand side we have a global vision with* $\theta \in [0, 2\pi]$ *and we can observe that* $\Re(F)$ *and* $\Re(\mathcal{L}_{-n,n-1}(F,))$ *are indistinguishable; in fact for this example the maximum error was* 3×10^{-9}.

Although $\Re(\mathcal{A}_{-n,n-1}(F,))$ *has a similar shape, see that it can drive us to catastrophic errors. On the right hand side we present a detail of the previous one, which give us an idea of the error. Notice that in general we cannot know the details of the nodal distribution.*

We have done the same with the imaginary part and we obtain the same results.

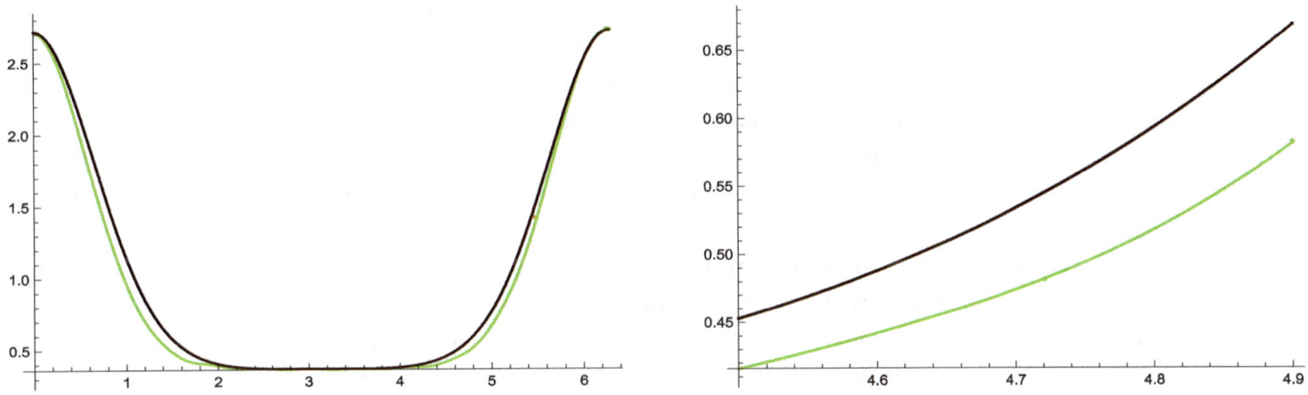

Figure 4. $\Re(F(z))$, $\Re(\mathcal{L}_{-n,n-1}(F,z))$ and $\Re(\mathcal{A}_{-n,n-1}(F,z))$ with $F(z) = e^z$, $z = e^{i\theta}$, $\theta \in [0, 2\pi]$, $\theta \in [4.5, 4.9]$ and $n = 24$.

Example 4. *Finally we choose* $F(z) = \chi_S(z)$ *defined on* $\mathbb{T}$ *as the characteristic function of the superior arc* S *of* $\mathbb{T}$, *we take* $n = 2000$, $A = 2$ *and we use expression (8) to obtain* $\mathcal{L}_{-n,n-1}(F,)$. *We know the behavior when the nodal system is related to para-orthogonal polynomials with respect to an analytic positive measure (see [10]), but we do not have a theory about the behavior of* $\mathcal{L}_{-n,n-1}(F,)$ *in our situation. We plot the results in Figure 5. Notice that the basic ideas of the Gibbs–Wilbraham phenomenon are present in this graphic, that is, the convergence of the interpolator to the function in regions which are far enough from the discontinuities and a heavy oscillation near the discontinuities. A representation of the oscillation and its amplitude, maybe, deserves a detailed study.*

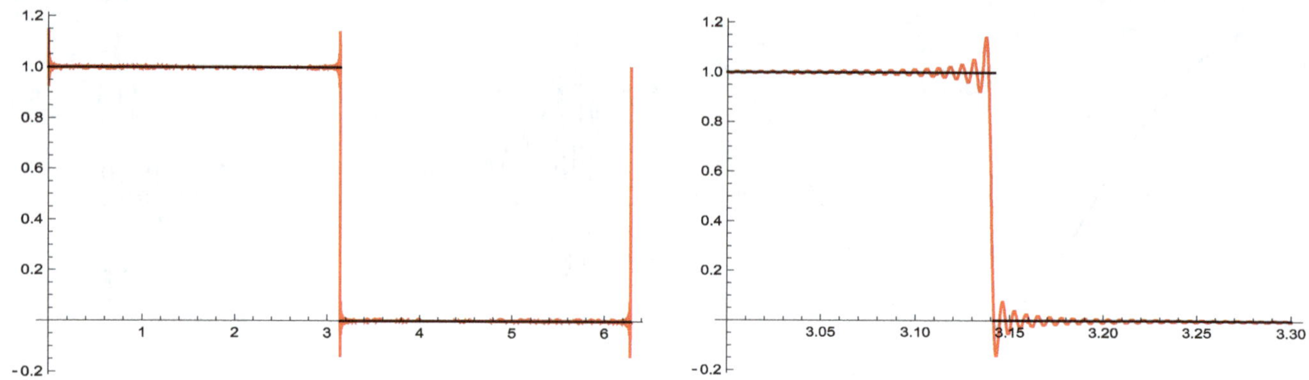

Figure 5. $F(e^{i\theta})$ and $\Re(\mathcal{L}_{-n,n-1}(F,e^{i\theta}))$ with $F(z) = \chi_S(z), z = e^{i\theta}$, $\theta \in [0,2\pi], \theta \in [3,3.3]$ and $n = 2000$.

5. Discussion

Usually, the nodal systems used for interpolation problems are strongly connected with measures on the bounded interval and on the unit circle and their corresponding orthogonal or paraorthogonal polynomials. We must point out that these choices are very suitable to construct the whole theory but in some numerical applicatons it is possible that the nodal systems do not satisfy this requisite. So, the starting point of the paper is a distribution for the nodes that can be obtained through a perturbation of the uniform distribution and, in particular of the roots of the unity, and which is more general than that related to measures and orthogonality.

The results of this article contribute to elaborate a theory over these type of nodal systems, as well as to the Lagrange interpolation theory based on these interpolatory arrays. Moreover, a theory about the rate of convergence for some types of smooth functions is given. Finally, we translate the results to perturbed Chevyshev nodal systems and to Lagrange interpolation on the bounded interval.

We think that this research could be of interest for some mechanical models that generate these types of nodal systems. As an example we consider the next problem.

Let us suppose that we are studying a equatorial characteristic $F(e^{i\theta})$ of a planet which depends on the angle θ and we have a theory which establishes that $F(e^{i\theta})$ is an analytic function. We observe the phenomenon using an observatory in the boundary of a spatial station in an elliptic orbit of period T which rotates over itself with period T_1 (with $T = n\,T_1$ and n large enough). Moreover, we take our observations when the center of the planet, the observatory and the center of the station are aligned. We can translate the problem thinking that the planet is our Sun, the spatial station is our Earth and the observatory is our city. So the time between observations is the equivalent of a solar day. It is well known that the duration of a solar day is not constant (see https://en.wikipedia.org/wiki/Equation_of_time for a brief introduction about the so called **Equation of time**), in our case have a little oscillation on T_1 and our observations are taken on a nodal system which satisfies (1). Notice that in this case we do not have a equispaced distribution nor the support of the theory of Orthogonal Polynomials. Therefore, before this paper we did not know how to use our data to reconstruct $F(e^{i\theta})$and after this paper we can be confident about the use of Lagrange interpolation.

Some future research directions could be the study of other types of interpolation on the unit circle and on the bounded interval by using these general interpolatory arrays; as well as to study the correspondig Gibbs–Wilbraham phenomena.

6. Materials and Methods

The experiments given in the section Numerical examples were obtained by using personal codes elaborated with Mathematica® 12 (Wolfram Research Europe Ltd, Long Hanborough Oxfordshire, United Kingdom). These programs to obtain the nodal points and to compute the interpolation polynomials as well as the plots of the test functions and their interpolators are available at the public repository https://www.dropbox.com/sh/0cx9chq3jfzov2w/AAA_SvL2i7HlC7ChMGpuG-Ata?dl=0 There one can find the program related to Example 2. To obtain the other examples some minor changes must be done.

Author Contributions: Conceptualization, E.B., A.C. (Alicia Cachafeiro), A.C. (Alberto Castejón) and J.M.G.-A.; Investigation, E.B., A.C. (Alicia Cachafeiro), A.C. (Alberto Castejón) and J.M.G.-A.; Software, E.B., A.C. (Alicia Cachafeiro), A.C. (Alberto Castejón) and J.M.G.-A.; Writing—original draft, E.B., A.C. (Alicia Cachafeiro), A.C. (Alberto Castejón) and J.M.G.-A. All authors have read and agreed to the published version of the manuscript.

References

1. Szili, L.; Vértesi, P. On the theorem of Géza Grünwald and Józef Marcinkiewicz. *Banach Cent. Publ.* **2011**, *95*, 251–259. [CrossRef]
2. Erdös, P.; Vértesi, P. On the almost everywhere divergence of Lagrange interpolation of polynomials for arbitrary systems of nodes. *Acta Math. Acad. Sci. Hung.* **1980**, *36*, 71–89. [CrossRef]
3. Trefethen, L.N. Six Myths of Polynomial Interpolation and Quadrature. *Math. Today (Southend-on-Sea)* **2011**, *47*, 184–188.
4. Grünwald, G. On the theory of interpolation. *Acta Math.* **1942**, *75*, 219–245. [CrossRef]
5. Erdös, P.; Turán, P. On Interpolation II: On the Distribution of the Fundamental Points of Lagrange and Hermite Interpolation. *Ann. Math.* **1938**, *9*, 703–724. [CrossRef]
6. Walsh, J.L. *Interpolation and Approximation by Rational Functions in the Complex Plane*, 5th ed.; American Mathematical Society Colloquium Publications: Providence, RI, USA, 1969; Volume 20.
7. Darius, L.; González-Vera, P. Some results about interpolation with nodes on the unit circle. *Indian J. Pure Appl. Math.* **2000**, *31*, 1273–1296.
8. Berriochoa, E.; Cachafeiro, A.; García-Amor, J.M. About nodal systems for Lagrange interpolation on the circle. *J. Appl. Math.* **2012**, *2012*. [CrossRef]
9. Szegő, G. *Orthogonal Polynomials*, 4th ed.; American Mathematical Society Colloquium Publications: Providence, RI, USA, 1975; Volume 23.
10. Berriochoa, E.; Cachafeiro, A.; García-Amor, J.M. Gibbs–Wilbraham phenomenon on Lagrange interpolation based on analytic weights on the unit circle. *J. Comput. Appl. Math.* **2020**, *365*, 112376. [CrossRef]
11. Chui, C.K.; Shen, X.-C.; Zhong, L. On Lagrange interpolation at disturbed roots of unity. *Trans. Am. Math. Soc.* **1993**, *336*, 817–830. [CrossRef]
12. Trefethen, L.N. *Approximation Theory and Approximation Practice*; Society for Industrial and Applied Mathematics (SIAM): Philadelphia, PA, USA, 2013.
13. Milovanovic, G.V.; Mitrinovic, D.S.; Rassias, T.M. *Topics in Polynomials: Extremal Problems, Inequalities, Zeros*; World Scientific Publishing Co.: Singapore, 1994.
14. Berriochoa, E.; Cachafeiro, A.; Martínez, E. About measures and nodal systems for which the Hermite interpolants uniformly converge to continuous functions on the circle and interval. *Appl. Math. Comput.* **2012**, *218*, 4813–4824. [CrossRef]
15. Berriochoa, E.; Cachafeiro, A.; García-Amor, J.M. Gibbs–Wilbraham oscillation related to an Hermite interpolation problem on the unit circle. *J. Comput. Appl. Math.* **2018**, *344*, 657–675. [CrossRef]
16. Simon, B. Fine structure of the zeros of orthogonal polynomials, I. A tale of two pictures. *Electron. Trans. Numer. Anal.* **2006**, *25*, 328–368.
17. Corless, R.M.; Fillion, N. *A Graduate Introduction to Numerical Methods*; Springer: New York, NY, USA, 2013.
18. Higham, N.J. The numerical stability of barycentric Lagrange interpolation. *IMA J. Numer. Anal.* **2004**, *24*, 547–556. [CrossRef]
19. Berrut, J.-P.; Mittelmann, H.D. Lebesgue constant minimizing linear rational interpolation of continuous functions over the interval. *Comput. Math. Appl.* **1997**, *33*, 77–86. [CrossRef]
20. Szabados, J.; Vértesi, P. *Interpolation of Functions*; World Scientific: Singapore, 1990.
21. Rivlin, T.H. *The Chebyshev Polynomials*; John Wiley & Sons: New York, NY, USA, 1974.

5

On the Performance of Variable Selection and Classification via Rank-Based Classifier

Md Showaib Rahman Sarker [†], **Michael Pokojovy** [†] **and Sangjin Kim** [*,†]

Department of Mathematical Sciences, The University of Texas at El Paso, El Paso, TX 79968, USA; msarker@miners.utep.edu (M.S.R.S.); mpokojovy@utep.edu (M.P.)

* Correspondence: skim10@utep.edu

† These authors contributed equally to this work.

Abstract: In high-dimensional gene expression data analysis, the accuracy and reliability of cancer classification and selection of important genes play a very crucial role. To identify these important genes and predict future outcomes (tumor vs. non-tumor), various methods have been proposed in the literature. But only few of them take into account correlation patterns and grouping effects among the genes. In this article, we propose a rank-based modification of the popular penalized logistic regression procedure based on a combination of ℓ_1 and ℓ_2 penalties capable of handling possible correlation among genes in different groups. While the ℓ_1 penalty maintains sparsity, the ℓ_2 penalty induces smoothness based on the information from the Laplacian matrix, which represents the correlation pattern among genes. We combined logistic regression with the BH-FDR (Benjamini and Hochberg false discovery rate) screening procedure and a newly developed rank-based selection method to come up with an optimal model retaining the important genes. Through simulation studies and real-world application to high-dimensional colon cancer gene expression data, we demonstrated that the proposed rank-based method outperforms such currently popular methods as lasso, adaptive lasso and elastic net when applied both to gene selection and classification.

Keywords: gene-expression data; ℓ_2 ridge; ℓ_1 lasso; adapative lasso; elastic net; BH-FDR; Laplacian matrix

MSC: 62F03; 62F07; 62P10

---

1. Introduction

Microarrays are an advanced and widely used technology in genomic research. Tens of thousands of genes can be analyzed simultaneously with this approach [1]. Identifying the genes related to cancer and building high-performance prediction models of maximal accuracy (tumor vs. non-tumor) based on gene expression levels are among central problems in genomic research [2–4]. Typically, in high-dimensional gene expression data analysis, the number of genes is significantly larger than the sample size, i.e., $m \gg n$. Hence, it is particularly challenging to identify those genes that are relevant to cancer disease and put forth prediction models. The main problem associated with high-dimensional data ($m \gg n$) is that of overfitting or overparametrization which leads to poor generalizability from training to test data.

Therefore, various researchers apply different types of regularization methods to overcome this "curse of dimensionality" in regression and other statistical and machine learning frameworks. These regularization approaches include, for example, the ℓ_1-penalty or lasso [5], which performs continuous shrinkage and feature selection simultaneously; smoothly clipped ℓ_1-penalty or SCAD [6], which is symmetric, non-concave and has singularities at the origin to produce sparse solutions; fussed lasso [7], which imposes the ℓ_1-penalty on the absolute difference of regression coefficients in order to enforce some smoothness of coefficients; or the adaptive lasso [8], etc. Unfortunately, ℓ_1-regularization sometimes perform inconsistently when used for variable selection [8]. In some situations, it introduces a major bias in estimated parameters in the logistic regression [9,10]. In contrast, the elastic net regularization procedure [11] as a combination of ℓ_1- and ℓ_2-penalties can successfully handle the highly correlated variables which are grouped together. Among the procedures mentioned above, elastic net and fussed lasso penalized methods are appropriate for gene expression data analysis. Unfortunately, when some prior knowledge needs to be utilized, e.g., when studying complex diseases such as cancer, those methods are not appropriate [4]. To account for a regulatory relationship between the genes and a priori knowledge about these genes, network-constrained regularization [4] is known to perform very well by incorporating a Laplacian matrix into the ℓ_2-penalty from the enet procedure. This Laplacian matrix represents a graph-structure of genes which are linked with each other. To select significant genes in high-dimensional gene expression data for classification, the graph-constrained regularization method is extended to logistic regression model [12].

Using penalized logistic regression methods [12,13] and graph-constrained procedures [4,12], we would build rank-based logistic regression method with variable screening procedure to improve the power of detecting most promising variables as well as classification capability.

The rest of this article is organized as follows. In Section 2, we describe variable screening procedure with adjusted p-values and regularization procedure for grouped and correlated predictors and present the computational algorithm. Further, we state the ranking criteria of four models and summarize the result of ranking procedure. In Section 3, we compare the proposed procedure with existing cutting edge regularization methods on simulation studies. Next, we apply four penalized logistic regression methods to the high dimensional gene expression data of colon cancer carcinoma to evaluation and comparison of the performance. Finally, we present a brief discussion of results and future research direction.

2. Materials and Methods

2.1. Adjusted p-Values: Benjamini and Hochberg False Discovery Rate (BH-FDR)

Multiple hypohtesis testing methods have been playing an important role in selecting most promising features while controlling type I error in high-dimensional settings. One of the most popular methods is BH-FDR [14,15] which is concerned with the expected proportion of incorrect number of rejections among a total number of rejections. The formula is mathematically expressed as $E\left(\frac{V}{R}\middle|R > 0\right)$, where V is the number of false positives and R is the total number of rejections. In this paper, the FDR method is used both for the purpose of prelimary variable screening both in the simulation studies and real data analysis to be presented later. The procedure of the method is as follows:

(1) Let $p_1, p_2, \ldots, p_m$ be the p-values of m hypothesis tests and sort them with the increasing oder: $p_{(1)}, p_{(2)}, \ldots, p_{(m)}$.
(2) Let $\hat{i} = \max\{i \,|\, p_{(i)} \leq \frac{iq}{m}, i = 1, \ldots, m\}$ for a given threshold q. If $\hat{i} > 1$, then reject the null hypotheses associated with $p_{(1)}, p_{(2)}, \ldots, p_{(i)}$. Otherwise, no hypotheses are rejected.

2.2. Regularized Logistic Regression

In the following, we present the regularized logistic regression model used in this paper (cf. [12]). Since this model is an integral part of our computational algorithm to be outlined in the section to follow, presenting the formula with all appropriate notations is necessary for our purposes.

Let the $n \times (m+1)$ matrix

$$X = \begin{pmatrix} 1 & x_{11} & x_{12} & \cdots & x_{1j} & \cdots & x_{1m} \\ 1 & x_{21} & x_{22} & \cdots & x_{2j} & \cdots & x_{2m} \\ \vdots & \vdots & \vdots & \ddots & \vdots & \ddots & \vdots \\ 1 & x_{i1} & x_{i2} & \cdots & x_{ij} & \cdots & x_{2m} \\ \vdots & \vdots & \vdots & \ddots & \vdots & \ddots & \vdots \\ 1 & x_{n1} & x_{n2} & \cdots & x_{nj} & \cdots & x_{nm} \end{pmatrix}$$

denote the design matrix, where n is the sample size and m is the total number of predictor variables. Without loss of generality, we assume the data are standardized with respect to each variable. This step is also performed by the `pclogit` R-package used in the present paper. Define the parameter vector $\boldsymbol{\eta} = (\beta_0, \boldsymbol{\beta})$ comprised of an intercept β_0 and m "slopes", $\beta_1, \ldots, \beta_m$. The objective function then is written as

$$f(\boldsymbol{\eta}) = -L(\boldsymbol{\eta}) + p(\boldsymbol{\beta}) \tag{1}$$

with the log-likelihood function

$$L(\boldsymbol{\eta}) = \frac{1}{n} \sum_{i=1}^{n} \left[y_i \log \pi(\boldsymbol{x}_i) + (1 - y_i) \log(1 - \pi(\boldsymbol{x}_i)) \right]$$

and resulting probabilities

$$\pi(\boldsymbol{x}_i) = \frac{\exp(\beta_0 + \boldsymbol{x}_i^T \boldsymbol{\beta})}{1 + \exp(\beta_0 + \boldsymbol{x}_i^T \boldsymbol{\beta})}.$$

Here, $p(\boldsymbol{\beta})$ is the penalty function and the response variable y_i takes the value 1 for cases and 0 for controls. The i-th individual is deemed case or control based on the probability π_i. Following [4], statistical dependence among the m explanatory variables can be modeled by a graph, which, in turn, can be described by its m-dimensional Laplacian matrix $\boldsymbol{L} = \big(L(u, v) \mid u, v \text{ vertices}\big)$ with the entries

$$L(u, v) = \begin{cases} 1, & \text{if } u = v \text{ and } d_u \neq 0, \\ -(d_u d_v)^{-\frac{1}{2}}, & \text{if } u \text{ and } v \text{ are adjacent}, \\ 0, & \text{otherwise}. \end{cases}$$

Here, d_v is the degree of a vertex v, i.e., the number of edges through this vertex. If there is no link in v (i.e., v is isolated), then $d_v = 0$. The martix L is symmetric, positive semi-definite and has 0 as the smallest eigenvalue and 2 as the largest eigenvalue. In the following, we will write $u \sim v$ to refer to adjecent vertices. The penalty term in equation (1) can is defined as

$$p(\boldsymbol{\beta}) = \lambda_1 \|\boldsymbol{\beta}\|_1 + \lambda_2 \boldsymbol{\beta}^T \boldsymbol{L} \boldsymbol{\beta} = \lambda_1 \sum_{j=1}^{m} |\beta_j| + \lambda_2 \sum_{u=1}^{m} \sum_{u \sim v} \left(\frac{\beta_u}{\sqrt{d_u}} - \frac{\beta_v}{\sqrt{d_v}} \right)^2. \tag{2}$$

Here, λ_1 and λ_2 are tuning parameters meant to control the sparsity and smoothness, $\|\boldsymbol{\beta}\|_1$ is the ℓ_1-norm and $\sum_{u \sim v}(\ldots)$ denotes the summation over all adjacent vertex pairs. When $\lambda_2 = 0$, the penalty reduces to that of lasso [5], and if $\boldsymbol{L}$ is replaced by the $m \times m$-identity matrix $\boldsymbol{I}$, the penalty corresponds to that of an elastic net [11]. If $\lambda_1 = 0$ and $\boldsymbol{L} = \boldsymbol{I}$, we arrive at ridge regression. In Equation (2), the penalty consists of ℓ_1- and ℓ_2-components. The ℓ_2-penalty is a degree-scaled difference of coefficients between linked predictors. According to [4], the predictor variables with more connections have larger coefficients. That is why small change of expression in the variables can lead to large change in response. Thus, this imposes sparsity and smoothness as well as correlation and grouping effects among variables. In case-control DNA methylation data analysis, ring networks and fully connected networks (cf. Figure 1) are typically used to describe correlation pattern of CpG sites within genes [12]. The Laplacian matrix is sparse and tri-diagonal (except for two corner elements) for ring networks and has all non-zero elements for fully connected networks. Those variables with more links produce strong grouping effects and are more likely to be selected in both networks [12].

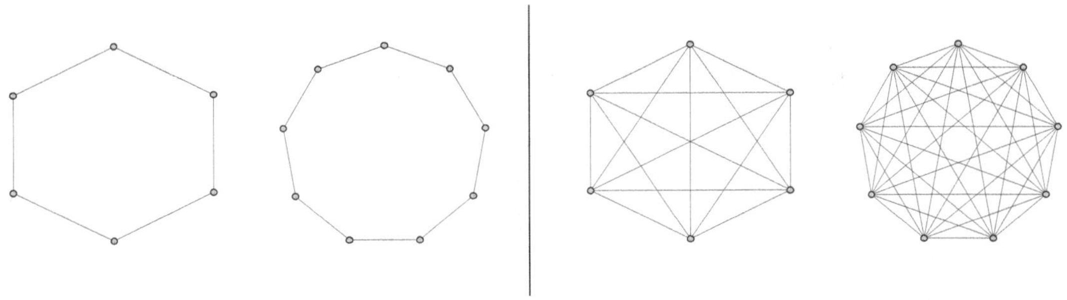

Figure 1. The ring network (left) and F.con network (right) are shown for the case there are two genes consisting of 6 and 9 CpG sites, respectively.

2.3. *Computational Algorithm*

Li & Li (2010) [16] developed an algorithm for graph-constrained regularization motivated by a coordinate descent algorithm from [17] for solving the unconstrained minimization problem for the objective in Equation (1). The algorithm implementation from the `pclogit` R-package [12,13] replaced the identity matrix by Laplacian matrix in the elastic net algorithm from the `glmnet` R-package [18]. According to Equation (1), the objective function is

$$f(\boldsymbol{\eta}) = -L(\boldsymbol{\eta}) + p(\boldsymbol{\beta}),$$

where

$$p(\boldsymbol{\beta}) = \lambda\alpha \sum_{i=1}^{m} |\beta_i| + \frac{1}{2}\lambda(1-\alpha) \sum_{u=1}^{m} \sum_{u\sim v} \left(\frac{\beta_u}{\sqrt{d_u}} - \frac{\beta_v}{\sqrt{d_v}}\right)^2 \tag{3}$$

with $\lambda = \lambda_1 + 2\lambda_2$ and $\alpha = \frac{\lambda_1}{\lambda_1+2\lambda_2}$ for some $\lambda_1, \lambda_2 > 0$.

Following [18], we perform a second-order Taylor expansion of $L(\cdot)$ around the current estimate $(\beta_0^*, \boldsymbol{\beta}^*)$ to approximate the objective $L(\cdot)$ in Equation (1) via

$$f^*(x) = -\frac{1}{2n} \sum_{i=1}^{n} q_i (t_i - \beta_0 - x_i^T \boldsymbol{\beta})^2 + p(\boldsymbol{\beta}),$$

where

$$\begin{aligned} t_i &= \beta_0^* + \boldsymbol{x}_i^T \boldsymbol{\beta}^* + q_i^{-1}(y_i - \pi^*(\boldsymbol{x}_i)), \\ q_i &= \pi^*(\boldsymbol{x}_i)(1 - \pi^*(\boldsymbol{x}_i)), \\ \pi^*(\boldsymbol{x}_i) &= 1 - (1 + \exp(\beta_0 + \boldsymbol{x}_i^T \boldsymbol{\beta}^*))^{-1}. \end{aligned}$$

Now, if all other estimates for all $v = u$ are fixed, $\beta_u = \beta_u^*$ can be computed. To update the estimate from β_u^*, we have to set the gradient of $f^*(\cdot)$ equal zero (strictly speaking, zero has to be included in the subgradient of $f^*(\cdot)$) and then solve for β_u to obtain

$$\beta_u^* = \frac{s\left(\frac{1}{n}\sum_{i=1}^{n} q_i x_{iu}(t_i - t_i^{(\tilde{u})}) + \lambda(1-\alpha)g(u), \lambda\alpha\right)}{\frac{1}{n}\sum_{i=1}^{n} q_i \boldsymbol{x}_{iu}^2 + \lambda(1-\alpha)},$$

where

$$t_i^{(\bar{u})} = \beta_0^* + \sum_{j \neq u} x_{ij}\beta_j^*,$$

$$g(u) = \sum_{u \sim v} \frac{\beta_v^*}{\sqrt{d_u d_v}} \tag{4}$$

and $s(z,r)$ denotes the "soft threshholding" operator given by

$$s(z,r) = \text{sign}(z)(|z| - r)_+ = \begin{cases} z - r, & \text{if } z > 0 \text{ and } r < |z|, \\ z + r, & \text{if } z < 0 \text{ and } r < |z|, \\ 0, & \text{otherwise.} \end{cases}$$

If the u-th predictor has no links to other predictors, then $g(u)$ in Equation (4) becomes zero, while Equation (3) takes the form

$$p(\boldsymbol{\beta}) = \lambda\alpha \sum_{i=1}^{m} |\beta_i| + \frac{1}{2}\lambda(1-\alpha) \sum_{u=1}^{m} \beta_u^2.$$

Thus, the regularization reduces to that of the elastic net (enet) procedure. In general, when the linkage is nontrivial, the term $\lambda(1-\alpha)g(u)$ is added to the elastic net to get the desired grouping effect.

2.4. *Adaptive Link-Constrained Regularization*

When there is a link between two predictors but their regression coefficients have different signs, the coefficients cannot be expected to be smooth [16]—even locally. To resolve this problem, we first need to estimate the sign of the coefficients and then refit the model with estimated signs. When the number of predictor variables is smaller than that of sample points, ordinary least squares are performed, while ridge estimates are computed, otherwise. We have to modify the Laplacian matrix in the penalty function:

$$L^*(u,v) = \begin{cases} 1, & \text{if } u = v \text{ and } d_u \neq 0, \\ -s_u s_v (d_u d_v)^{-\frac{1}{2}}, & \text{if } u \text{ and } v \text{ are adjacent,} \\ 0, & \text{otherwise} \end{cases}$$

and then update the $g(\cdot)$-function in Equation (4) via

$$g^*(u) = \sum_{u \sim v} \frac{s_u s_v \beta_v^*}{\sqrt{d_u d_v}}.$$

2.5. *Accuracy, Sensitivity, Specificity and Area under the Receiver Operating Curve (AUROC)*

We evaluated four metrics of binary classification for each of lasso, adaptive lasso, elastic net and the proposed rank based logistic regression methods to compare the performance. These metrics are accuracy, sensitivity, specificity and AUROC.

Based on the notations in Table 1, we define

$$\text{Accuracy} = \frac{a+b}{a+b+c+d}, \qquad \text{Specificity} = \frac{d}{n-m}, \qquad \text{Sensitivity} = \frac{a}{m}$$

as well as

$$\text{TPR (true positive rate)} = \frac{a}{k}, \qquad \text{FPR (false positive rate)} = \frac{b}{n-m}.$$

The last metric AUROC is related to the probability that the classifier under consideration will rank a randomly selected positive case higher than a randomly selected negative case [19]. The values of all these fours metrics—accuracy, sensitivity, specificity and AUROC—range from 0 to 1. The value of 1 represents a perfect model whereas the value of 0.5 corresponds to "coin tossing". The class prediction for each individual in binary classification is made based on a continuous random variable z. Given a threshold k as a tuning parameter, an individual is classified as "positive" if $z > k$ and "negative", otherwise. The random variable z follows a probability density $f_1(z)$ if the individual belongs to "positives" and $f_0(z)$, otherwise. So, the true positive and true negative rates are given by

$$\mathrm{TPR}(k) = \int_k^{\infty} f_1(z)\mathrm{d}z \quad \text{and} \quad \mathrm{FPR}(k) = \int_k^{\infty} f_0(z)\mathrm{d}z, \text{ respectively.}$$

Now, the AUROC statistic can be expressed as

$$A = \int_0^1 \mathrm{TPR}(\mathrm{FPR}^{-1}(z))\mathrm{d}z = \int_{-\infty}^{\infty}\int_{-\infty}^{\infty} \mathbb{1}\{k' > k\} f_1(k') f_0(k)\mathrm{d}k'\mathrm{d}k = \mathbb{P}(z_1 > z_0),$$

where z_1 and z_0 are the values of positive or negative instances, respectively.

Table 1. Confusion table: a is the number of true positives, b the number of false positives, c the number of false negatives and d the number of true negatives.

Predicted Condition	**True Condition**		
	Positive	Negative	Total
Positive	a	b	k
Nnegative	c	d	$n-k$
Total	m	$n-m$	$n = a+b+c+d$

2.6. Ranking and Best Model Selection

The penalty function in Equation (3) has two tuning parameters, namely, $\alpha \in [0,1]$ and $\lambda > 0$. The "limiting" cases $\alpha = 0$ and $\alpha = 1$ correspond to ridge and lasso regression, respectively. For a fixed value of α, the model selects more variables for smaller λ's and fewer variables for larger λ's. Theoretically, the result continuosly depends on α and should not significantly change under small perturbations of the latter [12,13]. Empirically, however, we discovered that the results produced by `pclogit` significantly vary with α. In `pclogit`, the Laplacian matrix determines the group effects of predictors and is calculated from adjacency matrix via

$$\boldsymbol{L} = \boldsymbol{D} - \boldsymbol{A},$$

where $\boldsymbol{D}$ is the degree matrix and $\boldsymbol{A}$ is the adjacency matrix. The degree-scaled difference of predictors in Equation (3) is computed from the normalized Laplacian matrix

$$\boldsymbol{L} = \boldsymbol{I} - \boldsymbol{D}^{-\frac{1}{2}}\boldsymbol{A}\boldsymbol{D}^{-\frac{1}{2}}.$$

We computed the adjacency matrix by using the information from the correlation matrix obtaining

$$A(u,v) = \begin{cases} 1, & \text{if } u \neq v \text{ and } |\mathrm{cor}(u,v)| \geq \epsilon, \\ 0, & \text{if } u = v \text{ or } |\mathrm{cor}(u,v)| < \epsilon. \end{cases}$$

Here, $\epsilon \in (0,1)$ is a specific cut-off value for correlation. So, ϵ is another tuning parameter in our model which needs to be optimally selected. In summary, to find an optimal combination of parameters α and ϵ, we make the combination of tuning parameter α and ϵ, where the total number of combinations is given by

$$C = K \times L$$

with K and L being the number of ϵ and α values, respectively. We compared the performance for each of different combinations with T resamplings. The (negative) measure of performance for each combination is the misclassification or error rate. The pair (α, ϵ) producing the smallest misclassification rate is declared optimal and used in the next step. The sparse coefficient matrix with dimensions $m \times$ nlam (nlam = number of λ's) is used in `pclogit` (cf. [12,13]). By default, nlam = 100. We extracted all predictors with non-zero coefficients for each of λ values. Then we built 100 logistic regression models. Given estimated parameter values $\boldsymbol{\beta}$, we have the estimated class probability for a predictor vector $\boldsymbol{x}$ at each of λ values.

$$\pi(\boldsymbol{x}) = \frac{\exp(\boldsymbol{x}^T\boldsymbol{\beta})}{1 + \exp(\boldsymbol{x}^T\boldsymbol{\beta})}$$

Using the "naïve" Bayesian approach, we infer $y = 1$ if $\pi \geq 0.5$ and $y = 0$, otherwise. The values of accuracy, sensitivity, specificity and AUROC statistics are computed for each of 100 models and ranked in an increasing order by their values. Note that AUROC method does not use a fixed cut-off value, e.g., $c = 0.5$, but rather describes the overall performance with all possible cut-off values in the decision rule. Let R_{ij}, $i = 1, 2, 3, 4$, $j = 1, 2, \ldots, 100$, comprise the ranking matrix $\boldsymbol{R}$. The first row, i.e., $i = 1$, displays the ranking of models with respect to their accuracy. Similarly, $i = 2$ ranks the models with respect to their sensitivity, $i = 3$, in terms of specificity and $i = 4$ by AUROC. Suppose, $R_{1,5} > R_{1,8}$. Then in the 1st row (i.e., in terms of accuracy), model 5 outperforms model 8. We calculate the column means ($\bar{R}_{.j}$) of the $\boldsymbol{R}$ matrix. The column with the highest overall mean value of accuracy, sensitivity, specificity and AUROC will be chosen as the resulting optimal model. Note that there is a one-to-one correspondence between columns and the 100 competing models. In (the unlikely) case of two or more columns producing the same mean, the column with a smaller index j is selected since the model represented by such column is more parsimonious. Formally, suppose p and q, $p > q$, are two column indices in the $\boldsymbol{R}$ matrix. If $\bar{R}_{.p} = \bar{R}_{.q} = \max_r \bar{R}_{.r}$, the q-th column will be selected and the associated model becomes our proposed rank-based penalized logistic regression model.

3. Results

3.1. Analysis of Simulated Data

We conducted extensive simulation studies to compare the performance in terms of accuracy, sensitivity, specificity and AUROC as well as the power of detecting true important variables by the proposed method with the performance of such three prominent regularized logistic regression methods as lasso, adaptive lasso and elastic net. We decided to focus on these (meanwhile) classical methods due to their popularity both in the literature and applications. Some of their very recently developed comptetitors such as [20] (R-package `SelectiveInference`) and [21] (R-package `islasso`) are currently gaining attention from the community and will be used as benchmarks in our future research.

Continuing with the description of our simulation study, all predictors $\boldsymbol{x}$ were generated from a multivariate normal distribution with the following probability density function

$$f(\boldsymbol{x}) = \left(\frac{1}{2\pi}\right)^{\frac{m}{2}} \frac{1}{\sqrt{\det(\boldsymbol{\Sigma})}} \exp\left(-\frac{1}{2}(\boldsymbol{x} - \boldsymbol{\mu})^T \boldsymbol{\Sigma}^{-1} (\boldsymbol{x} - \boldsymbol{\mu})\right)$$

with an m-dimensional mean vector $\boldsymbol{\mu}$ and an $(m \times m)$-dimensional covariance matrix $\boldsymbol{\Sigma}$. Writing out the covariance matrix

$$\boldsymbol{\Sigma} = (\sigma_{ij}) = \begin{pmatrix} \sigma_{11} & \sigma_{12} & \sigma_{13} & \cdots & \sigma_{1m} \\ \sigma_{21} & \sigma_{22} & \sigma_{23} & \cdots & \sigma_{2m} \\ \vdots & \vdots & \vdots & \ddots & \vdots \\ \sigma_{m1} & \sigma_{m2} & \sigma_{m3} & \cdots & \sigma_{mm} \end{pmatrix} \quad \text{with} \quad \sigma_{ii} = \sigma_i^2,$$

the correlation matrix $\boldsymbol{M}$ can be expressed as

$$\boldsymbol{M} = (\rho_{ij}) = \begin{pmatrix} \rho_{11} & \rho_{12} & \rho_{13} & \cdots & \rho_{1m} \\ \rho_{21} & \rho_{22} & \rho_{23} & \cdots & \rho_{2m} \\ \vdots & \vdots & \vdots & \ddots & \vdots \\ \rho_{m1} & \rho_{m2} & \rho_{m3} & \cdots & \rho_{mm} \end{pmatrix} \quad \text{with} \quad \rho_{ij} = \frac{\sigma_{ij}}{\sqrt{\sigma_{ii}^2 \sigma_{jj}^2}}.$$

The binary response variable is generated using Bernoulli distribution with individual probability (π) defined as

$$\pi(\boldsymbol{x}) = \frac{1}{1 + \exp(-\boldsymbol{x}^T \boldsymbol{\beta})},$$

$\boldsymbol{x}$ is the matrix of true important variables and $\boldsymbol{\beta}$ is the associated preassigned regression coefficients. Next, we present the details of the three different simulation scenarios considered.

- Under scenario 1, each of the simulated datasets has 200 observations and 1000 predictors. Here, for all $\boldsymbol{x}$ vectors, we let $\boldsymbol{\mu} = \boldsymbol{0}$ and $\mathbf{Var}(x_j) = 0.3$. Pairwise correlation of $\rho = 0.4$ was applied to the first eight variables, while the remaining 992 variables were left uncorrelated. The $\boldsymbol{\beta}$-vector was chosen as

$$\boldsymbol{\beta} = (\underbrace{2, 2, 2, 2, 2,}_{\text{5 entries}} \underbrace{3, 3, 3}_{\text{3 entries}}, \underbrace{0, 0, 0, \ldots, 0}_{\text{992 entries}}).$$

  Each of the datasets was split into training and test sets with equal proportions.

- The datasets under scenario 2 also have 200 observations and 1000 predictors. Again, $\boldsymbol{\mu} = \boldsymbol{0}$ and $\mathbf{Var}(x_j) = 0.3$. Now, the first five variables were assumed to have a correlation of $\rho = 0.4$. The remaining 995 variables were independent. The $\boldsymbol{\beta}$-vector was selected as

$$\boldsymbol{\beta} = (\underbrace{2.0, 2.0, 2.0, 2.7, 2.0, 2.0, 2.5, 2.7, -2.8, 3.0, 2.6, 3.0, 3.0, 3.0, 3.0,}_{\text{15 entries}} \underbrace{0, 0, 0, \ldots, 0}_{\text{985 entries}}).$$

  Each of the datasets was split into training and test sets with equal proportions.

- Under the last scenario 3, each of the datasets has 150 observations and 1000 predictors. We let $\boldsymbol{\mu} = \boldsymbol{0}$ and $\mathbf{Var}(x_j) = 0.4$. The first five variables were assigned into a correlation value of $\rho = 0.3$, while the variables with indices from 11 to 30 were chosen to have the correlation value of $\rho = 0.6$. Outside of these two blocks, the variables were assumed uncorrelated. The $\boldsymbol{\beta}$-vector was chosen

$$\boldsymbol{\beta} = (\underbrace{2.0, 2.0, 2.0, 2.0, 2.0, 2.5, -2.6, 2.7, 3.0, -2.9, 2.0, 2.0, 2.0, 2.0, 2.0,}_{\text{15 entries}}$$
$$\underbrace{2.5, -2.0, 2.7, 3.0, -2.5, 2.0, 2.0, 2.0, 2.0, 2.0, 2.5, -2.0, 2.7, 3.0, -2.5,}_{\text{15 entries}} \underbrace{0, 0, 0, \ldots, 0}_{\text{970 entries}}).$$

  The dataset was split into training and test sets with ratio of 70 to 30.

We compared the proposed rank-based penalized logistic regression method with lasso, adaptive lasso and elastic net methods from the glmnet R-package [11]. Algorithm 1 summarizes the procedure to calculate the average value of accuracy, sensitivity, specificity and AUROC based on a given number of iterations for each of the three simulation scenarios.

Algorithm 1 Calculation of overall mean and standard deviation on simulation studies

Step 1: Generate the data on each of the three simulation scenarios.
Step 2: Split the data into training and test sets randomly with the ratio of 70 to 30.
Step 3: Screen the variables using BH-FDR based on the training dataset.
Step 4: Plug the screened variables to each of the four methods.
Step 5: Calculate the values of Accuracy, Sensitivity, Specificity and AUROC for each of the methods.
Step 6: Repeats Step 1–5 to achieve a given number of replications.
Step 7: Calculate the means and standard deviations for each of the methods.

In Table 2, we compare the estimated mean and standard deviation of accuracy, sensitivity, specificity and AUROC values based on 200 iterations under correlation structure of $\rho = 0.4$ in the simulation of scenario 1. The proposed rank-based penalized method shows the highest accuracy of 0.963 with the standard deviation of 0.02, sensitivity of 0.961 with standard deviation of 0.03, specificity of 0.965 with standard deviation of 0.03. In addition, it yields the same AUROC of 0.995 with standard deviation of 0.01 as elastic net and adaptive lasso.

Table 2. Comparison of the performance among the four methods over 200 replications under simulation scenario 1. The values in parentheseses are the standard deviations.

Method	Accuracy	Sensitivity	Specificity	AUROC
rank-based	0.963 (0.02)	0.961 (0.03)	0.965 (0.03)	0.995 (0.01)
lasso	0.953 (0.03)	0.952 (0.04)	0.955 (0.03)	0.993 (0.01)
alasso	0.957 (0.03)	0.955 (0.04)	0.960 (0.03)	0.995 (0.01)
enet	0.961 (0.02)	0.959 (0.04)	0.962 (0.03)	0.995 (0.01)

In Table 3, we compare estimated mean and standard deviation of accuracy, sensitivity, specificity and AUROC values using 200 iterations under correlation structure of $\rho = 0.4$ in the simulation of scenario 2. The proposed rank-based method also shows highest accuracy of 0.831 with standard deviation 0.04, sensitivity of 0.833 with standard deviation of 0.06, specificity of 0.829 with standard deviation of 0.05. In addition, the proposed method produces AUROC of 0.913 with standard deviation of 0.03. This is the second highest value which is slightly lower than the AUROC value of the elastic net.

Table 3. Comparison of the performance among the four methods over 200 replications under simulation scenario 2. The values in parentheseses are the standard deviations.

Method	Accuracy	Sensitivity	Specificity	AUROC
rank-based	0.831 (0.04)	0.833 (0.03)	0.829 (0.05)	0.913 (0.03)
lasso	0.826 (0.05)	0.827 (0.07)	0.825 (0.09)	0.910 (0.04)
alasso	0.815 (0.04)	0.814 (0.07)	0.815 (0.07)	0.902 (0.04)
enet	0.826 (0.04)	0.828 (0.07)	0.825 (0.07)	0.915 (0.03)

In Table 4, we compare estimated mean and standard deviation of accuracy, sensitivity, specificity and AUROC values with 150 iterations under correlation structure of $\rho = 0.3$ and $\rho = 0.6$ in simulation of scenario 3. The proposed method shows highest accuracy of 0.916 with standard deviation of 0.04, sensitivity of 0.919 with standard deviation of 0.06, specificity of 0.912 with standard deviation of 0.06 and AUROC of 0.977 with standard deviation of 0.02.

Table 4. Comparison of the performance among the four methods over 150 replications under simulation scenario 3. The values in parentheseses are the standard deviations.

Method	Accuracy	Sensitivity	Specificity	AUROC
rank-based	0.916 (0.04)	0.919 (0.06)	0.912 (0.06)	0.977 (0.02)
lasso	0.888 (0.04)	0.898 (0.06)	0.880 (0.07)	0.963 (0.02)
alasso	0.866 (0.04)	0.877 (0.07)	0.855 (0.07)	0.949 (0.03)
enet	0.909 (0.04)	0.916 (0.06)	0.903 (0.06)	0.975 (0.02)

Furthermore, we compared the performance in terms of selecting the number of true important variables by each of the four methods under three different simulation scenarios. First, we performed multiple hypothesis testing with BH-FDR [15] to reduce the dimensionality of the data. After performing a screening step to retain the relevant variables, we used them as input for the proposed rank-based penalized method with the regularization step outlined in Section 2.3. We illustrate the performance of variable selection with boxplots in Figures 2–4 for simulation scenarios 1, 2, and 3. Each figure displays two boxplots, which, in turn, depict the distribution of the number of variables selected (NVS) and the number of true important variables (NTIV) within the number of variables selected (NVS) with each of the four methods computed based on the given number of iterations in each of the three simulation scenarios.

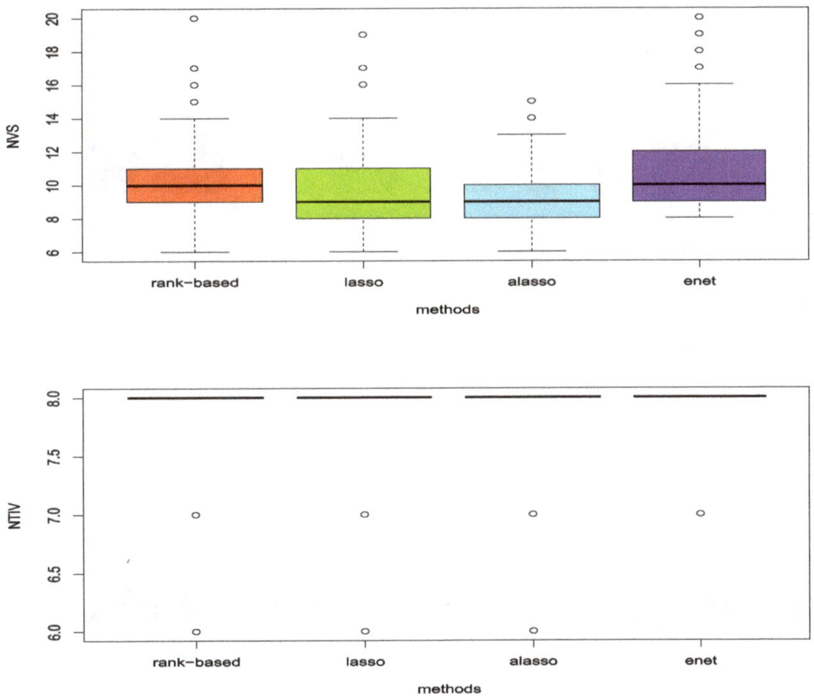

Figure 2. Boxplots of total number of variables (NVS) selected and the number of true important variables (NTIV) within the number of variables selected with four different models under scenario 1 based on 200 replications.

Figure 2 reports that the proposed rank-based method has a slightly higher median number of variables selected (displayed as a thick line in the upper boxplots) than lasso, adaptive lasso and elastic net under scenario 1. The lower boxplots show that all four methods performed head-to-head for selection of true important variables under scenario 1 with 200 replications. Table 5 compares the mean and the standard deviation (in parentheses) of the number of variables (NVS) selected and the number of true important variables (NTIV) in NVS for each of the four methods over 200 replications. The proposed rank-based method and elastic net performed head-to-head while slightly outperforming lasso and adaptive lasso.

Table 5. Estimated mean and standard deviation of number of variables (NVS) selected and the number of true important variables (NTIV) among NVS with four different models under simulated scenario 1 with 200 replications. The values in parentheseses are standard deviations.

Method	NVS	NTIV
rank-based	10.465 (2.24)	7.975 (0.19)
lasso	9.885 (1.98)	7.880 (0.37)
alasso	9.475 (1.47)	7.970 (0.20)
enet	10.805 (2.37)	7.975 (0.16)

Figure 3 suggests the proposed method has a marginally higher median number of variable selected compared to the other three methods in the upper boxplot. It is also clear that the proposed method has a slightly higher median number of true important variables in the lower boxplot on scenarios 2 computed with 200 replications. Table 6 confirms that the rank-based penalized method has the highest mean both for selecting the number of variables and important variables.

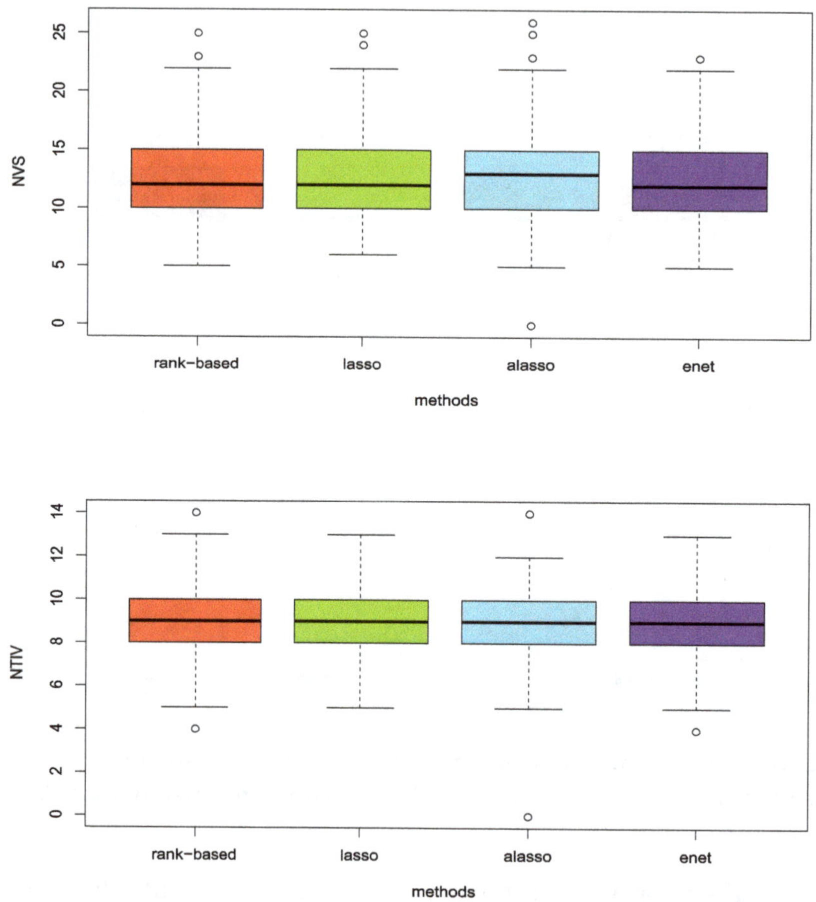

Figure 3. Boxplots of total number of variables (NVS) selected and the number of true important variables (NTIV) within the number of variables selected with four different models on scenario 2 based on 200 replications.

Table 6. Estimated mean and standard deviation of number of variables (NVS) selected and the number of true important variables (NTIV) among NVS in four different models under simulation scenario 2 with 200 replications. The values in parentheseses are standard deviations.

Method	NVS	NTIV
rank-based	13.675 (3.95)	9.345 (1.62)
lasso	12.750 (3.50)	8.905 (1.81)
alasso	11.965 (3.18)	8.720 (1.73)
enet	13.115 (3.92)	9.105 (1.73)

In Figure 4, the upper boxplot demonstrates that the proposed rank-based method has the highest median number of variables selected, elastic net has second highest median, lasso has third largest median and adaptive lasso has the smallest median under scenario 3 based on 150 replications. The lower boxplots also show that the proposed rank-based method has the highest median number of true important variables selected. However, adaptive lasso has a higher median number of true important variables than lasso unlike the upper boxplots. Thus, the proposed rank based-method clearly outperforms other three methods under high-correlation settings among variables.

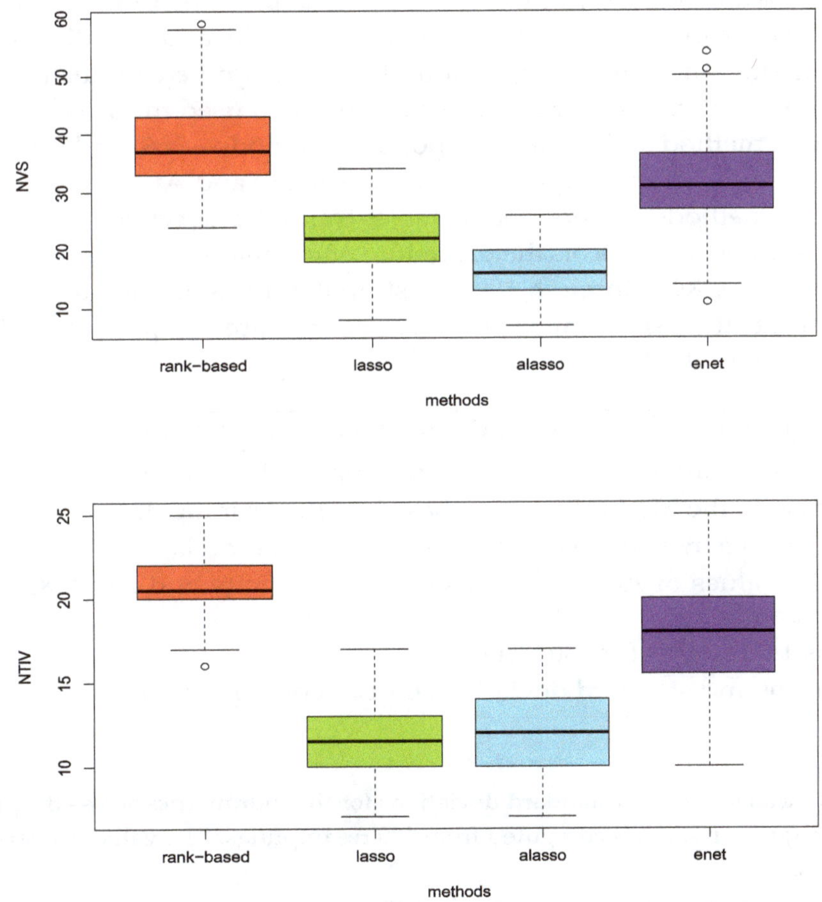

Figure 4. Boxplots of total number of variables (NVS) selected and the number of true important variables (NTIV) within the number of variables selected with four different models under scenario 3 based on 150 replications.

Table 7 summarizes the number of variables selected and true important variables selected across the four methods under the high-correlation setting among variables computed from 150 replications.

The proposed rank-based method has the highest mean number of overall variables selected and true important variables selected.

Table 7. Estimated mean and standard deviation of number of variables (NVS) selected and the number of true important variables (NTIV) among NVS in four different models under simulation scenario 3 with 150 replications. The values in parentheseses are standard deviations.

Method	NVS	NTIV
rank-based	37.830 (7.14)	20.770 (1.88)
lasso	22.010 (4.96)	11.780 (2.21)
alasso	16.430 (4.39)	11.920 (2.28)
enet	32.270 (8.55)	17.600 (3.15)

3.2. *Real Data Example*

We applied four logistic regression methods to select differentially expressed genes and assess their discrimination capability between colon cancer cases and healthy controls using high-dimensional gene expression data [22]. The colon cancer gene expression dataset is available at [23]. It contains 2000 genes with the highest minimal intensity across 62 tissues. The data were measured on 40 colon tumor samples and 22 normal colon tissue samples. We split the data set into training and testing sets with proportions 70% and 30%, respectively. To detect significantly differentially expressed genes for high-dimensional colon cancer carcinoma and measure classification prediction, we adapted two step procedures of filtering and variable selection. First, we applied BH-FDR [15] to select most promising candidates of genes as a preprocessing step and then used the screened genes as input to the proposed rank-based method and three other popular methods—lasso, adaptive lasso and elastic net. The performance in terms of accuracy, sensitivity, specificity and AUROC as well as the selection probabilities for the four methods are reported in Tables 8 and 9, respectively.

Algorithm 2 outlines above protocols the procedure of calculating the average values of accuracy, sensitivity, specificity and AUROC through 100 bootstrap iterations applied to the colon cancer gene expression data. In Table 8, the performance of all four metrics are computed based on 100 iterations of resampled subsets of individuals.

Algorithm 2 Calculation of mean with standard deviation on colon cancer data

Step 1: Split the data into training and test sets randomly with the ratio of 70 to 30.

Step 2: Screen genes with the BH-FDR method based on the training data.

Step 3: Plug the screened genes as the input to each of four methods.

Step 4: Calculate the values of Accuracy, Sensitivity, Specificity and AUROC across each of the methods on the test data.

Step 5: Repeat Steps 1 through 4 for 100 times.

Step 6: Calculate means and standard deviations for each of the methods.

Table 8. Estimated mean values and standard deviations for the four metrics across the four competing penalized logistic regression models computed from 100 resamplings. The values in parentheseses are standard deviations.

Colon Cancer Data Analysis Based on 100 Times Resmpling				
Method	**Accuracy**	**Sensitivity**	**Specificity**	**AUROC**
rank-based	0.853 (0.08)	0.860 (0.13)	0.840 (0.13)	0.917 (0.06)
lasso	0.801 (0.09)	0.911 (0.07)	0.637 (0.21)	0.897 (0.08)
adaptive lasso	0.804 (0.09)	0.869 (0.09)	0.719 (0.21)	0.877 (0.08)
elastic net	0.802 (0.09)	0.917 (0.07)	0.640 (0.22)	0.903 (0.07)

The average AUROC of 0.853 with standard deviation of 0.06 in the proposed rank-based method has the highest value compared to other three methods. Also the accuracy of 0.853 with standard deviation of 0.08 are optimal among the four methods. The values of sensitivity (0.860) and specificity (0.840) are also better than those of the other three methods. In summary, it is fair to conclude that the proposed rank-based method outperforms the other three popular penalized logistic regression methods. Table 9 shows top 5 ranked genes with highest selection probabilities for the proposed rank-based method, lasso, adaptive lasso and elastic net. An expressed sequence tag (EST) of Hsa.1660 associated with colon cancer carcinoma is found by all four methods. Hsa.36689 [24,25] is shown and top ranked by the proposed method, lasso and elastic net. Hsa692 also appeared and is second ranked by the proposed method, lasso and elastic net. In addition, Hsa.37937 is shown and is third and second ranked by the proposed method and elastic net, respectively.

Table 9. List of top 5 ranked genes across rank-based, lasso, adaptive and elastic net. An extra asterix (*) sign is put next to a gene each time the gene is selected by one of four methods.

EST Name	Gene ID	Gene Description	Selection Probability
		Rank-Based	
***Hsa.36689	Z50753	H.sapiens mRNA for GCAP-II/uroguanylin precursor	1.00
***Hsa.692.2	M76378	Human cysteine-rich protein (CRP) gene, exons 5 and 6	0.99
**Hsa.37937	R87126	Myosin heavy chain,nonmuscle(Gallus gallus)	0.97
****Hsa.1660	H55916	Peptidyl-prolyl cis-trans isomerase, mitrochondrial precursor(human)	0.91
Hsa.1832	R44887	nedd5 protein (Mus musculus)	0.90
		Lasso	
***Hsa.36689	Z50753	H.sapiens mRNA for GCAP-II/uroguanylin precursor	0.87
Hsa.692.2	M76378	Human cysteine-rich protein (CRP) gene, exons 5 and 6	0.82
*****Hsa.1660	H55916	Peptidyl-prolyl cis-trans isomerase, mitrochondrial precursor(human)	0.66
Hsa.6814	H08393	Collagen alpha 2(XI) chain(Homo sapiens)	0.52
Hsa.8147	M63391	Human desmin gene, complete cds	0.50
		Adaptive Lasso	
Hsa.1454	M82919	H. gamma amino butyric acid(GABAA)receptor beta3 subunit mRNA,cds	0.83
Hsa.6814	H08393	Collagen alpha 2(XI) chain(Homo sapiens)	0.77
****Hsa.1660	H55916	Peptidyl-prolyl cis-trans isomerase, mitrochondrial precursor(human)	0.77
Hsa.14069	T67077	Sodium/Potasssium-transporting atpase gamma chain(Ovis aries)	0.69
Hsa.2456	U25138	Human MaxiK potassium channel beta subunit mRNA, complete cds	0.55
		Elastic Net	
***Hsa.36689	Z50753	H.sapiens mRNA for GCAP-II/uroguanylin precursor	0.98
**Hsa.37937	R87126	Myosin heavy chain,nonmuscle(Gallus gallus)	0.94
***Hsa.692.2	M76378	Human cysteine-rich protein (CRP) gene, exons 5 and 6	0.94
Hsa.8147	M63391	Human desmin gene, complete cds	0.91
****Hsa.1660	H55916	Peptidyl-prolyl cis-trans isomerase, mitrochondrial precursor(human)	0.84

4. Discussion

In this paper, we proposed a new rank-based penalized logistic regression method to improve classification performance and the power of variable selection in high-dimensional data with strong correlation structure.

Our simulation studies demonstrated that the proposed method improves not only the performance of classification or class prediction but also the detection of true important variables under various correlation settings among features when compared to existing popular regularization methods such as lasso, adaptive lasso, and elastic net. As demonstrated by simulation studies, if the true important variables are not passed through the filtering method such as BH-FDR, their chance of being selected in the final model decreases signficantly, thus, leading to reduction in variable selection and classification performance. Therefore, effective filtering methods which are likely to retain as many most promising variables as possible are indispensable.

Applied to high-dimensional colon gene expression data, the proposed rank-based logistic regresson method with BH-FDR screening produced the highest average AUROC value of 0.917 with standard deviation of 0.06 and accuracy of 0.853 with standard deviation of 0.08 using 100 resampling steps. The proposed method produced a good balance between sensitivity and specificity in contrast to other methods. Elastic net demonstrated the second best peformance with an average AUROC value of 0.903 with standard deviation of 0.07. A probable reason is that elastic net accounts for group correlation effects. In addition, we compared top 5 ranked ESTs across the proposed method, lasso, adaptive lasso and elastic net [12]. They had a common EST of Hsa.1660 associated to colon cancer data. We also found that Hsa.36689 was both deemed important and top ranked by the proposed method, lasso and elastic net. This also applied to Hsa.692, which was deemed important and second top ranked by the proposed method and lasso, whereas it was only third-ranked by the elastic net. Hsa.37937 was detected by both the proposed method and the elastic net. Hence, the four ESTs mentioned appear to be promising candidate biomarkers associated with colon cancer carcinoma. The function of the genes corresponding to ESTs is summarized in Table 9.

5. Conclusions

In this study the proposed rank-based classifier demonstrated the superiority in not only classification prediction but also the power of detecting true important variables when compared to lasso, adaptive lasso, and elastic net through the extensive simulation studies. Besides, in the application of high-dimensional colon cancer gene expression data, the proposed classifier showed the best performance in terms of accuracy and AUROC among the four classifiers considered in the paper. As a future research, we would develop the methodology of variable selection and compare the performance with those of most recent competitors such as [20,21,26,27], etc.

Author Contributions: All authors have equally contributed to this work. All authors wrote, read, and approved the final manuscript.

Acknowledgments: We would like to thank the reviewers for their valuable comments.

References

1. Houwelingen, H.C.V.; Bruinsma, T.; Hart, A.A.M.; Veer, L.J.V.; Wessels, L.F.A. Cross-validated Cox regression on microarray gene expression data. *Stat. Med.* **2006**, *25*, 3201–3216. [CrossRef] [PubMed]
2. Lofti, E.; Keshavarz, A. Gene expression microarray classification using PCA–BEL. *Comput. Biol. Med.* **2014**, *54*, 180–187.
3. Algamal, Z.Y.; Lee, M.H. Penalized logistic regression with the adaptive LASSO for gene selection in high-dimensional cancer classification. *Expert Syst. Appl.* **2015**, *42*, 9326–9332. [CrossRef]
4. Li, C.; Li, H. Network-constrained regularization and variable selection for analysis of genomic data. *Bioinformatics* **2008**, *24*, 1175–1182. [CrossRef] [PubMed]
5. Tibshirani, R. Regression shrinkage and selection via the Lasso. *J. R. Stat. Soc. B* **1996**, *58*, 267–288. [CrossRef]
6. Fan, J.; Li, R. Variable selection via nonconcave penalized likelihood and its oracle properties. *J. Am. Stat. Assoc.* **2001**, *96*, 1175–1182. [CrossRef]
7. Tibshirani, R.; Saunders, M. Sparsity and smoothness via the fused lasso. *J. R. Stat. Soc. B* **2005**, *67*, 91–108. [CrossRef]
8. Zou, H. The adaptive Lasso and its oracle properties. *J. Am. Stat. Assoc.* **2006**, *101*, 1418–1429. [CrossRef]
9. Meinshausen, N.; Yu, B. Lasso-type recovery of sparse representations for high-dimensional data. *Ann. Stat.* **2009**, *37*, 246–270. [CrossRef]
10. Huang, H.; Liu, X.Y.; Liang, Y. Feature selection and cancer classification via sparse logistic regression with the hybrid $L_{1/2+2}$ regularization. *PLoS ONE* **2009**, *11*, e0149675. [CrossRef]
11. Zou, H.; Hastie, T. Regularization and variable selection via the elastic net. *J. R. Stat. Soc. B* **2005**, *67*, 301–320. [CrossRef]

12. Sun, H.; Wang, S. Penalized logistic regression for high-dimensional DNA methylation data with case-control studies. *Bioinformatics* **2012**, *28*, 1368–1375. [CrossRef] [PubMed]
13. Sun, H.; Wang, S. Network-based regularization for matched case-control analysis of high-dimensional DNA methylation data. *Stat. Med.* **2012**, *32*, 2127–2139. [CrossRef] [PubMed]
14. Reiner, H.; Yekutieli, D.; Benjamin, Y. Identifying differentially expressed genes using false discovery rate controlling procedures. *Bioinformatics* **2003**, *19*, 368–375. [CrossRef]
15. Benjamini, Y.; Hochberg, Y. Controlling the false discovery rate: A practical and powerful approach to multiple testing. *J. R. Stat. Soc. B* **1995**, *57*, 289–300. [CrossRef]
16. Li, C.; Li, H. Variable selection and regression analysis for graph-structured covariates with an application to genomics. *Ann. Appl. Stat.* **2010**, *4*, 1498–1516. [CrossRef]
17. Friedman, J.; Hastie, T.; Hofling, H.; Tibshirani, R. Pathwise coordinate optimization. *Ann. Appl. Stat.* **2007**, *1*, 302–332. [CrossRef]
18. Friedman, J.; Hastie, T.; Tibshirani, R. Regularization paths for generalized linear models via coordinate descent. *J. Stat. Softw.* **2010**, *33*, 1–22. [CrossRef]
19. Fawcett, T. An introduction to ROC analysis. *Pattern Recognit. Lett.* **2006**, *27*, 861–874. [CrossRef]
20. Lee, J.D.; Sun, D.L.; Sun, Y.; Taylor, J.E. Exact post-selection inference, with application to the lasso. *Ann. Stat.* **2016**, *44*, 907–927. [CrossRef]
21. Cilluffo, G.; Sottile, G.; La Grutta, S.; Muggeo, V.M.R. The Induced Smoothed lasso: A practical framework for hypothesis testing in high dimensional regression. *Stat. Methods Med. Res.* **2019**, doi:10.1177/0962280219842890. [CrossRef] [PubMed]
22. Alon, U.; Barakai, N.; Notterman, D.A.; Gish, K.; Ybarra, S.; Mack, D.; Levine, A.J. Broad patterns of gene expression revealed by clustering analysis of tumor and normal colon tissues probed by oligonucleotide arrays. *Proc. Natl. Acad. Sci. USA* **1999**, *96*, 6745–6750. [CrossRef]
23. Available online: http://genomics-pubs.princeton.edu/oncology/affydata/index.html (accessed on 25 April 2019).
24. Ding, Y.; Wilkins, D. A simple and efficient algorithm for gene selection using sparse logistic regression. *Bioinformatics* **2003**, *19*, 2246–2253.
25. Li, Y.; Campbell, C.; Tipping, M. Bayesian automatic relevance determination algorithms for classfifying gene expression data. *Bioinformatics* **2002**, *18*, 1332–1339. [CrossRef]
26. Frost, H.R.; Amos, C.I. Gene set selection via LASSO penalized regression (SLPR). *Nucleic Acids Res.* **2017**, doi:10.1093/nar/gkx291. [CrossRef]
27. Boulesteix, A.L.; De, B.R.; Jiang, X.; Fuchs, M. IPF-LASSO: Integrative L_1-Penalized Regression with Penalty Factors for Prediction Based on Multi-Omics Data. *Comput. Math. Methods Med.* **2017**. [CrossRef] [PubMed]

6

A Functional Data Analysis for Assessing the Impact of a Retrofitting in the Energy Performance of a Building

Miguel Martínez Comesaña [1,*], Sandra Martínez Mariño [1], Pablo Eguía Oller [1], Enrique Granada Álvarez [1] and Aitor Erkoreka González [2]

[1] Department of Mechanical Engineering, Heat Engines and Fluid Mechanics, Industrial Engineering School, University of Vigo, Maxwell s/n, 36310 Vigo, Spain; samartinez@uvigo.es (S.M.M.); peguia@uvigo.es (P.E.O.); egranada@uvigo.es (E.G.Á.)

[2] ENEDI Research Group, Department of Thermal Engineering, University of the Basque Country, 48013 Bilbao, Spain; aitor.erkoreka@ehu.eus

* Correspondence: migmartinez@uvigo.es

Abstract: There is an increasing interest in reducing the energy consumption in buildings and in improving their energy efficiency. Building retrofitting is the employed solution for enhancing the energy efficiency in existing buildings. However, the actual performance after retrofitting should be analysed to check the effectiveness of the energy conservation measures. The aim of this work was to detect and to quantify the impact that a retrofitting had in the electrical consumption, heating demands, lighting and temperatures of a building located in the north of Spain. The methodology employed is the application of Functional Data Analyses (FDA) in comparison with classic mathematical techniques such as the Analysis of Variance (ANOVA). The methods that are commonly used for assessing building refurbishment are based on vectorial approaches. The novelty of this work is the application of FDA for assessing the energy performance of renovated buildings. The study proves that more accurate and realistic results are obtained working with correlated datasets than with independently distributed observations of classical methods. Moreover, the electrical savings reached values of more than 70% and the heating demands were reduced more than 15% for all floors in the building.

Keywords: retrofitting; refurbishment; functional data analysis; vectorial analysis; energy efficiency

---

1. Introduction

The building sector is considered the largest energy consumer in the European Union, representing 40% of the final energy consumption [1]. Globally, the energy consumption of this sector accounts for 20% of the total delivered energy [2]. Thus, there is increasing interest in improving the energy efficiency of buildings [3–5]. Furthermore, the potential of saving energy by renovation in Europe is considerable as two-thirds of European buildings were constructed before 1980 [6]. Building retrofitting can contribute to reduce the energy consumption of existing buildings with lower energy efficiencies. In this context, it is important to develop methodologies that can evaluate the actual impact of refurbishment on renovated buildings in terms of energy consumption, thermal comfort and lighting.

Therefore, building retrofitting is essential to prove the effectiveness of the applied energy conservation measures to check if the energy efficiency of the building has certainly improved. However, there are very few studies that actually evaluate the retrofit of buildings [7]. Most of the studies evaluate measures for building refurbishment based on energy simulation [8–10], mathematical models [11], artificial neural networks (ANN) [12] and building information modelling (BIM) [13]. Thus, most studies analyse the energy conservation measures based on model outputs and not proving the real effect on the building with monitored data. There is a performance gap between simulated and measured energy consumption. Some studies have found that the calculated heating energy consumption levels in the design phase were much lower than the measured values [14].

Some authors have evaluated the effect of refurbishment and renovations made in buildings with real data. Ardente et al. [15] presented the results of an energy and environmental assessment of retrofit actions implemented in six public buildings by using life cycle analysis (LCA). It is a common approach to evaluate the decrease of energy consumption, operational cost and environmental impact in building retrofitting [16–18]. Another approach for energy diagnosing of existing buildings is U-value in-situ measurement [19] that characterises the heat losses through the building envelope and that can be used to evaluate retrofitting actions [20]. Zavadskas et al. [21] proposed an approach to assess indoor environmental conditions before and after retrofitting of dwellings with multiplicative optimality criteria and experimental data. Hamburg et al. [14] analysed how well the energy performance targets of building refurbishment are reached by collecting energy consumption and indoor measurements after renovation and constructing simulations.

All of the previous presented methodologies to evaluate the retrofitting made in buildings use vector-based data approaches. The presented methodology in this work for evaluating the impact of a building retrofitting is based on functional analysis of monitored data before and after retrofitting. By comparison of vector and functional analysis, we demonstrate that functional analysis provides more realistic and accurate evaluations of the studied variables.

The methods for assessing building renovation have been applying vectorial analysis to the data. These methods do not take into account the observations within a day as a set when evaluating the daily behaviour of the data. As a consequence, the correlation between observations is missed. In this context, Functional Data Analysis (FDA) can be useful because it is able to detect days that do not have individual outliers, but may be far from the mean behaviour [22–26]. A proof of its application is that FDA has expanded to a great number of scientific fields related with continuous-time monitoring processes such as the environment [24,26–29], health and medical research [30,31], industrial processes [32,33], sensor technology [34,35] or even econometrics [36]. Moreover, it has also been applied with machine learning techniques in optimisation and classification problems [37,38]. Nowadays, FDA continues to expand its applications in more fields such as quality control or sports [39,40]

In this work, we propose the use of FDA for assessing the impact that building retrofitting has in the energy performance, indoor temperatures and lighting conditions of a building. The methodology was applied to a case study of the renovated building of the Rectorate of the University of the Basque Country (Spain). The novelty of this work is the application of FDA to statistically contrast the differences in the energy performance of a building before and after a retrofitting. The literature review shows that just few studies actually evaluated the retrofit of buildings with monitored data and functional analysis was not used for this application.

The samples are composed of daily curves of variables such as electricity, heating demands and temperatures. FDA allows making the contrast between samples taking into account the average behaviour of the group throughout all day [24,37,41], which would not be possible with a vectorial approach. In vectorial analysis, the data of a whole day have to be summarised in a single value to work with daily observations. To classify a day as outlier, it has to move away from the sample mean, in this case calculated with simplified daily observations [42,43].

On the one hand, to carry out the functional analysis, a functional ANOVA (FANOVA) was used to evaluate whether there are differences between monitored data in the building before and after retrofitting. On the other hand, a classical analysis of variance (ANOVA) was also used to study the differences between the samples before and after the retrofitting [43–46]. To complement the vectorial analysis, Kruskal's non-parametric test was applied to contrast if the two samples come from the same initial distribution [47–49]. In addition, some variables such as the rate of change in sample variance or the functional $\mathcal{L}_2$ distance between curves are also presented to measure the impact of the refurbishment. The vectorial method is based on the differences between the medians, and the functional method consists of measuring the distance between the curves that represent the mean functions [50–52].

The results show that FDA efficiently demonstrates that the heating demands in the building were reduced thanks to the envelope insulation, although ventilation was increased, indoor temperatures were increased and internal lighting loads were reduced. The results also show that a significant reduction in lighting consumption was achieved with the installation of LED lighting. Moreover, it is demonstrated that taking into account the correlation of the data from a functional approach is a more realistic and informative way to study how different two or more samples are.

2. Materials and Methods

2.1. Functional Data Analysis (FDA)

Functional Data Analysis (FDA) studies observations which form functions defined over a determined set T. The infinite-dimensional structure of the data enlarges the possibilities of research [23,27,53]. A random variable $\mathcal{X}$ is defined such a functional variable if it takes values in a complete metric or semi-metric space and is observed in a discrete set of points $\{t_j\}_{j=1}^{n_p} \in [a,b]$ (not necessarily equispaced) for each of the n individuals studied [54,55]. Thus, the data consist of a **X** matrix with n rows representing the different individuals and n_p columns representing the different discrete points where the functions are evaluated [55].

The functional model, through a process known as smoothing, converts the initial discrete values into a set of continuous functions over time $x(t) \in \mathcal{X} \subset \mathcal{F}$, being $\mathcal{F}$ a functional space. To estimate these functions, $\mathcal{F}$ is $\mathcal{F} = span\{\phi_1, ..., \phi_{n_b}\}$, where ϕ_k is a base function and n_b the number of basis functions necessary to build a functional sample. Although there are other types, the basis functions used commonly are spline or Fourier functions [56], and the expansion considered is [24,25,27,57]:

$$x(t) = \sum_{k=1}^{n_b} c_k \phi_k(t) \tag{1}$$

where $\{c_k\}_{k=1}^{n_b}$ represent the coefficients that shape the function $x(t)$ with respect to the chosen set of basis functions. In this way, the smoothing process consist on solving the following regularisation problem [24,25,27,57]:

$$\min_{x \in F} \sum_{j=1}^{n_p} \{z_j - x(t_j)\}^2 + \lambda \Gamma(x) \tag{2}$$

where $z_j = x(t_j) + \epsilon_j$ (being ϵ_j a value of the zero-mean random noise) is the result of observing x at the point t_j, λ a regularisation parameter that controls the intensity of the regularisation, and Γ an operator that penalises the complexity of the solution. Taking into account the expansion, Equations (1) and (2) can be expressed as [24,25,27,57]:

$$\min_{\mathbf{c}} \{(\mathbf{z} - \mathbf{\Phi c})^T (\mathbf{z} - \mathbf{\Phi c}) + \lambda \mathbf{c}^T \mathbf{R c}\} \tag{3}$$

where $\mathbf{z} = (z_1, ..., z_{n_p})^T$ is the observation vector, $\mathbf{c} = (c_1, ..., c_{n_b})^T$ the vector coefficients of the functional expansion, $\mathbf{\Phi}$ the $n_p \times n_b$ matrix of $\mathbf{\Phi}_{jk} = \phi_k(t_j)$ elements, and **R** the matrix formed by $n_b \times n_b$ elements [24,25,57,58]:

$$R_{kl} = \langle D^2\phi_k, D^2\phi_l \rangle_{\mathcal{L}_2(1)} = \int_{\mathbf{T}} D^2\phi_k(t) D^2\phi_l(t) dt \tag{4}$$

where $D^n\phi_k(t)$ represents the nth-order differential operator of the function ϕ_k. After this, it is easy to know that the solution can be calculated as follows:

$$c = (\mathbf{\Phi}^t\mathbf{\Phi} + \lambda\mathbf{R})^{-1}\mathbf{\Phi}^T\mathbf{z} \tag{5}$$

2.1.1. Functional Depths

Initially, the depth concept appeared in the multivariate statistics for measuring the centrality of a point $x \in \mathbb{R}^d$ within a specific dataset: giving greater value to the points near the center [22]. Then, some authors extended this measured to FDA [59,60]. The functional depth give us a centrality measure of a specific curve x_i with respect a set of curves $x_1, ..., x_n$ that comes from a stochastic process $\mathcal{X}(\cdot)$ in a defined interval $[a, b] \in \mathbb{R}$.

There are several functional depths in the statistical literature, but the three main are: *Fraiman–Muniz* [59], *h-modal* [61] and *Random Projections* [60]. The most used is the *h-modal* depth because it has a better frequency of correct detection than the others [22]. The *h-modal* depth defines the functional mode as the curve most densely surrounded by other curves of the dataset. In this manner, the functional depth of a curve x_i with respect the other curves in the sample is given by:

$$MD_n(x_i, h) = \sum_{k=1}^{n} K\left(\frac{||x_i - x_k||}{h}\right) \tag{6}$$

with $||\cdot||$ being a norm in the functional space, $K : \mathbb{R}^+ \to \mathbb{R}^+$ a kernel function, and h a bandwidth parameter [61]. Thus, the curve that gets the maximum value in Equation (6) is considered the functional mode. Moreover, some authors [60,61] recommended the use of $\mathcal{L}^2(I)$ norms and a truncated Gaussian kernel:

$$||x_i - x_k||_2 = \left(\int_a^b (x_i(t) - x_k(t))^2 dt\right)^{1/2} \qquad K(t) = \frac{2}{\sqrt{2\pi}} exp\left(-\frac{t^2}{2}\right), t > 0 \tag{7}$$

The principal aim of functional depths, viewed as functional dispersion measure, is the detection of outliers. As in the classical analysis, detecting and examining these curves is important because they may bias our functional estimations and because it allows us to discover the reasons that make these curves deviate from the mean. Furthermore, from a functional approach, it is essential because it may occur that the individual values of a curve are not outliers vectorially, but, instead, the complete curve is a functional outlier [22,58]. If we assume that every curve in the data come from the same stochastic process, a curve would be considered such an outlier for two reasons: it is at a significant distance from the expected function of the stochastic process or its shape represents a very different behaviour from the other curves. Therefore, the curves with functional depth below a specific C value would be considered atypical and would be removed from the sample (see [23–26]). On the other hand, it would be convenient to choose a C that provides a controlled type I error level. It should be a value that, in absence of outliers, the probability of mislabelling a correct data as outlier would be approximately a 1% [23–26]:

$$P(D_n(x_i) \leq C) = 0.01, \quad i = 1, ..., n \tag{8}$$

In this way, the chosen C will be first percentile of the depths distribution chosen. Since this distribution is unknown this, percentile must be estimated using the sample data. For this purpose, there are two different bootstrap techniques: trimming bootstrap [62] and weighting bootstrap [63]. Some studies demonstrate that, despite having a larger incorrect outlier detection, the trimming bootstrap has a better performance detecting the curves that are actually outliers [22,64].

2.1.2. Functional Test ANOVA (FANOVA)

Any test or contrast that can be made in a vectorial analysis can have a functional version that usually provides more relevant information. An example of this is the classical ANOVA. Its functional version, although also contrasting the mean levels of a variable, is based on k independent samples $X_{ij}(t), j = 1, ..., n_i\ t \in [a, b]$ drawn from $\mathcal{L}^2(l)$ processes $X_i,\ i = 1, ..., k$ such that $E(X_i(t)) = m_i(t)$ [65–68]. If we have a functional sample classified in several groups such as $\{\mathcal{X}_i, \mathcal{G}_i\}_{i=1}^n \in \mathcal{F} \times \mathcal{G} = \{1, ..., G\}$, where $\mathcal{G}$ is a discrete variable that tell us the membership group, the contrast will be:

$$\begin{cases} H_0 : \overline{X}_1 = \overline{X}_2 = ... = \overline{X}_G \\ H_1 : \exists k, j \quad s.t. \quad \overline{X}_k \neq \overline{X}_j \end{cases} \tag{9}$$

After a few operations, as shown in [50,51], it is possible to go from classic (F_n) to functional statistic (V_n).

$$F_n = \frac{\sum_i^G n_{i.}(\overline{Y}_{i.} - \overline{Y})^2/(G-1)}{\sum_i^G \sum_j^{n_i} (Y_{ij} - \overline{Y}_{i.})^2/(n-G)} \qquad \Longrightarrow \qquad V_n = \sum_{i<j}^{n_i} ||\overline{Y}_{i.} - \overline{Y}_{j.}||^2 \tag{10}$$

In addition, according to Cuevas et al. [50] and Tarrío-Saavedra et al. [51], the asymptotic distribution of V_n under H_0 is the same as the following statistic:

$$V := \sum_{i<j}^{G} ||Z_i(t) - C_{ij} Z_j(t)||^2 \tag{11}$$

where $C_{ij} = (p_i/p_j)^{1/2}$ and $Z_1(t), ..., Z_G(t)$ are independent Gaussian processes with mean 0 and covariance functions $K_i(s, t)$.

Finally, H_0 will be rejected, at a level α, whenever $V_n > V$ where $P_{H_0}\{V > V_\alpha\} = \alpha$ [52]. Because in practise it is not easy to estimate the distribution of V, usually, it is necessary to implement a Monte Carlo procedure through which we get for each $i = 1, ..., G$, N iid observations

$$Z_{il}^* = (Z_{il}^*(t_1), ..., Z_{il}^*(t_m)), \qquad l = 1, ..., N \tag{12}$$

from a *m-dimensional* Gaussian random variable with mean 0 and covariance matrix $(\hat{K}_i(t_p, t_q))_{1 \le p,\ q \le m}$. The functional $\mathcal{L}^2(l)$-distances $||Z_i(t) - C_{ij} Z_j(t)||^2$ are approximated by the $\mathbb{R}^m$-Euclidean distances $||Z_{il}^* - C_{ij} Z_{jl}^*(t)||^2$. Ultimately, the replications $\tilde{V}_l$ of V are

$$\tilde{V}_l = \sum_{i<j}^{G} ||Z_{il}^* - C_{ij} Z_{jl}^*|| \tag{13}$$

and the distribution of V is approximated from the empirical distribution to a sample $\tilde{V}_1, ..., \tilde{V}_N$ [50,51].

2.1.3. Functional Strengths

FDA has numerous advantages but the following are the ones that make FDA suitable for studying daily behaviours in energy variables [24,28,58]:

- It is not mandatory to have prior information on data distribution. The study does not depend on or is not limited to certain distributions.
- The analysis takes into account time intervals as a unit. The sample analysed focusses on complete time units such as days, months or years.
- Analysis of homogeneity. The definition of outliers is different; it is based on the idea that, even though data do not surpass the cut-off, if they show constant deviations, they will be identified as outlier.

- Possibility of study trends. Besides calculating mean functions or detect outliers, it is also possible to study slight variations from the normal data behaviour of the data without outliers.
- Complete analysis of the time spectrum. Before this approach, most analyses were based on the values obtained in a given grid of discrete points. On the contrary, with FDA, it is possible to work with the entire time set in a continuous way.

2.2. Building Description

The case study of this work is the Rectorate building of the University of the Basque Country (Spain) constructed in the 1970s. It is a large building divided in three blocks (west, central and east), as shown in Figure 1. For the purpose of this work, only the west block was considered, which is an office building that includes a nursery. Four storeys form the block: floor 0 (Ground Floor) and Floor 2 (2F) consist of rooms and offices while Floor 1 (1F) and Floor 3 (3F) are mainly open spaces (see Figure 2). As the building use is office work, the occupation takes place only during weekdays, being reduced during summer. There is lighting consumption from 7 a.m. to 8 p.m., except on the second floor, where lighting consumption ends at 6 p.m. The building has a centralised heating system with hot water radiators powered by a campus district heating. The heating works the whole year following a control program according to a schedule from 6 a.m. to 7 p.m. except in July and August, when the heating remains off. A refurbishment of the building was made in the summer of 2016 on the ground floor and in the summer of 2017 in the rest of the building to reduce its energy consumption.

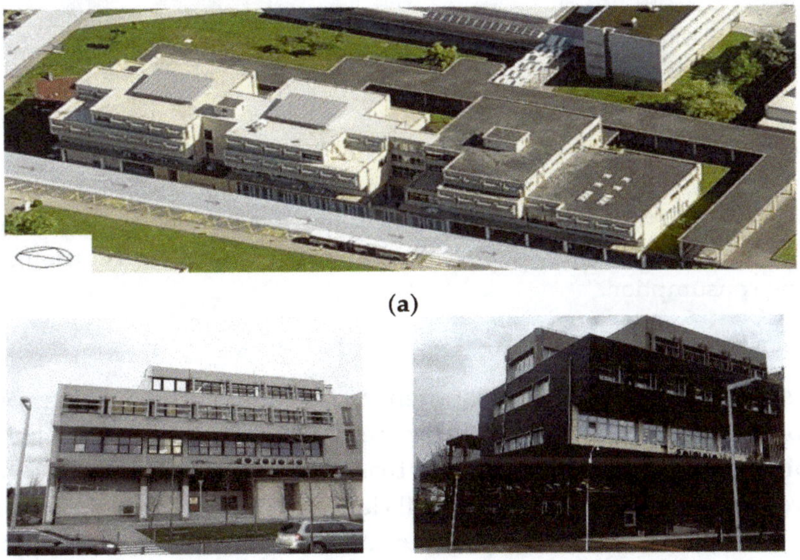

(a)

(b) (c)

Figure 1. (**a**) View of the whole rectorate building that is formed by three blocks (west, central and east). This work focussed in the study of the retrofitting measures carried out in the west block. (**b**) West block before retrofitting. (**c**) West block after retrofitting.

2.2.1. Building Description before Retrofitting

Before the retrofitting, the building did not have any insulation. Most of the façade was built with precast reinforced concrete panels with non-ventilated air gap. Regarding the windows, some were single-glazed with wooden frame, and others were double-glazed but with aluminium frame without thermal break. Before retrofitting, there was no air conditioning or mechanical ventilation system.

2.2.2. Building Description after Retrofitting

The refurbishment consisted in several measures to improve the building's envelope and energy system. The façade was insulated with vacuum insulated panels (VIPs) to reduce heat losses through the envelope. Some windows were replaced by high-performance windows. To improve the lighting consumption of the building, LED lighting was installed. A ventilation system with heat recovery was added on each floor. Furthermore, thermostatic valves were installed to improve the control of the hot water radiators. A detailed description of the refurbishment and an assessment of the heat loss coefficient of the building is provided in [20].

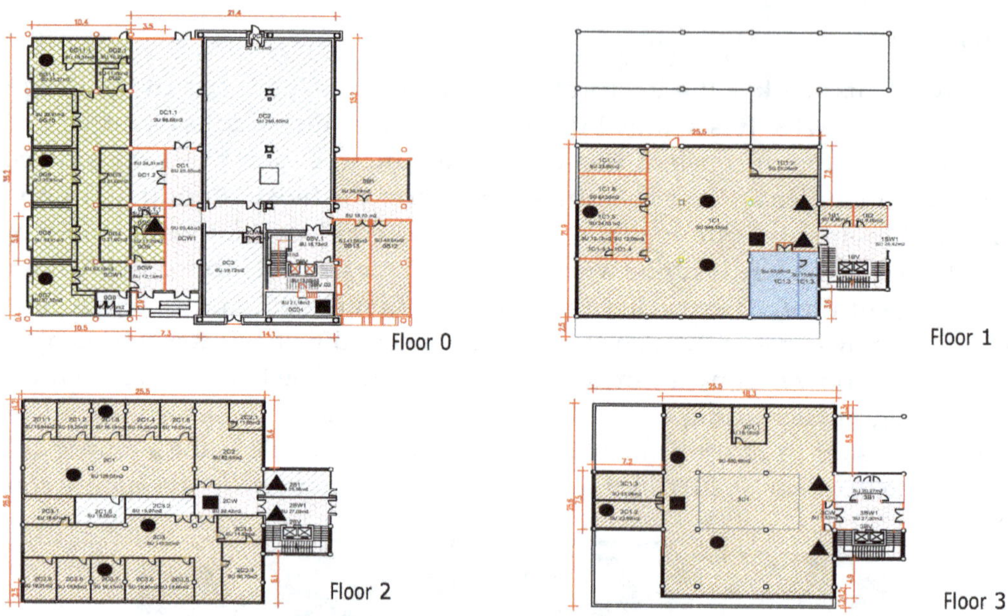

Figure 2. Plans of the four floors in the west block of the building with their space distribution and dimensions. The use of each space is represented with colours: offices (brown), nursery (green), storage rooms (grey), corridors (pink) and server room (blue). In addition, the distribution of the sensors is also displayed: circles, illuminance, temperature and relative humidity; triangles, calorimeters; and square, lighting consumption.

2.2.3. Monitoring Description of the Building

The monitoring started before the refurbishment to investigate the necessary energy conservation measures. Thus, monitored data before and after retrofitting from sensors located all around the building are available. Monitored data before retrofitting correspond to 2016 and 2017, and data after retrofitting correspond to s 2018 and 2019; all data are minute-by-minute values. In particular, on the ground floor, the renovation started a year earlier with the insulation of a false ceiling and the installation of LED lighting. In this case, the years considered before retrofitting are 2015 and 2016. The monitored variables include outdoors conditions, indoors conditions and building heating and lighting consumption. Table 1 presents all the monitored variables in the building used to assess the impact that the retrofitting had on the building's energy performance, comfort and lighting. Indoor condition sensors are located in several points at each floor, as shown in Figure 2. The electrical consumption and the heating demand is provided per floor. Further information about the monitoring of the building can be found in the work carried out by Erkoreka et al. [69].

2.3. Pre-Processing Data

Before presenting the results, it is necessary to explain the specific smoothing process performed in this study. On the one hand, because there are parts of the day in which the variables are constant, the basis functions chosen are splines. On the other hand, to select the optimal number of basis, the determination coefficient R^2 was taken into account to measure the smoothing adjustment in relation to the raw data. As shown in Figure 3, the criterion was to select the minimum number of basis (in a given grid) where the R^2 surpasses the value of 99%.

Table 1. Monitored variables in the building to assess the impact of the refurbishment in the energy performance, illuminance and comfort of the building.

Type of Measurement		Monitored Variable	Units	Sensor
Indoor conditions		Illuminance	[LUX]	Siemens 5WG1 255-4AB12
		Temperature	[°C]	ARCUS SK04-S8-CO2-TF
Electrical Consumption	Before retrofitting	Lighting + elec. equipment	W	Power meters ABB EM/S and ABB a41/43 per floor
	After retrofitting	Lighting	W	
		Ventilation + elec. equipment	W	Power meters ABB EM/S and ABB a41/43 per floor
Heating consumption		Thermal energy of the heating water	W	Calorimeter: Kamstrup Multical 602 and ZENNER Zelsius (DN20)

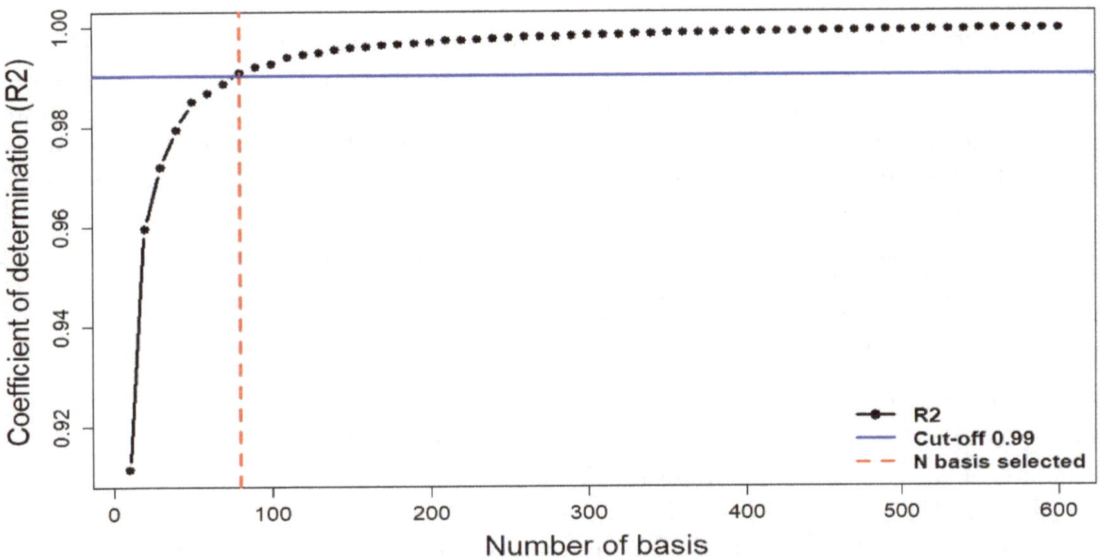

Figure 3. Example of the process of selecting the optimum number of basis for smoothing the original sample. The red line represents the minimum number of basis where the R^2 is higher than 0.99.

In addition, it is important highlight the *data cleaning* that was carried out using a functional approach. It is known that throughout all the analysed years there were many days when the building was closed and the information that these days provide is not only not useful, but can also distort the final results. To solve this problem, an algorithm that searches for these days and deletes them from the sample was developed (see Algorithm 1).

After an exploratory analysis of the data, the values of the chosen parameters for Algorithm 1 are: $\beta_1 = 500$, $\beta_2 = 250$, $\alpha = 0.25$, and $\theta = 0.5$. Moreover, because there are parts of the day in which the variables are constant, the chosen basis functions are spline. With the application of the algorithm to the data, the sample becomes smaller but only with relevant days that take into account the normal behaviour of the building. Figure 4 illustrates the performance of the algorithm. It can be seen that it is capable of deleting the days with an abnormal behaviour without affecting the bulk of the data. The days when the building was unoccupied and closed, and therefore with very small or no electrical consumption, are detected and eliminated in the picture on the right of Figure 4.

Figure 4 also presents the mean functions (in form of curves) and the change that they suffer after deleting non-representative days. Working with non-representative days will produce erroneous results in any study; for example, if an ANOVA test is performed, the test may not reject the equality of means even if the groups are different.

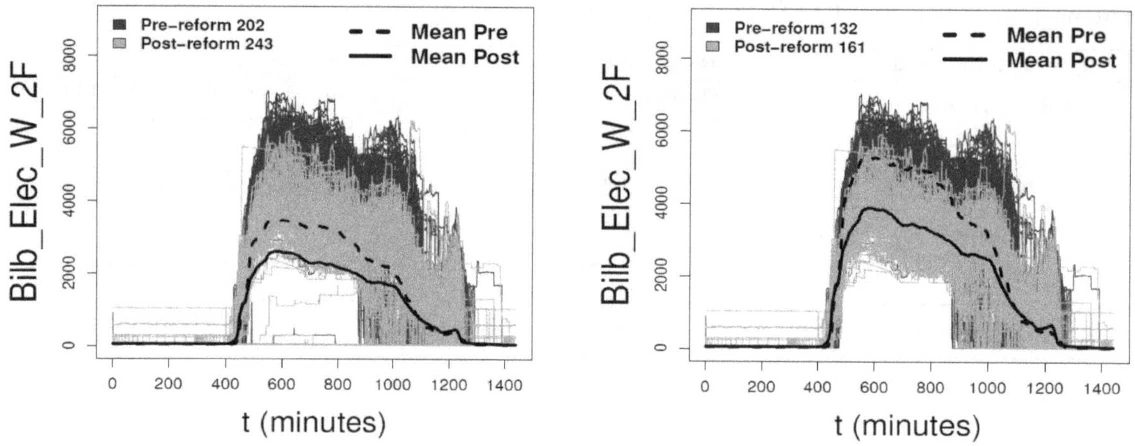

Figure 4. Performance of the *Functional cleaning* algorithm with an example variable: Electrical demand on the second floor of the building in the representative months (in dark gray the data before the refurbishment, in light gray after). The number of days taken into account in each sample is also shown: (**left**) the raw data and the mean functions separated between before and after the refurbishment; and (**right**) the data after applying the algorithm and the mean functions separated between before and after the refushbishment.

Algorithm 1: Functional cleaning.

Input: Data divided in groups and the parameters: β_1, β_2, α, θ.
Output: Data without inappropriate days.

Transform the data to funcional format: 1440 minute data each day.
Searching for missing values (NAs). The daily limits are:
β_1 NAs per day $\qquad$ β_2 consecutive NAs per day
Delete the days that exceeded the daily limits.
Approximation, with an interpolation technique, of the remaining missing values.
Calculate the variability of every daily curve in the sample.
Delete the curves that:

- Have a variability less than or equal to a percentage $\alpha \in \mathbb{R}$ of the average sample variability.
- Be below the sample mean function for at least a percentage $\theta \in \mathbb{R}$ of the day.

---

3. Results and Discussion

The effects of the refurbishment carried out in 2017 in the Rectorate building of the University of the Basque Country were analysed. In this analysis, lighting consumption, illuminance, indoor temperatures and heating demand were studied. These variables were measured every minute between 2016 and 2019 (changing 2016 for 2015 in the case of the ground floor) and only taking into account those months in which the heating systems operate significantly (October to March). As the retrofitting started in the summer of 2017, the months of this year after the summer are not suitable for analysis. In this way, data were divided into nine months before retrofitting (six months in 2016 and three months in 2017) and nine months after retrofitting (six months in 2018 and three months in 2019).

Section 3.1 presents the lighting analysis of the study and Section 3.2 the same analysis for heating demands. The numerical results were based on the p-values of the ANOVA and Kruskal tests in the vectorial analysis, and on the p-values of the FANOVA in the functional analysis. Different measures are also shown to illustrate the differences between the sample groups: $\mathrm{D_{vec}}$, the difference between the medians from the vectorial approach; $\mathrm{D_{func}}$, the average minute difference between the mean functions; and D_{dist}, the $\mathcal{L}_2(I)$ distance between the mean functions. Additionally, to measure the functional smoothing adjustment to the raw data, the coefficient of determination R^2 is shown in the tables. The change in the variance of monitored data ($\triangle$Var) and the savings obtained with the retrofitting, calculated in relation to the initial energy demands, are also shown from both approaches. Lastly, all figures presented here were made with the R-programming software.

3.1. Lighting Analysis

The results of the electrical lighting consumption are shown in Figure 5. On the one hand, the vectorial analysis by means of box plots is shown in the first row. It can be seen that, after the refurbishment, the lighting demands of the building decreased and became more homogeneous on all floors. This is demonstrated in Table 2, where the change is quantified, and the statistical tests used (ANOVA and Kruskal) corroborate the change. In this case, the floor with the highest reduction was the third floor (2004 W less) and the floor with the lowest reduction was the second floor (489 W less). On the other hand, in the second row of Figure 5, the functional analysis, through the daily curve graph, is presented. In this case, thanks to this approach, it can be seen that, despite the decrease in the consumption, the retrofitting hardly affects the daily behaviour of the lighting consumption. As it is observable in Figure 5 that the curves have very similar shapes, with the exception on the third floor where a change in the lighting schedule was implemented. Table 2 presents the FANOVA results. The similarity of the samples is rejected on every floor. However, the impact varies among floors, as shown in Table 2. The first floor was the most benefited (2032 W less per minute on average) and the third floor the least benefited (400 W less per minute on average). The differences between the two analyses, in absolute terms, have been noticed: with the vectorial analysis the third floor obtained the highest reduction, while with functional analysis it obtained the lowest reduction. As shown in the second row of Figure 5, on the third floor, the reduction was concentrated on the last hours of the day; however, on average during the day, this reduction was lower. The vectorial approach does not take this fact into account because it distorts the sample with the calculation of daily averages.

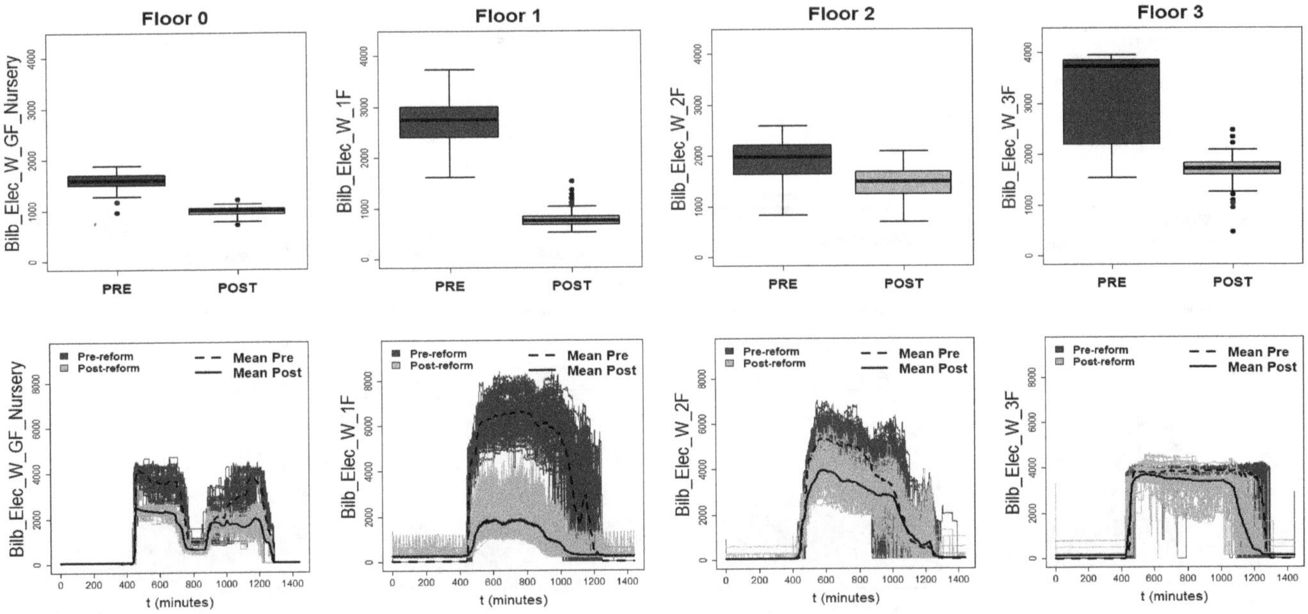

Figure 5. Analysis of the electrical lighting consumption on each floor measured in W. In the first row, the vectorial results (in form of box plots) are presented. In the second row, instead, the functional data are represented with the respective mean functions. The data are divided into winter days before and after the refurbishment.

Table 2. Numerical results on each floor for lighting consumption. The vectorial results are presented with D_V measuring the difference between medians. The functional results, accompanied with the average minute difference (D_{func}), the $\mathcal{L}_2(l)$ distance between the mean functions (D_{dist}) and the smoothing adjustment (R^2), are also presented. For both analyses, the change in the variability of the data ($\triangle$Var) and the electrical savings are displayed. Lastly, the *p*-values for the tests, both vectorial and functional, are displayed in this table.

	Electrical Consumption										
	Vectorial Analysis					Functional Analysis					
	p_{anova}	$p_{kruskal}$	D_{vec} (W)	$\triangle$Var	Savings	p_{fanova}	D_{func} (W)	D_{dist} ($\mathcal{L}_2(l)$)	$\triangle$Var	Savings	R^2
Floor 0	≈0	≈0	−587.48	−72.63%	36.96%	≈0	−609.69	34536.02	−71.80 %	38.11%	0.9909
Floor 1	≈0	≈0	−1987.68	−94.47%	72.68%	≈0	−2032.31	116400.04	−89.54%	73.15%	0.9903
Floor 2	≈0	≈0	−489.05	−43.49%	24.83%	≈0	−434.68	29537.58	−55.54%	22.50%	0.9929
Floor 3	≈0	≈0	−2004.80	−95.32%	53.82%	≈0	−400.82	35603.07	+17.29%	18.18%	0.9915

On the other hand, the effects of the refurbishment on the illuminance conditions were studied. Figure 6 shows the analysis for the illuminance levels. Through vectorial analysis, an impact is also detected, but is different on each floor. In general, the illuminance level was improved. Table 3 indicates that the floors with the biggest increase in illuminance were the first and third floors (356 lx and 414 lx more, respectively). Ground floor was the only one with an illuminance reduction from this point of view (270 lx less). In this case, this reduction is related to the shading that was installed to have a better protection against natural light (see Table 3). Furthermore, observing the functional illuminance curves floor by floor shown in the second row of Figure 6, the conclusion is the same: there was also an improvement in the illuminance levels, and the biggest increase of illuminance took place on the first and third floors. From this point of view, Table 3 shows that the increments were about 174 lx on the first floor and 204 lx on the third floor, on average, every minute. Once again, FDA makes it possible to see that the behaviour of the illuminance was maintained on all floors. As shown in Table 3, FDA detects that the illuminance changes on the second floor, while the vectorial approach fails in this detection (the *p*-values obtained are bigger than 0.05).

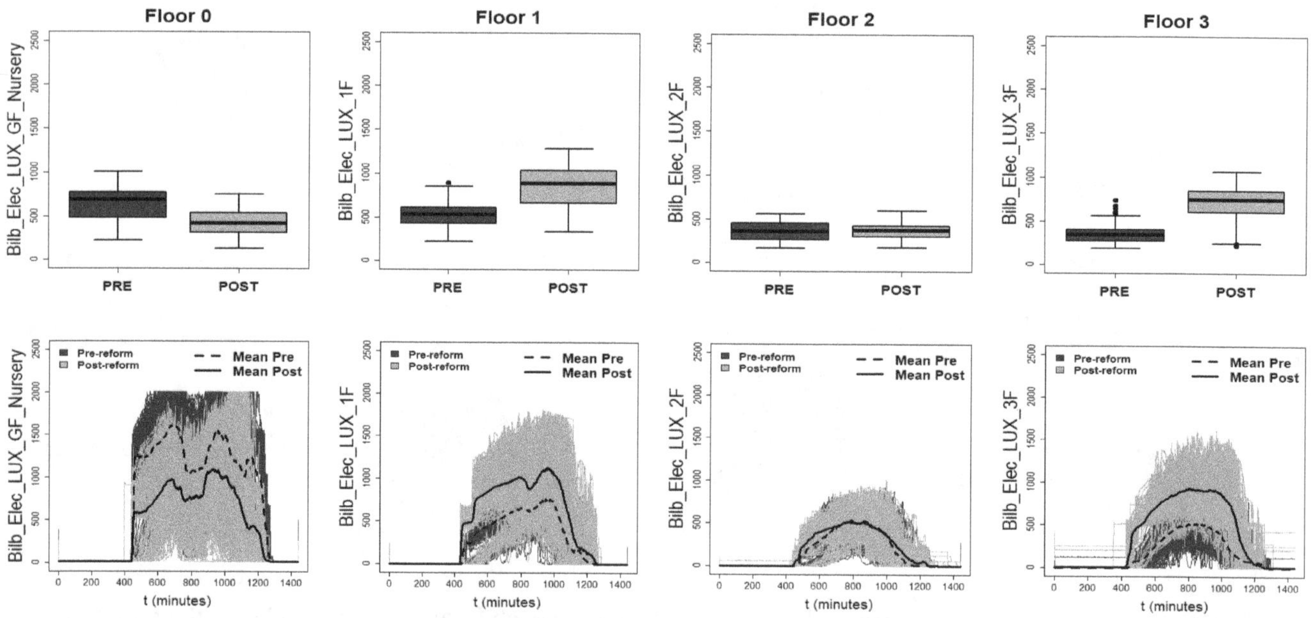

Figure 6. Analysis of the illuminance on each floor measured in lx. In the first row, the vectorial results (in form of box plots) are presented. In the second row, instead, the functional data represented with the respective mean functions. The data are divided into winter days before and after the refurbishment.

Table 3. Numerical results on each floor for illuminance. The vectorial results are presented with D_{vec} measuring the difference between medians. The functional results, accompanied with the average minute difference (D_{func}), the $\mathcal{L}_2(I)$ distance between the mean functions (D_{dist}) and the smoothing adjustment (R^2), are also presented. For both analyses, the change in the variability of the data is displayed ($\triangle$Var). Lastly, the p-values for the tests, both vectorial and functional, are displayed in this table.

	Illuminance								
	Vectorial Analysis				Functional Analysis				
	p_{anova}	$p_{kruskal}$	D_{vec} (lx)	$\triangle$Var	p_{fanova}	D_{func} (lx)	D_{dist} ($\mathcal{L}_2(I)$)	$\triangle$Var	R^2
Floor 0	≈0	≈0	−266.22	−42.27%	≈0	−286.81	15274.23	−12.51%	0.9909
Floor 1	≈0	≈0	+356.12	+62.88%	≈0	+174.78	9379.46	+59.44%	0.9913
Floor 2	0.217	0.278	+9.46	−27.78%	≈0	+23.33	1608.27	−9.29%	0.9904
Floor 3	≈0	≈0	+414.86	+81.24%	≈0	+204.32	11140.84	+66.30%	0.9914

Analysing the graphs shown in Figures 5 and 6, it can be seen that the retrofitting reduced the electrical consumption of lighting while the illuminance levels were improved or maintained in proper levels. The level of illuminance on the first and third floors was improved mainly due to the replacement by LED technology, as shown in Figure 6 and Table 3. In the case of the second floor, although it obtained an average reduction of 435 W, the illuminance level was maintained according to European illuminance regulations. This regulation states that the optimal lighting level in offices is 500 lx, a level already reached before the renovation. In the case of the ground floor, as mentioned above, it had to be analysed individually. The mean function of illuminance before retrofitting on this floor had peaks above 1500 lx (see Figure 6), which indicates a high influence of natural light. After the retrofitting, the shading that was installed to protect from sunlight achieved a reduction, on average, of 286 lx each minute, as it can be seen in Table 3 (266 lx, observing the vectorial results). Furthermore, the electrical consumption associated to lighting on the ground floor is also reduced. Table 2 shows that the vectorial reduction was 587 W, and the functional reduction 609 W, on average, every minute.

Generally, monitored lighting consumption data after retrofitting have less dispersion (see $\triangle$Var in Table 2). This means that the monitored lighting data are more homogeneous. In this case, the reduction of the data variability reaches values higher than 30% from both methods. However, the functional analysis detects an increase of 17% in the data dispersion of the third floor, as shown in Table 2. The functional graph of this floor in Figure 5 supports this result. The vectorial approach, which summarises the days with the mean, distorts the results by considering false conclusions such as, in this case, that the data variance decreased for all floors (see Table 2). On the other hand, the homogeneity of the monitored illuminance data after the renovation is different depending on the floor. Table 3 shows that the floors where the data dispersion was reduced are the ground and the second floors.

From an energetic point of view, the relative electrical consumption savings after retrofitting are also calculated. Table 2 shows that savings of more than 18% were obtained on every floor with the functional analysis and more than 24% with the vectorial analysis. The first floor is the most benefited floor with savings around 73% from the two approaches applied. However, there are differences between the results of the vectorial analysis and the functional analysis. This is shown in Figures 5 and 6 and, specifically, in Table 2 where the savings are presented. The savings calculation of functional analysis are more accurate because it is based on the areas under the curves representing the mean functions, taking into account the entire daily behaviour. The vectorial results, instead, only quantify the differences between the medians of the samples.

3.2. Heating Analysis

The results show that the refurbishment had a significant impact in the heating demands of the building. First, observing the box plots from the vectorial analysis, as shown in the first row of Figure 7, it is clear that the heating demand was reduced on all floors. This fact is supported by vectorial tests that reject similarity between samples on all floors (see Table 4). After that, with the functional graphs, as shown in the second row of Figure 7, it is also appreciated that the heating demand curves after refurbishment are below the initial curves. This is also proved by the functional tests that reject all the sample similarity hypothesis (see Table 4). The results shown in Table 4 change from vectorial to functional analysis. Both analyses come to the same conclusions, but the magnitude of change is different for each one. With vectorial approach, the reduction per floor ranges from 3667 W as the highest reduction on the third floor to 1057 W as the lowest on the first floor. In contrast, with the functional analysis, although the highest and lowest reduction were on the same floors, the values of the reduction are not the same. The heating demand, each minute, was reduced on average 3918 W on third floor and 1456 W on first floor. Therefore, the relative heating savings were significant (see Table 4). The floor most benefited was the third floor; both analysis obtained savings higher than a 30% on this floor. Again, the ground floor had to be analysed individually because, in addition to the installation of shading in 2017 that prevents solar gains, a false ceiling was already insulated on this floor in 2015. Thus, it was possible to obtain a saving of about 12% of the initial heating demands from vectorial results and about 17% from functional results (see Table 4).

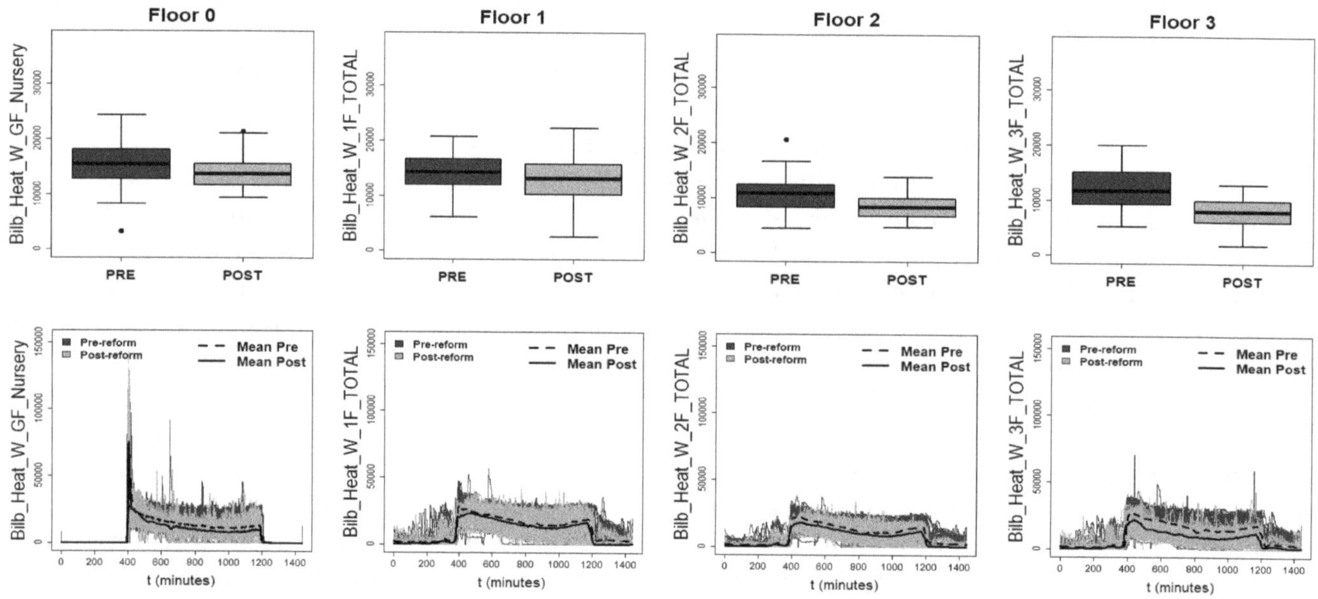

Figure 7. Analysis of the heating demands of each floor measured in W. In the first row, the vectorial results (in form of box plots) are presented. In the second row, instead, the functional data are represented with the respective mean functions. The data are divided into winter days before and after the refurbishment.

Table 4. Numerical results on each floor results for heating demands. The vectorial results are presented with D_{vec} measuring the difference between medians. The functional results, accompanied with the average minute difference (D_{func}), the $\mathcal{L}_2(l)$ distance between the mean functions (D_{dist}) and the smoothing adjustment (R^2), are also presented. For both analyses, the change in the variability of the data ($\triangle$Var) and the heating savings are displayed. Lastly, the p-values for the tests, both vectorial and functional, are calculated.

	Heating Demand										
	Vectorial Analysis					Functional Analysis					
	p_{anova}	$p_{kruskal}$	D_{vec} (W)	$\triangle$Var	Savings	p_{fanova}	D_{func} (W)	$D_{dist}(\mathcal{L}_2(l))$	$\triangle$Var	Savings	R^2
Floor 0	1.761×10^{-6}	7.047×10^{-6}	−1838.12	−35.68%	11.86%	≈0	−1455.83	158722.42	−30.66%	17.36%	0.9901
Floor 1	0.018	0.02	−1057.67	+24.71%	7.46%	≈0	−1975.15	95981.90	−23.89%	16.97%	0.9908
Floor 2	≈0	≈0	−2457.87	−43.99%	22.73%	≈0	−2158	96002.01	−43.48%	23.60%	0.9914
Floor 3	≈0	≈0	−3667.06	−60.22%	31.49%	≈0	−3917.87	185968.48	−51.68%	35.51%	0.9911

In the case of the vectorial analysis, Table 4 shows that, in general, the measurements are less variable in general (between 35% and 60% less). On the first floor, instead, this method detects an increase in data dispersion after the retrofitting. However, this fact is not supported by the functional approach (see Table 4). Figure 7 shows in its second row that the heating curves on this floor are similar or even less variable. In the case of functional analysis results, as shown in Table 4, the variation of the measurements is lower on all floors (between 23% and 51% less). The functional method is demonstrated to be more accurate and provides more information. For instance, functional analysis detects the demand peaks on ground floor in the morning when the heating starts, as it can be seen in second row of Figure 7. With the vector analysis, this information is lost as shown in first row of Figure 7.

The possible consequences of the retrofitting on the building temperatures are also studied. Both the vectorial approach and the functional approach conclude that the temperatures increased on every floor except on the third floor (see Figure 8 and Table 5). The tests do not detect any change in the average indoor temperature on this floor. This is probably because this floor is a large space with low occupancy where the temperatures were stable before and after refurbishment. This is supported, on the one hand, by the functional analysis in Figure 8 where the curves are almost overlapping. On the other hand, Table 5 shows that the temperature on the third floor increased very slightly and the FANOVA test does not detect a significant change (p-value = 0.17). On the contrary, the ground, first and second floors had higher temperatures after retrofitting, as shown in Figure 8 and Table 5. The increase, depending on the floor and method, was between 0.5 and 2 °C. The reason is that after the refurbishment the building is more insulated, heat losses are reduced and it is easier to keep it warmer. Moreover, the temperature set point have been increased in the common zones. Only with FDA it is detected that the retrofitting succeeds to reduce the influence of natural light on the ground floor temperatures. In the second row of Figure 8, it is observed that temperatures on the ground floor after refurbishment do not have peaks at the end of the day due to solar radiation. Finally, both analyses show that the homogeneity of monitored data is significantly improved. Table 5 shows that, on each floor and from both approaches, there is more homogeneity in the measurements related to the building's indoor temperatures.

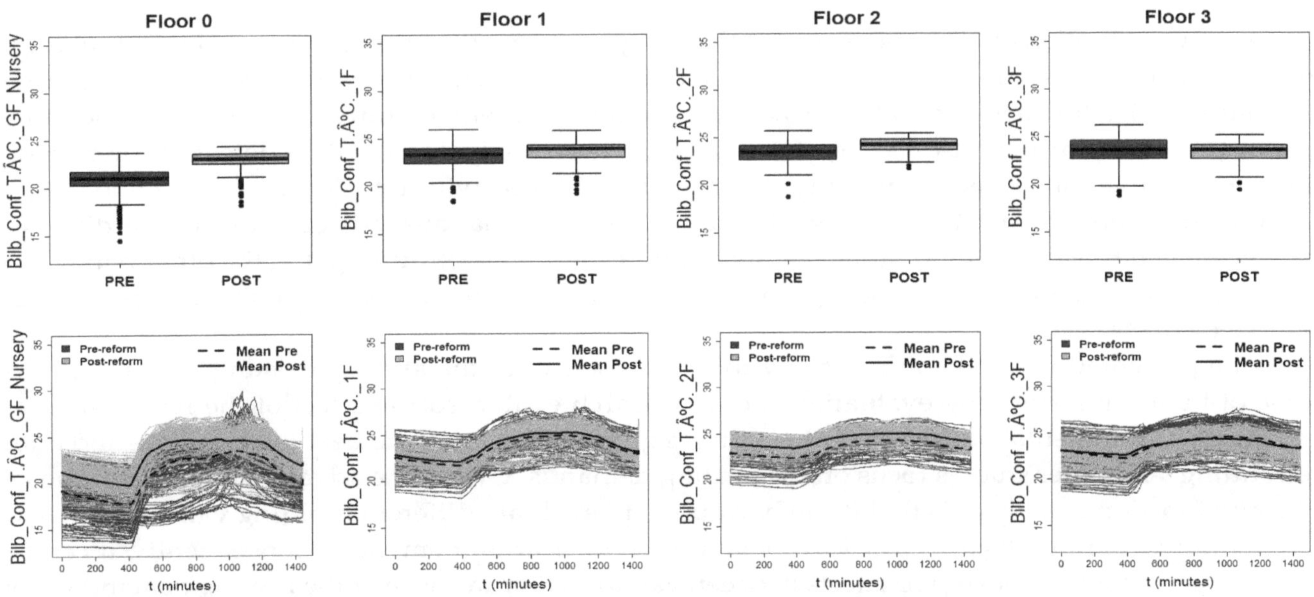

Figure 8. Analysis of the temperatures on each floor measured in °C. In the first row, the vectorial results (in form of box plots) are presented. In the second row, instead, the functional data are represented with the respective mean functions. The data are divided into winter days before and after the refurbishment.

Table 5. Numerical results on each floor indoor temperatures. The vectorial results are presented with D_{vec} measuring the difference between medians. The functional results, accompanied with the average minute difference (D_{func}), the $\mathcal{L}_2(l)$ distance between the functional mean (D_{dist}) and the smoothing adjustment (R^2), are also presented. For both analyses, the change in the variability of the data is displayed ($\triangle$Var). Lastly, the p-values for the tests, both vectorial and functional, are calculated.

	Indoor Temperatures								
	Vectorial Analysis				Functional Analysis				
	P_{anova}	$P_{kruskal}$	D_{vec} (W)	$\triangle$Var	P_{fanova}	D_{func} (W)	D_{dist} ($\mathcal{L}_2(l)$)	$\triangle$Var	R^2
Floor 0	≈0	≈0	+1.92	−62.75%	≈0	+1.95	75.09	−40.10%	0.9918
Floor 1	0.001	0.004	+0.55	−45.33%	≈0	+0.35	13.43	−37.77%	0.9970
Floor 2	≈0	≈0	+0.76	−53.31%	≈0	+0.92	35.29	−60.68%	0.9976
Floor 3	0.39	0.274	−0.072	−50.65%	0.17	+0.002	7.97	−63.15%	0.9959

As in the electrical analysis, the relative savings obtained in the heating analysis were calculated, as shown in Table 4. These savings reached values of more than 30% (in particular on third floor), and, in this case, with the functional approach are higher than with the vectorial approach. The floor with the smallest relative saving, observing the results of both methods, was the first floor, but with the vectorial method the reduction was almost 8% and with the functional method almost 17%. In this case, the results are more homogeneous among floors; there is not much difference from floor to floor. As expected, the form of the heating demands curves, before and after retrofitting, is the same although the values after retrofitting decreased (see Figure 7). Similar behaviour is appreciated in the temperature curves (see Figure 8).

After retrofitting the indoors temperature increased on all floors, a ventilation system with exterior air was installed and the internal gains were reduced with LED lighting. These changes should contribute to an increase of the heating demand. However, Figure 7 and Table 4 show that the heating demand on each floor has been reduced, demonstrating the effectiveness of façade insulation. Besides the decrease on the heating demands, indoor temperatures in the building are maintained or even increased, as demonstrated in Figure 8 and Table 5.

4. Conclusions

A new application of FDA and a new methodology to assess the impact of retrofitting in buildings are presented in this paper. The study was conducted by analysing monitored data of lighting consumption, heating demands, illuminance levels and indoor temperatures of the Rectorate building of the University of the Basque Country (Spain). These analyses aimed to detect an impact on the measured variables and to quantify the changes achieved with retrofitting. The methodology used in this study is based, on the one hand, in the functional analysis contrasting the distance between mean functions of monitored samples before and after retrofitting. On the other hand, as a comparison, the classical or vectorial approach was carried out measuring the dissimilarities between sample medians.

The presented method contributes with advantages over the already existing research in the topic of building retrofitting evaluation. Some research evaluated the effect of the refurbishment based on environmental indicators. Other studies give a measure of the heat losses before and after retrofitting, and other studies focus on the heating demands. One of the advantages of the proposal of using functional analysis is that it can be applied to evaluate different building variables such as temperatures, lighting levels, electrical consumption or heating demands. Consequently, it can be used to search for relationship or effects between variables. For instance, if the heating demand is not reduced as expected, the evolution of other variables can be observed to look for the cause. Moreover, a daily based analysis can be done, evaluating the peculiarities of some days in the performance of the studied variables.

The research contribution of this paper is the application of a mathematical method such as functional analysis to evaluate the impact that a building retrofitting had in its energy performance. There are few studies that present evaluations of retrofitting actions in buildings with monitored data and the employed methods are based in vector-based data approaches. The benefits of applying FDA to contrast and measure the similarity between samples of monitored data before and after a retrofitting were demonstrated. An advantage of FDA is that it is not restricted to certain characteristics of the data distribution, thus it is not necessary to test its normality. It considers complete time units, working with the time set in a continuous way, without having to summarise them, which is beneficial to evaluate monitored building variables. In addition, the outlier detection takes into account constant deviations as a reason to identify an outlier, even though it does not surpass the cut-off criterion.

Furthermore, the variables used in this study are commonly analysed in different types of buildings. The methodology presented here can be applied to assess the energy and thermal performance of different buildings, such as industrial, residential, educational or office buildings. Functional analysis can be applied, with variables or tools different from those used here, to evaluate different aspects of the studied building. Thus, this method could also be used to evaluate monitored variables in thermal facilities within an energy efficiency framework. The effectiveness and usefulness of the functional approach to evaluate variables that affect the energy efficiency was demonstrated in this study.

The results illustrate the greater accuracy of FDA in detecting if there was a significant change in the studied variables in comparison with vectorial analyses results. In the illuminance analysis of the second floor, the FANOVA detected a change (p-value ≈ 0) while the vectorial ANOVA did not (p-value $= 0.22$). Additionally, FDA could detect that most of the lighting consumption reduction on the third floor was concentrated in the last hours of the day. In addition, it has been demonstrated that FDA can provide trustable information about the dispersion of the data. In the lighting consumption analysis of the third floor, the greater dispersion of data after retrofitting was only identified with FDA. Moreover, the representation of the monitored variables in a continuous way throughout the day shows that the FDA allows a greater accuracy and a better adjustment to reality than the vectorial methods. It allows identifying average patterns of the studied variables and it has the potential for detecting anomalous behaviours of monitored variables. In the analysis of the heating demand of the ground floor, the FDA reported a peak in the early hours of the day (around 7 p.m.), both before and after the retrofitting, which the vectorial analysis did not notice.

From an energetic point of view, the conclusion is that the refurbishment carried out in the building under study had a significant impact on its energy performance. The main aim of the retrofitting was to reduce the heat losses. This goal was achieved, as demonstrated by the decrease in the heating demands of the entire building even though the temperatures were increased, a ventilation system being installed and the LED lighting reducing the internal heat gains. The analysis also shows that the illuminance levels were improved on all floors, only decreasing on the ground floor as a shading was installed to prevent from direct sunlight. According to FDA results, the heating demands were reduced 17% on the ground and first floor, 24% on the second floor and 36% on the third floor. Moreover, the reduction in lighting consumptions were 38% for the ground floor, 73% for the first floor and around 20% for the second and third floors.

Author Contributions: Mathematical methodology, M.M.C.; engineering knowledge, S.M.M. and E.G.Á.; and data collection, P.E.O. and A.E.G. All authors have read and agreed to the published version of the manuscript.

Acknowledgments: This paper was supported by the Spanish Government (Science, Innovation and Universities Ministry) under the project RTI2018-096296-B-C21.

References

1. Foucquier, A.; Robert, S.; Suard, F.; Stéphan, L.; Jay, A. State of the art in building modelling and energy performances prediction: A review. *Renew. Sustain. Energy Rev.* **2013**, *23*, 272–288. [CrossRef]
2. Energy Information Administration. International Energy Outlook 2019. Available online: https://www.eia.gov/ieo (accessed on 21 March 2020).
3. RCP policy: public health and health inequality. Every breath we take: the lifelong impact of air pollution. Royal College of Physicians (RCPCH). 2016. Available online: https://www.rcplondon.ac.uk/projects/outputs/every-breath-we-take-lifelong-impact-air-pollution (accessed on 21 March 2020).
4. Wang, L.; Pereira, N.; Hung, Y. *Air Pollution Control Engineering*; Humana Press: New York, NY, USA, 2004.
5. EEA. *Air Quality in Europe*; European Environmental Agency: Copenhagen, Denmark, 2019.
6. Frances Bean, J.; Dorizas, V.; Bourdakis, E.; Staniaszek, D.; Pagliano, L.; Roscetti, A. *Future-Proof Buildings for all Europeans—A Guide to Implement the Energy Performance of Buildings Directive*; Buildings Performance Institute Europe (BPIE): Brussels, Belgium, 2019.
7. Cabeza, L.F.; Rincón, L.; Vilariño, V.; Castell, A. Life cycle assessment (LCA) and life cycle energy analysis (LCEA) of buildings and the building sector: A review. *Renew. Sustain. Energy Rev.* **2014**, *29*, 394–416. [CrossRef]
8. Santamouris, M.; Pavlou, C.; Doukas, P.; Mihalakakou, G.; Synnefa, A.; Hatzibiros, A.; Patargias, P. Investigating and analysing the energy and environmental performance of an experimental green roof system installed in a nursery school building in Athens, Greece. *Energy* **2007**, *32*, 1781–1788. [CrossRef]
9. Chidiac, S.E.; Catania, E.J.; Morofky, E.; Foo, S. Effectiveness of single and multiple energy retrofit measures on the energy consumption of office buildings. *Energy* **2011**, *36*, 5037–5052. [CrossRef]
10. Zmeureanu, R. Assessment of the energy savings due to the building retrofit. *Build. Environ.* **1990**, *25*, 95–103. [CrossRef]
11. Asadi, E.; Da Silva, M.G.; Antunes, C.H.; Dias, L. Multi-objective optimization for building retrofit strategies: A model and an application. *Energy Build.* **2012**, *44*, 81–87. [CrossRef]
12. Yalcintas, M. Energy-savings predictions for building-equipment retrofits. *Energy Build.* **2008**, *40*, 2111–2120. [CrossRef]
13. Tobias, L.; Vavaroutsos, G. *Retrofitting Buildings to be Green and Energy-Efficient: Optimizing Building Performance, Tenant Satisfaction, and financial Return*; Urban Land Institute: Washington, DC, USA, 2012.
14. Hamburg, A.; Kalamees, T. How well are energy performance objectives being achieved in renovated apartment buildings in Estonia? *Energy Build.* **2019**, *199*, 332–341. [CrossRef]
15. Ardente, F.; Beccali, M.; Cellura, M.; Mistretta, M. Energy and environmental benefits in public buildings as a result of retrofit actions. *Renew. Sustain. Energy Rev.* **2011**, *15*, 460–470. [CrossRef]
16. Mohammadpourkarbasi, H.; Sharples, S. Eco-retrofitting very old dwellings: current and future energy and carbon performance for two UK cities, Plea 2013. In Proceedings of the 29th Conference, Munich, Germany, 10–12 September 2013.
17. Beccali, M.; Cellura, M.; Fontana, M.; Longo, S.; Mistretta, M. Energy retrofit of a single-family house: Life cycle net energy saving and environmental benefits. *Renew. Sustain. Energy Rev.* **2013**, *27*, 283–293. [CrossRef]
18. Famuyibo, A.A.; Duffy, A.; Strachan, P. Achieving a holistic view of the life cycle performance of existing dwellings. *Build. Environ.* **2013**, *70*, 90–101. [CrossRef]
19. Ficco, G.; Iannetta, F.; Ianniello, E.; dAmbrosio Alfano, F.R.; DellIsola, M. U-value in situ measurement for energy diagnosis of existing buildings. *Energy Build.* **2015**, *104*, 108–121. [CrossRef]
20. Uriarte, I.; Erkoreka, A.; Giraldo-Soto, C.; Martin, K.; Uriarte, A.; Eguia, P. Mathematical development of an average method for estimating the reduction of the Heat Loss Coefficient of an energetically retrofitted occupied office building. *Energy Build.* **2019**, *192*, 101–122. [CrossRef]
21. Zavadskas, E.; Kaklaukas, A.; Turskis, Z.; Kalibatas, D. An approach to multi-attribute assessment of indoor environment before and after refurbishment of dwellings. *J. Environ. Eng. Landsc. Manag.* **2009**, *17*, 5–11. [CrossRef]
22. Febrero, M.; Galeano, P.; Wenceslao, G. Outlier detection in functional data by depth measures, with application to identify abnormal NOx levels. *Environmetrics* **2008**, *19*, 331–345. [CrossRef]

23. Horváth, L.; Kokoszka, P. *Inference for Functional Data with Applications*; Springer: Berlin/Heidelberg, Germany, 2012. [CrossRef]
24. Martínez, J.; Saavedra, A.; García, P.; Piñeiro, J.; Iglesias, C.; Taboada, J.; Sancho, J.; Pastor, J. Air quality parameters outliers detection using functional data analysis in the Langreo urban area (Northern Spain). *Appl. Math. Comput.* **2014**, *241*, 1–10. [CrossRef]
25. Pi neiro, J.; Torres, J.; García, P.; Alonso, J.; Mu niz, C.; Taboada, J. Analysis and detection of functional outliers in waterquality parameters from different automated monitoring stationsin the Nalón River Basin (Northern Spain). *Environ. Sci. Pollut. Res.* **2015**, *22*, 387–396. [CrossRef]
26. Martínez, J.; Pastor, J.; Sancho, J.; McNabola, A.; Martínez, M.; Gallagher, J. A functional data analysis approach for the detection of air pollution episodes and outliers: A case study in Dublin, Ireland. *Mathematics* **2020**, *8*, 225. [CrossRef]
27. Sancho, J.; Martínez, J.; Pastor, J.; Taboada, J.; Piñeiro, J.; García Nieto, P. New methodology to determine air quality in urban areas based on runs rules for functional data. *Atmos. Environ.* **2014**, *83*, 185–192. [CrossRef]
28. Sancho, J.; Iglesias, C.; Piñeiro, J.; Martínez, J.; Pastor, J.; Araújo, M.; Taboada, J. Study of water quality in a spanish river based on statistical process control and functional data analysis. *Math. Geosci.* **2016**, *48*, 163–186. [CrossRef]
29. Iglesias, C.; Sancho, J.; Piñeiro, J.I.; Martínez, J.; Pastor, J.J.; Taboada, J. Shewhart-type control charts and functional data analysis for water quality analysis based on a global indicator. *Desalin. Water Treat.* **2016**, *57*, 2669–2684. [CrossRef]
30. Dombeck, D.; Graziano, M.; Tank, D. Functional clustering of neurons in motor cortex determined by cellular resolution imaging in awake behaving mice. *J. Neurosis.* **2009**, *29*, 13751–13760. [CrossRef]
31. Martínez, J.; Ordoñez, C.; Matìas, J.M.; Taboada, J. Determining noise in an aggregates plant using functional statistics. *Hum. Ecol. Risk Assess.* **2011**, *17*, 521–533. [CrossRef] [PubMed]
32. Ordóñez, C.; Martínez, J.; Saavedra, A.; Mourelle, A. Intercomparison Exercise for Gases Emitted by a Cement Industry in Spain: A Functional Data Approach. *J. Air Waste Manag. Assoc. 1995* **2011**, *61*, 135–141. [CrossRef]
33. Sancho, J.; Pastor, J.; Martínez, J.; García, M. Evaluation of Harmonic Variability in Electrical Power Systems through Statistical Control of Quality and Functional Data Analysis. In *Procedia Engineering*; The Manufacturing Engineering Society: Southfield, MI, USA, 2013; Volume 63, pp. 295–302. [CrossRef] [PubMed]
34. Wu, D.; Huang, S.; Xin, J. Dynamic compensation for an infrared thermometer sensor using least-squares support vector regression (LSSVR) based functional link artificial neural networks (FLANN). *Meas. Sci. Technol.* **2008**, *19*, 105202.1–105202.6. [CrossRef]
35. Ordoñez, C.; Martínez, J.; Cos Juez, J.; Sánchez Lasheras, F. Comparison of GPS observations made in a forestry setting using functional data analysis. *Int. J. Comput. Math.* **2012**, *89*, 402–408. [CrossRef]
36. Müller, H.G.; Sen, R.; Stadtmüller, U. Functional data analysis for volatility. *J. Econometr.* **2011**, *165*, 233–245. [CrossRef]
37. López, M.; Martínez, J.; Matías, J.M.; Taboada, J.; Vilán, J.A. Functional classification of ornamental stone using machine learning techniques. *J. Comput. Appl. Math.* **2010**, *234*, 1338–1345. [CrossRef]
38. López, M.; Martínez, J.; Matías, J.M.; Taboada, J.; Vilán, J.A. Shape functional optimization with restrictions boosted with machine learning techniques. *J. Comput. Appl. Math.* **2010**, *234*, 2609–2615. [CrossRef]
39. Flores, M.; Naya, S.; Fernández-Casal, R.; Zaragoza, S.; Raña, P.; Tarrío-Saavedra, J. Constructing a Control Chart Using Functional Data. *Mathematics* **2020**, *8*, 58. [CrossRef]
40. Warmenhoven, J.; Harrison, A.; Robinson, M.A.; Vanrenterghem, J.; Bargary, N.; Smith, R.; Cobley, S.; Draper, C.; Donnelly, C.; Pataky, T. A force profile analysis comparison between functional data analysis, statistical parametric mapping and statistical non-parametric mapping in on-water single sculling. *J. Sci. Med. Sport* **2018**, *21*, 1100–1105. [CrossRef]
41. Ordoñez, C.; Martínez, J.; Matías, J.M.; Reyes, A.N.; Rodríguez-Pérez, J.R. Functional statistical techniques applied to vine leaf water content determination. *Math. Comput. Model.* **2010**, *52*, 1116–1122. [CrossRef] [PubMed]

42. Eisenhart, C. The Assumptions Underlying the Analysis of Variance. *Int. Biometr. Soc.* **1947**, *3*, 1–21. [CrossRef]
43. Montgomery, D. *Design and Analysis of Experiments*, 8th ed.; John Wiley & Sons, Inc.: Hoboken, NJ, USA, 2013. [CrossRef]
44. Kotz, S.; Johnson, N.L. *Breakthroughs in Statistics. Volume II. Methodology and Distribution*; Springer: Berlin/Heidelberg, Germany, 1993; Volume 2.
45. Vikneswaran. *An R companion to "Experimental Design"*; Vikneswaran: Duxbury, MA, USA, 2005.
46. Crawley, M. *The R Book*, 2nd ed.; John Wiley & Sons: Hoboken, NJ, USA, 2013.
47. Theodorsson-Norheim, E. Kruskal-Wallis test: BASIC computer program to perform nonparametric one-way analysis of variance and multiple comparisons on ranks of several independent samples. *Comput. Meth. Prog. Biomed.* **1986**, *23*, 57–62.
48. Ostertagová, E.; Ostertag, O.; Kovác, J. Methodology and application of the Kruskal-Wallis test. *Mech. Mater.* **2014**, *611*, 115–120. [CrossRef]
49. Van Hecke, T. Power study of anova versus Kruskal Wallis test. *Stat. Manag. Syst.* **2012**, *15*, 241–247. [CrossRef]
50. Cuevas, A.; Febrero, M.; Fraiman, R. An anova test for functional data. *Comput. Stat. Data Anal.* **2004**, *47*, 111–122.
51. Tarrío-Saavedra, J.; Naya, S.; Francisco-Fernández, M.; Artiaga, R.; Lopez-Beceiro, J. Application of functional ANOVA to the study of thermal stability of micro–nano silica epoxy composites. *Chemometr. Intell. Lab. Syst.* **2011**, *105*, 114–124. [CrossRef]
52. Kaufman, C.G.; Sain, S.R. Bayesian functional (ANOVA) modeling using Gaussian process prior distributions. *Bayes. Anal.* **2010**, *5*, 123–149. [CrossRef]
53. Wang, J.L.; Chiou, J.M.; Müller, H.G. Functional Data Analysis. *Ann. Rev. Stat. Appl.* **2016**, *3*, 257–295. [CrossRef]
54. Kramosil, I.; Michálek, J. Fuzzy metrics and statistical metric spaces. *Kybernetika* **1975**, *11*, 336–344. [CrossRef]
55. Ramsay, J.; Silverman, B. *Applied Functional Data Analysis: Methods and Cae Studies*; Springer: Berlin/Heidelberg, Germany, 2002.
56. Walz, M.; Zebrowski, T.; Küchenmeister, J.; Busch, K. B-spline modal method: A polynomial approach compared to the Fourier modal method. *Opt. Express* **2013**, *21*, 14683–14697. [CrossRef]
57. Muñiz, C.; García, P.; Alonso, J.; Torres, J.; Taboada, J. Detection of outliers in water quality monitoring samples using functional data analysis in San Esteban estuary (Northern Spain). *Sci. Total Environ.* **2012**, *439*, 54–61. [CrossRef] [PubMed]
58. Martínez, J.; García, P.; Alejano, L.; Reyes, A. Detection of outliers in gas emissions from urban areas using functional data analysis. *J. Hazard. Mater.* **2011**, *186*, 144–149. [CrossRef] [PubMed]
59. Fraiman, R.; Muniz, G. Trimmed means for functional data. *TEST Off. J. Spanish Soc. Stat. Operat. Res.* **2001**, *10*, 419–440. [CrossRef]
60. Cuevas, A.; Febrero, M.; Fraiman, R. Robust estimationand classification for functional data via projection-based notions. *Comput. Stat.* **2007**, *22*, 481–496. [CrossRef]
61. Cuevas, A.; Febrero, M.; Fraiman, R. On the use of bootstrap for estimating functions with functional data. *Comput. Stat. Data Anal.* **2006**, *51*, 1063–1074. [CrossRef]
62. Léger, C.; Romano, J. Boostrap adaptive estimation. The trimmed-mean example. *Can. J. Stat.* **1990**, *18*, 297–314. [CrossRef]
63. Hall, P.; Maesono, Y. A Weighted Bootstrap Approach to Bootstrap Iteration. *J. R. Stat. Soc. Ser. B Stat. Methodol.* **2000**, *62*, 137–144. [CrossRef]
64. Millán-Roures, L.; Epifanio, I.; Martínez, V. Detection of Anomalies in Water Networks by Functional Data Analysis. *Math. Problems Eng.* **2018**, *2018*, 13. [CrossRef]
65. Dette, H.; Derbort, S. Analysis of Variance un Non Parametric regression Models. *J. Multivar. Anal.* **2001**, *76*, 110–137. [CrossRef]
66. Maldonado, Y.; Staniswalis, J.; Irwin, L.; Byers, D. A similarity analysis of curves. *Can. J. Stat.* **2002**, *30*, 373–381. [CrossRef]

67. Cuesta-Albertos, J.A.; Febrero, M. A simple multiway ANOVA for functional data. *TEST* **2010**, *19*, 537–557. [CrossRef]
68. Zhang, J.T. Analysis of Variance for Functional Data. In *A Chapman & Hall Book*; Press, C., Ed.; Taylor & Francis Group: Abingdon, UK, 2013; Chapter 5, p. 412. [CrossRef]
69. Erkoreka, A.; García, K.; Teres-Zubiaga, J.; Del Portillo, L. In-use office building energy characterization through basic monitoring and modelling. *Energy Build.* **2016**, *119*, 256–266. [CrossRef]

7

Combination of Ensembles of Regularized Regression Models with Resampling-Based Lasso Feature Selection in High Dimensional Data

Abhijeet R Patil [1] and Sangjin Kim [2,*]

[1] Computational Science, University of Texas at El Paso, El Paso, TX 79968, USA; arpatil@miners.utep.edu
[2] Department of Mathematical Sciences, University of Texas at El Paso, El Paso, TX 79968, USA
* Correspondence: skim10@utep.edu

Abstract: In high-dimensional data, the performances of various classifiers are largely dependent on the selection of important features. Most of the individual classifiers with the existing feature selection (FS) methods do not perform well for highly correlated data. Obtaining important features using the FS method and selecting the best performing classifier is a challenging task in high throughput data. In this article, we propose a combination of resampling-based least absolute shrinkage and selection operator (LASSO) feature selection (RLFS) and ensembles of regularized regression (ERRM) capable of dealing data with the high correlation structures. The ERRM boosts the prediction accuracy with the top-ranked features obtained from RLFS. The RLFS utilizes the lasso penalty with sure independence screening (SIS) condition to select the top k ranked features. The ERRM includes five individual penalty based classifiers: LASSO, adaptive LASSO (ALASSO), elastic net (ENET), smoothly clipped absolute deviations (SCAD), and minimax concave penalty (MCP). It was built on the idea of bagging and rank aggregation. Upon performing simulation studies and applying to smokers' cancer gene expression data, we demonstrated that the proposed combination of ERRM with RLFS achieved superior performance of accuracy and geometric mean.

Keywords: ensembles; feature selection; high-throughput; gene expression data; resampling; lasso; adaptive lasso; elastic net; SCAD; MCP

MSC: 62P10; 62F40; 62F07

---

1. Introduction

With the advances of high throughput technology in biomedical research, large volumes of high-dimensional data are being generated [1–3]. Some of the examples of what produces such data are microarray gene expression [4–6] data sequencing, RNA-seq [7], genome-wide association studies (GWASs) [8,9], and DNA-methylation studies [10,11]. These data are high dimensional in nature, where the total count of features is significantly larger than the number of samples ($p >> n$)—termed the curse of dimensionality. Although this is one of the major problems, there are many other problems, such as noise, redundancy, and over parameterization. To deal with these problems, many two-stage approaches of feature selection (FS) and classification algorithms have been proposed in machine learning over the last decade.

The FS methods are used to reduce the dimensionality of data by removing noisy and redundant features that help in selecting the truly important features. The FS methods are classified into rank-based and subset methods [12,13]. Rank-based methods rank all the features with respect to their importance based on some criteria. Although there is a lack of threshold to select the optimal number of top-ranked features, this can be solved using the sure independence screening (SIS) [14] conditions. Some of the popular rank-based FS methods used in bioinformatics are information gain [15], Fisher score [16], chi-square [17], and minimum redundancy maximum relevance [18]. These rank-based FS methods have several advantages, such as that they avoid overfitting and are computationally faster because they do not depend on the performances of classification algorithms. However, these methods do not consider joint importance because they focus on marginal significance. To overcome this issue, feature subset section methods were introduced. The subset methods [19] are the ones where the subsets of features are selected with some predetermined threshold based on some criteria, but these methods need more computational time in a high-dimensional data setting and lead to an NP-hard problem [20]. Some of the popular subset methods include Boruta [21] and relief [22].

For the classification of gene expression data, there are non-parametric-based popular algorithms, such as random forests [23], Adaboost [24], and support vector machines [25]. The support vector machines are known to perform well in highly correlated gene expression data compared to the random forests [26]. The random forests and Adaboost are based on the concept of decision trees, and the support vector machines are based on the idea of hyperplanes. In addition to the above, there are parametric machine learning algorithms, such as penalized logistic regression (PLR) models, that have five different penalties which are predominantly popular in high-dimensional data. The first two classifiers are Lasso [27] and ridge [28] that are based on L1 and L2 penalties. The third classifier is a combination of these and is termed as elastic net [29]. The other two PLR classifiers are SCAD [30] and MCP [31], which are based on non-concave and concave respectively. All these individual classifiers are very common in machine learning and bioinformatics [32]. However, in highly correlated gene expression data, the individual classifiers do not perform well in terms of prediction accuracy. To overcome the issue of individual classifiers, ensemble classifiers are proposed [33,34]. The ensemble classifiers are bagging and aggregating methods [35,36] that are employed to improve the accuracy of several "weak" classifiers [37]. The tree-based method of classification by ensembles from random partitions (CERP) [38] showed good performance but is computer-intensive. The ensembles of logistic regression models (LORENS) [39] for high-dimensional data were proven to be better for classification. However, there was a decrease in performance when there were a smaller number of true, important variables in the high-dimensional space because of random partitioning.

To address these issues, there is a need to develop a novel combination of FS with a classification method and compare the proposed method with the other combinations of popular FS with the classifiers through extensive simulation studies and a real data application. In a high dimensional data set, it is necessary to filter out the redundant and unimportant features using the FS methods. This helps in reducing the computational time and helps in boosting the prediction accuracy with the help of significant features.

In this article, we introduce the combination of an ensemble classifier with an FS method—the resampling-based lasso feature selection (RLFS) method for ranking features, and ensemble of regularized regression models (ERRM) for classification purposes. The resampling approach was proven to be one of the best FS screening steps in a high-dimensional data setting [13]. The RLFS uses the selection probability with lasso penalty, and the threshold for selecting the top-ranked features is set using b-SIS condition; and these select features were applied to the ERRM to achieve the best prediction accuracy. The ERRM uses five individual regularization models, lasso, adaptive lasso, elastic net, SCAD, and MCP.

2. Materials and Methods

The FS method includes the proposed RLFS method, information gain, chi-square, and minimum redundancy maximum relevance. The classification methods include support vector machines, penalized regression models, and tree-based methods, such as random forests and adaptive boosting. The programs for all the experiments were written using R software [40]. The FS and classification were performed with the packages [41–46] obtained from CRAN. The weighted rank aggregation was evaluated with the RankAggreg package obtained from [47]. The codes for implementing the algorithms are available at [48]. The SMK-CAN-187 data were obtained from [49]; some of the applications of the data can be found in the articles [50,51] where the importance of screening approach in high dimensional data is elaborated.

2.1. Data Setup

To assess the performances of the models, we developed simulation study and also considered a real application of gene expression data.

2.1.1. Simulation Data Setup

The data were generated based on a random multivariate normal distribution where the mean was assigned as 0, and the variance-covariance matrix $\sum_x$ adapts a compound symmetry structure with the diagonal items set to 1 and the off-diagonal items being ρ values.

$$\sum_x = \begin{pmatrix} 1 & \rho & \cdots & \rho \\ \rho & 1 & \cdots & \rho \\ \vdots & \vdots & \ddots & \vdots \\ \rho & \rho & \cdots & 1 \end{pmatrix}_{p \times p} . \tag{1}$$

The class labels were generated using the Bernoulli trails with the following probability:

$$\pi_i(y_i = 1|x_i) = \frac{\exp(x_i\beta)}{1 + \exp(x_i\beta)}. \tag{2}$$

The data matrix $x_i \sim N_p(0, \sum_x)$ was generated using the random multivariate normal distribution, and the response variable y_i was generated by binomial distribution, as shown in Equations (1) and (2) respectively. For sufficient comparison of the performance of the model and subsidizing the effects of the data splits, all of the regularized regression models were built using the 10-fold cross-validation procedure, and the averages were taken over 100 partitioning times referred to as 100 iterations in this paper. The data generated are high-dimensional in nature with the number of samples, $n = 200$ and total features, $p = 1000$. The true regression coefficients were set to 25, which were generated using uniform distribution with the minimum and maximum values 2 and 4, respectively.

With this setup of high-dimensional data, we simulated three different types of data, each with correlation structures $\rho = 0.2$, 0.5, and 0.8 respectively. These values show the low, intermediate, and high correlation structures in the datasets which are significantly similar to what we usually see in the gene expression or others among many types of data in the field of bioinformatics [13,52]. At first, the data were divided randomly into training and testing sets with 75% and 25% of samples respectively; 75% of the training data was given to the FS methods, which ranked the genes concerning their importance, and then the top-ranked genes were selected based on b-SIS condition. The selected genes were applied in all the classifiers. For standard comparison and mitigating the effects of the data splitting, all of the regularized regression models were built using the 10-fold cross-validation; the models were assessed for testing the performance with the testing data using different evaluation metrics, and averages were taken over 100 splitting times referred to as 100 iterations.

2.1.2. Experimental Data Setup

To test the performance of the proposed combination of ERRM with RLFS, and compare it with the rest of the combinations of FS and classifiers, the gene expression data SMK-CAN-187 were analyzed. The data include 187 samples and 19,993 genes obtained from smokers, which included 90 samples from those with lung cancer and 97 samples from those without lung cancer. This data is high-dimensional, with the number of genes being 19,993. The preprocessing procedures are necessary to handle these high-dimensional data. At first, the data were randomly divided into training and testing sets with 75% and 25% of samples respectively. As the first filtering step, 75% of the training data were given to the marginal maximum likelihood estimator (MMLE), to overcome the redundant noisy features, and the genes were ranked based on their level of significance. The ranked significant genes were next applied to the FS methods along with the proposed RLFS method as the second filtering step, and a final list of truly significant genes was obtained. These significant genes were applied to all the classification models along with the proposed ERRM classifier. All of the models were built using the 10-fold cross-validation. The average classification accuracy and Gmean of our proposed framework were tested using the test data. The above procedure was repeated for 100 times and the averages were taken.

2.1.3. Data Notations

Let the expression levels of features in ith sample be represented as $x_i = (x_{i1}, x_{i2}, \ldots.., x_{ip})$ for $i = 1, \ldots.., n$, where n is the total number of samples and p is the total number of features. The response variables, $y_i \in \{0, 1\}$, where $y_i = 0$ means that ith individual is in the non disease group and $y_i = 1$ is disease group.

The original data x_i were split into 75% for the training set x_j and 25% for the testing set x_k. The training set $x_j = (x_{j1}, x_{j2}, \ldots.., x_{jp})$ for $j = 1, \ldots.., t$, where t is the number of training samples, the response variable y_j for the training set. The testing set $x_k = (x_{k1}, x_{k2}, \ldots.., x_{kp})$ for $k = 1, \ldots.., v$, where v is the number of testing samples; the response variable is y_k for the testing set. The classifiers are fitted on x_j, and the class labels y_j as training data set to predict the classification of y_k using x_k of the testing set.

The detailed procedure is as follows. The training data x_j were given to the FS methods, and the new reduced feature set $x_r = (x_{j1}, x_{j2}, \ldots.., x_{jf})$ for $j = 1, \ldots.., t$, where t was the samples included in training data, and f was the reduced number of features after the FS step. This reduced feature set x_r was used as new training data for building the classification models.

2.2. *Rank Based Feature Selection Methods*

With the gain in popularity of high dimensional data in bioinformatics, the challenges to deal with it also grew. In gene expression data, having large p and small n problems, the n represents the samples as patients and p represents the features as genes. Dealing with such a large number of genes that are generated by conducting large biological experiments involves computationally intensive tasks that become too expensive to handle. The performance drops when such a large number of genes are added to the model. To overcome this problem, employing FS methods becomes a necessity. In statistical machine learning, there are many FS methods developed to deal with the gene expression data. But most of the existing algorithms are not completely robust applications to the gene expression data. Hence, we propose an FS method that ranks the features based on some criteria explained in the next section. We also explain some other popular FS methods in classification problems, such as information gain, chi-square, and minimum redundancy maximum relevance.

2.2.1. Information Gain

The information gain (IG) method [15] is simple, and one of the widely used FS methods. This univariate FS method is used to assess the quantity of information shared between the training feature set $x_j = (x_{j1}, x_{j2}, \ldots.., x_{jp})$ for $j = 1, \ldots.., t$, where t is the number of training samples, for $g = 1, 2, \ldots.p$, where g is the feature in p number of features, and the response variable y_j. It provides an ordered ranking of all the features having a strong correlation with the response variable that helps to obtain good classification performance.

The information gain between the gth feature in x_j and the response variable y_j is given as follows:

$$\mathrm{IG}(x_j; y_j) = \mathrm{H}(x_j) - \mathrm{H}(x_j|y_j), \tag{3}$$

where $H(x_j)$ is entropy of x_j and $H(x_j|y_j)$ is entropy of x_j given y_j. The entropy [53] of x_j is defined by the following equation:

$$\mathrm{H}(x_j) = \sum_{g \in x_j} \pi(g) \log(\pi(g)), \tag{4}$$

where g indicates discrete random variable x_j and $\pi(g)$ gives the probability of g on all values of x_j. Given the random variable y_j, the conditional entropy of x_j is:

$$\mathrm{H}(x_j|y_j) = \sum_{y \in y_j} \pi(y) \sum_{g \in x_j} \pi(g|y) \log(\pi(g|y)), \tag{5}$$

where $\pi(y)$ is the prior probability of y_j; $\pi(g|y)$ is conditional probability of g in a given y that shows the uncertainty of x_j given y_j.

$$\mathrm{IG}(x_j; y_j) = \sum_{g \in x_j} \sum_{y \in y_j} \pi(g, y) \log \frac{\pi(g, y)}{\pi(g)\pi(y)}, \tag{6}$$

where $\pi(g, y)$ is the joint probability of g and y . IG is symmetric such that $\mathrm{IG}(x_j; y_j) = \mathrm{IG}(y_j; x_j)$, and is zero if the variables x_j and y_j are independent.

2.2.2. Chi-Square Test

The chi-square test (Chi2) belongs to the category of the non-parametric test, which is used mainly in determining the significant relation between two categorical variables. As part of the preprocessing step, we used the "equal interval width" approach to transform the numerical variables into categorical counterparts. The "equal interval width" algorithm first divides the data into q intervals of equal size. The width of each interval is defined as: $w = (max - min)/q$ and the interval boundaries are determined by: $min + w, min + 2w, ..., min + (q - 1)w$.

The general rule in Chi2 is that the features have a strong dependency on the class labels selected, and the features independent of the class labels are ignored.

From the training set, $x_j = (x_{j1}, ...x_{jp})$, $g = 1, 2, ...p$, where g is every feature in p number of features. Given a particular feature g with r different feature values [53], the Chi2 score of that particular feature can be calculated as:

$$\chi^2(g) = \sum_{j=1}^{r} \sum_{s=1}^{p} \frac{(O_{js} - E_{js})^2}{E_{js}}, \tag{7}$$

where O_{js} is the number of instances with the jth feature value given feature g. In addition, $E_{js} = \frac{O_{*s} O_{j*}}{O}$, where O_{j*} indicates the number of data instances with the feature value given feature g, O_{*s} denotes the number of data instances in r, and p is total number of features.

When two features are independent, the O_{js} is closer to the expected count E_{js}; consequently, we will have smaller Chi2 score. On the other hand, the higher Chi2 score implies that the feature is more dependent on the response and it can be selected for building the model during training.

2.2.3. Minimum Redundancy Maximum Relevance

The minimum redundancy and maximum relevance method (MRMR) is built on optimization criteria of mutual information (redundancy and relevance); hence, it is also defined under mutual information based methods. If a feature has uniformly of expressions or if they are randomly distributed in different classes, its mutual information with such classes is null [18]. If a feature is expressed differentially for different classes, it should have strong mutual information. Hence, we use mutual information as a measure of the relevance of features. MRMR also reduces the redundant features from the feature set. For a given set of features, it tries to measure both the redundancy among features and relevance between features and class vectors.

The redundancy and relevance are calculated based on mutual information, which is as follows: We know that, in the training set x_j, $g = 1, ..., p$ represents every feature in x_j and y_j is the response variable.

$$I(g,y) = \sum_{g \in x_j} \sum_{y \in y_j} \log \frac{\pi(g,y)}{\pi(g)\pi(y)}. \tag{8}$$

In the following equation, for simplicity, let us consider the training set x_j as X and response variable y_j as Y. The objective function is shown below:

$$J_{\mathrm{MRMR}}(X_S, Y) = \frac{1}{|S|} \sum_{i \in S} I(X_i, Y) - \frac{1}{|S|^2} \sum_{i,j \in S} I(X_i, X_j), \tag{9}$$

where S is the subset of selected features and X_i is the ith feature. The first term is a measure of relevance that is the sum of mutual information of all the selected features in the set S with respect to the output Y. The second term is measure of redundancy that is the sum of the mutual information between all the selected features in the subset S. By optimizing the Equation (9), we are maximizing the first term and minimizing the second term simultaneously.

2.3. *Classification Algorithms*

Along with gene selection, improving prediction accuracy when dealing with high-dimensional data has always been a challenging task. There is a wide range of popular classification algorithms used when dealing with high throughput data, such as tree-based methods [54], support vector machines, and penalized regression models [55]. These popular models are discussed briefly in this section.

2.3.1. Logistic Regression

Logistic regression (LR) is perhaps one of the primary and popular models used while dealing with binary classification problems [56]. Logistic regression for dealing with more than two classes is called multinomial logistic regression. The primary focus here is on the binary classification. Given the set of inputs, the output is a predicted probability that the given input point belongs to a particular class. The output is always within [0, 1]. Logistic regression is based on the assumption that the original input space can be divided into two separate regions, one for each class, by a plane. This plane helps to discriminate between the dots belonging to different classes and is called as linear discriminant or linear boundary.

One of the limitations is the number of parameters that can be estimated needs to be smaller and should not exceed the number of samples.

2.3.2. Regularized Regression Models

Regularization is a technique used in logistic regression by employing penalties to overcome the limitations of dealing with high-dimensional data. Here, we discuss the PLR models such as lasso, adaptive lasso, elastic net, SCAD, and MCP. These five methods are included in the proposed ERRM and also tested as independent classifiers for comparing performance with the ERRM.

The logistic regression equation:

$$\log\left(\frac{\pi(y_j = 1|x_j)}{1 - \pi(y_j = 1|x_j)}\right) = \beta_0 + \beta x_j, \tag{10}$$

where $j = 1....t$ and $\beta = (\beta_1 ... \beta_f)^T$.

From logistic regression Equation (10), the log-likelihood estimator is shown as below:

$$l(\beta, y_j) = \sum_{j=1}^{t} \{y_j \log(\pi(y_j = 1|x_j)) + (1 - y_j)\log(1 - \pi(y_j = 1|x_j))\}. \tag{11}$$

Logistic regression offers the benefit by simultaneous estimation of the probabilities $\pi(x_j)$ and $1 - \pi(x_j)$ for each class. The criterion for prediction is $I\{\pi(y_j = 1|x_j) \geq 0.5\}$, where $I(\cdot)$ is an indicator function.

The parameters for PLR are estimated by minimizing above function:

$$\hat{\beta}_{\text{PLR}} = \underset{\beta}{\text{argmin}} \left[-l(\beta, y_j) + p(\beta) \right], \tag{12}$$

where $p(\beta)$ is a penalty function, $l(\beta, y_j)$ is the log-likelihood function.

Lasso is a widely used method in variable selection and classification purposes in high dimensional data. It is one of the five methods used in the proposed ERRM for classification purposes. The LASSO penalized regression method is defined below:

$$\hat{\beta}_{\text{LASSO}} = \underset{\beta}{\text{argmin}} \left[-l(\beta, y_j) + \lambda \sum_{j=1}^{f} |\beta_j| \right], \tag{13}$$

where f is the reduced number of features; λ is the tuning parameter that controls the strength of the L1 penalty.

The oracle property [30] has consistency in variable selection and asymptotic normality. The lasso works well in subset selection; however, it lacks the oracle property. To overcome this, different weights are assigned to different coefficients: this describes a weighted lasso called adaptive lasso. The adaptive lasso (ALASSO) penalty is shown below:

$$\hat{\beta}_{\text{ALASSO}} = \underset{\beta}{\text{argmin}} \left[-l(\beta, y_j) + \lambda \sum_{j=1}^{f} w_j |\beta_j| \right], \tag{14}$$

where f is the reduced number of features, λ is the tuning parameter that controls the strength of the L2 penalty, and w_j is the weight vector based on ridge estimator. The ridge estimator [28] uses the L2 regularization method which obtains the size of coefficients by adding the L2 penalty.

The elastic net (ENET) [57] is the combination of lasso which uses the L1 penalty, and ridge which uses the L2 penalty. The sizable number of variables is obtained, which helps in avoiding the model turning into an excessively sparse model.

The ENET penalty is defined as:

$$\hat{\beta}_{\text{ENET}} = \underset{\beta}{\text{argmin}} \left[-l(\beta, y_j) + \lambda \left(\frac{1-\alpha}{2} \sum_{j=1}^{f} |\beta_j|^2 + \alpha \sum_{j=1}^{f} |\beta_j| \right) \right], \tag{15}$$

where λ is the tuning parameter that controls the penalty, f is the number of features, α is the mixing parameter between ridge $\alpha = 0$ and lasso $\alpha = 1$.

The smoothly clipped absolute deviation penalty (SCAD) [30] is a sparse logistic regression model with a non-concave penalty function. It improves the properties of the L1 penalty. The regression coefficients are estimated by minimizing the log-likelihood function:

$$\hat{\beta}_{\text{SCAD}} = \underset{\beta}{\text{argmin}} \left[-l(\beta, y_j) + \lambda \sum_{j=1}^{f} p_{\lambda}(\beta_j) \right]. \tag{16}$$

In Equation (16) the $p_\lambda(\beta_j)$ is defined by:

$$|\beta_i| I_{(|\beta_j| \le \lambda)} + \left(\frac{\{(c^2-1)\lambda^2 - (c\lambda - |\beta_j|)_+^2\} I(\lambda \le |\beta_j|)}{2(c-1)} \right), \; c > 2 \text{ and } \lambda \ge 0\,. \tag{17}$$

Minimax concave penalty (MCP) [31] is very similar to the SCAD. However, the MCP relaxes the penalization rate immediately, while for SCAD, the rate remains smooth before it starts decreasing. The MCP equation is given as follows:

$$\hat{\beta}_{\text{MCP}} = \underset{\beta}{\operatorname{argmin}} \left[-l(\beta, y_j) + \lambda \sum_{j=1}^{f} p_\lambda(\beta_j) \right]. \tag{18}$$

In Equation (18) the $p_\lambda(\beta_j)$ is defined as:

$$\left(\frac{2c\lambda|\beta_j| - \beta_j^2}{2c} \right) I(|\beta_j| \le c\lambda) + \left(\frac{c\lambda^2}{2} \right) I(|\beta_j| > c\lambda), \text{ for } \lambda \ge 0 \text{ and } c > 1. \tag{19}$$

2.3.3. Random Forests

The random forest (RF) [23] is an interpretive and straightforward method commonly used for classification purposes in bioinformatics. It is also known for its variable importance ranking in high dimensional data sets. RF is built on the concept of decision trees. Decision trees are usually more decipherable when dealing with binary responses. The idea of RF is to operate as an ensemble instead of relying on a single model. RF is a combination of a large number of decision trees where each tree has some random subset of features obtained from the data by allowing repetitions. This process is called bagging. The majority voting scheme is applied by aggregating all the tree models and obtaining one final prediction.

2.3.4. Support Vector Machines

Support vector machines (SVM) [25] are well known amongst most of the mainstream algorithms in supervised learning. The main goal of a SVM is to choose a hyperplane that can best divide the data in the high dimensional space. This helps to avoid overfitting. The SVM detects the maximum margin hyperplane, the hyperplane that maximizes the distance between the hyperplane, and the closest dots [58]. The maximum margin indicates that the classes are well separable and correctly classified. It is represented as a linear combination of training points. As a result, the decision boundary function for classifying points as to hyperplane only involves dot products between those points.

2.3.5. Adaboost

Adaboost is also known as adaptive boosting (AB) [24]. It improves the performance of a particular weak boosting classifier through an iterative process. This ensemble learning algorithm can be extensively applied to classification problems. The primary objective here is to assign more weights to the patterns that are harder to classify. Initially, the same weights are assigned to each training item. The weights of the wrongly classified items are incremented while the weights of the correctly classified items are decreased in each iteration. Hence, with the additional iterations and more classifiers, the weak learner is bound to cast on the challenging samples of the training set.

2.4. The Proposed Framework

We propose a combination of the FS method and classification method. For the filtering procedure, the resampling-based lasso feature selection method is introduced, and for the classification, the ensemble of regularized regression models is developed.

2.4.1. The Resampling-Based Lasso Feature Selection

From [13], we see that the resampling-based FS is relatively more efficient in comparison to the other existing FS methods in gene expression data. The RLFS method is based on the lasso penalized regression method and the resampling approach employed to obtain the ranked important features using the frequency.

The least absolute shrinkage and selection operator (LASSO) [27] estimator is based on L1-regularization. The L1-regularization method limits the size of coefficients pushes the unimportant regression coefficients to zero by using the L1 penalty. Due to this property, variable selection is achieved. It plays a crucial role in achieving better prediction accuracy along with the gene selection in bioinformatics.

$$\hat{\beta}_{\text{lasso}} = \underset{\beta}{\text{argmin}}\left[-\sum_{j=1}^{t}\{y_j \log(\pi(y_j = 1|x_j)) + (1-y_j)\log(1-\pi(y_j=1|x_j))\} + \lambda\sum_{j=1}^{p}|\beta_j|\right]. \quad (20)$$

The selection probability $S(f_m)$ of the features based on the lasso is shown in the below equation.

$$S(f_m) = \frac{1}{R}\sum_{i=1}^{R}\frac{1}{L}\sum_{j=1}^{L} I(\beta_{ijm} \neq 0), \text{ for } m = 1,2,...,p. \quad (21)$$

The b-SIS criteria to select the top k ranked features is defined by,

$$\left\lceil b \times \frac{n}{\log(n)} \right\rceil, \quad (22)$$

where R is defined by the total number of resampling, L is total number of λ values, f_m is the feature indexed as i, p is total number of features, n is total number of samples, and β_{ijm} is defined as regression coefficient of mth feature and $I()$ indicator variable. Each R number of resamples and L number of values of λ are considered to build the variable selection model. The 10-fold cross validation is considered while building the model.

After ranking the features using the RLFS method, we employ the b-SIS approach to select the top features based on Equation (22) where b is set to two. The number of true important variables selected among the top b-SIS ranked features is calculated in each iteration and the average of this is taken over 100 iterations.

2.4.2. The Ensembles of Regularized Regression Models

LASSO, ALASSO, ENET, SCAD, and MCP are the five individual regularized regression models included as base learners in our ERRM. The role of bootstrapped aggregation or bagging is to reduce the variance by averaging over an "ensemble" of trees, which will improve the performance of weak classifiers. $B = B_1^k, ..., B_M^k$ is the number of random bootstrapped samples obtained from reduced training set x_r with corresponding class label y_j. The five regularized regression models are trained on each bootstrapped sample B named sub-training data, leading to $5 \times B$ models. These five regularized models are then trained using the 10-fold cross-validation to predict the classes on the out of bag samples called sub-testing data where the best model fit in each of the five regularized regression model is obtained. Henceforth, in each of the five regularized models, the best model is selected and the testing data x_k is applied to obtain the final list of predicted classes for each of these models. For binary classification problems, in addition to accuracy, the sensitivity and specificity are primarily sought. The E evaluation metrics are computed for each of these best models of five regularized models. In order to get an optimized classifier using all the evaluation measures E is essential, and this is achieved using weighted rank aggregation. Here, each of the regularized models is ranked based on the performance of E evaluation metrics. The models are ranked based on the increasing order of performance; in the case of a matching score of accuracy for two or more models, other metrics such as sensitivity and specificity are considered. The best performing model among the five models is obtained based on these ranks. This procedure is repeated to obtain the best performing model in each of the tree T. Finally, the majority voting procedure is applied over the T trees to obtain a final list of predicted classes. The test class label is applied to measure the final E measures for assessing the performance of the proposed ensembles. The Algorithm 1 defines the proposed ERRM procedure.

The complete workflow of the proposed RLFS-ERRM framework is shown in Figure 1.

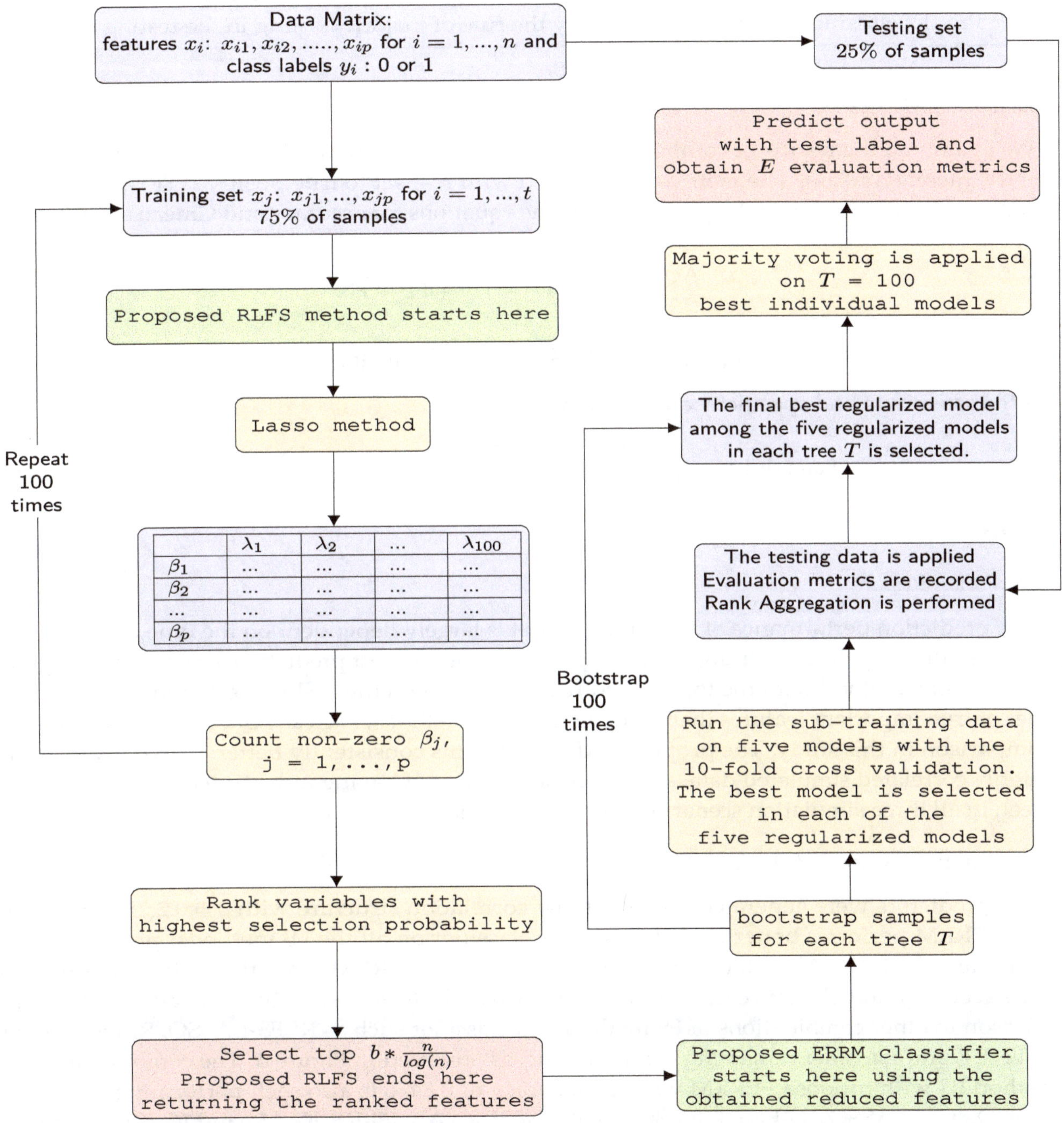

Figure 1. The complete workflow depicting the proposed combination of RLFS-ERRM framework.

Algorithm 1 Proposed ERRM

Step 1: Obtain new training data x_r with most informative features using the proposed RLFS method.
Step 2: Draw bootstrap samples from x_r and apply them to each of the regularized methods to be fitted with 10-fold cross validation.
Step 3: Apply out of bag samples (OOB) not used in bootstrap samples to the above fitted models to choose the best model using E evaluation metrics.
Step 4: Repeat steps 2 and 3 until getting 100 bootstrap models.
Step 5: Apply testing set x_k to each of 100 models to aggregate votes of classification.
Step 6: Predict classification of each sample by the rule of majority voting in the testing set.

2.5. Evaluation Metrics

We evaluated the results of combinations of FS methods with the classifier using accuracy and geometric mean (Gmean). The metrics are detailed with respect to true positive (TP), true negative (TN), false negative (FN), and false positive (FP). The equations for accuracy and Gmean are as follows:

$$\text{Accuracy} = \frac{\text{TP} + \text{TN}}{\text{TP} + \text{TN} + \text{FP} + \text{FN}} \tag{23}$$

$$\text{Gmean} = \sqrt{\text{Sensitivity} \times \text{Specificity}},$$

where the sensitivity and specificity are given by:

$$\text{Sensitivity} = \frac{\text{TP}}{\text{TP} + \text{FN}} \quad \text{and} \quad \text{Specificity} = \frac{\text{TN}}{\text{TN} + \text{FP}}. \tag{24}$$

3. Results

3.1. Simulation Results

The prediction performance of any given model is largely dependent on the type of the features. The features affecting the classification will help in attaining the best prediction accuracies. In Figure 2, we see the RLFS method with the top-ranked features based on the b-SIS criterion includes a higher number of true important features than other existing FS methods, such as IG, Chi2, and MRMR used for comparison in this study. The proposed RLFS performs consistently better across low, medium, and highly correlated simulated data, and the positive effect of having more true important variables was seen in all three simulation scenarios (further explained in detail).

3.1.1. Simulation Scenario (S1): Low Correlation 0.2

The predictors were generated, having a low correlation structure with $\rho = 0.2$. The proposed classifier ERRM performs better than the existing classifier on all the FS methods: proposed RLFS, IG, Chi2, and MRMR. Also, the proposed combination of ERRM classifier with RLFS method, with the accuracy and Gmean, each of which is 0.8606 and 0.8626 respectively, is relatively better in comparison to other combinations of FS method and classifier such as RLFS-LASSO, RLFS-ALASSO, RLFS-ENET, and the other remaining combinations, as observed in Figure 3. The combination of the FS method IG with proposed ERRM with an accuracy of 0.8476 is also seen performing better than IG-LASSO, IG-ALASSO, IG-ENET, IG-SCAD, IG-MCP, IG-AB, IG-RF, IG-LR, and IG-SVM. Similarly, the combination of Chi2-ERRM with an accuracy of 0.8538 is seen better than FS method Chi2 with the other remaining classifiers. The results are reported in Table 1. The combination of MRMR-ERRM has an accuracy of 0.8550 and Gmean of 0.8552 is better than the combination of FS method MRMR with the rest of the nine classifiers.

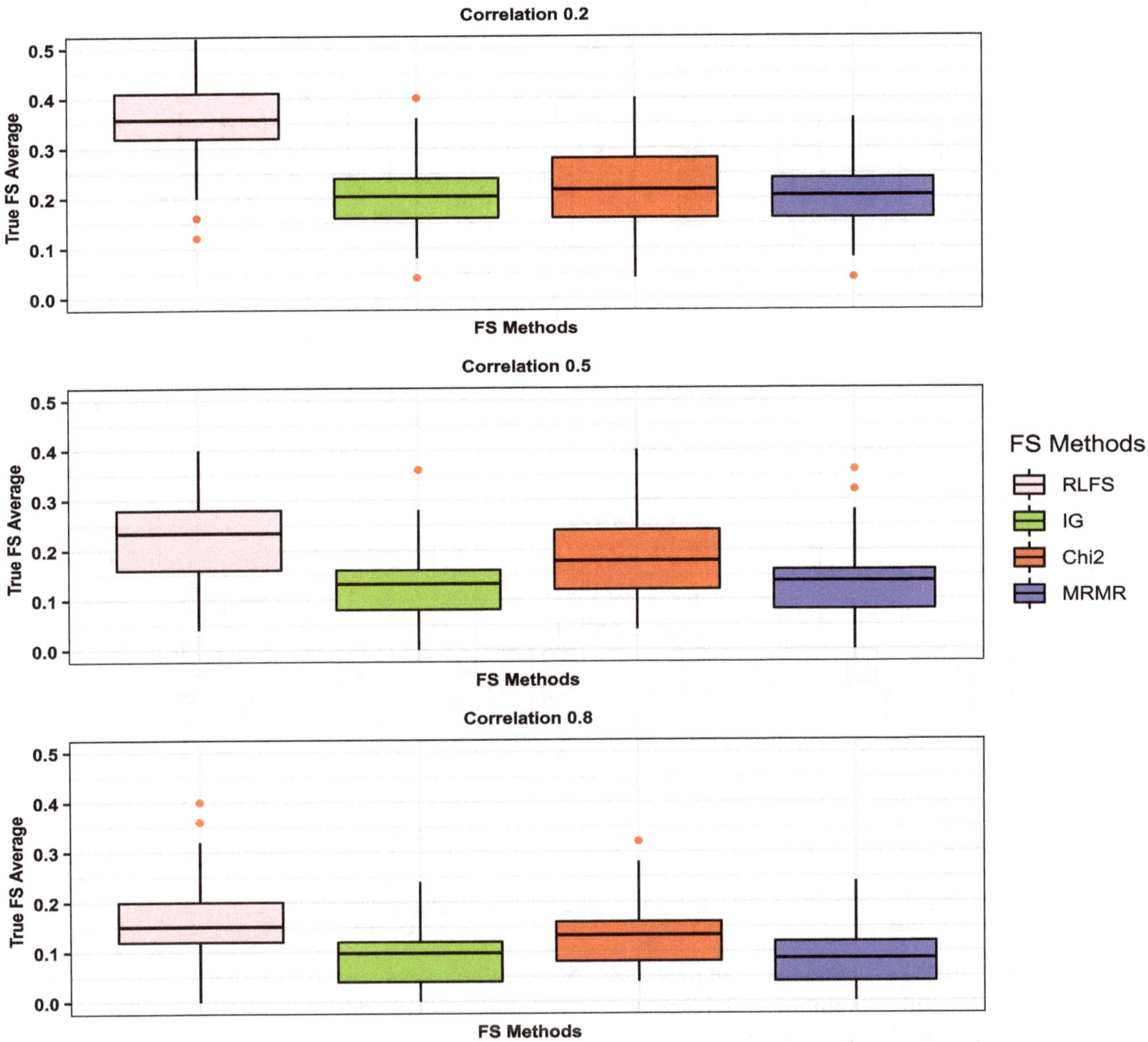

Figure 2. True number of features selected among top *b*-SIS ranked features, and the average of this taken over 100 iterations for three different scenarios. The first horizontal line in the box shows the first quartile and the second horizontal dark line which usually represents the median values are shown as the mean values in this article. The third horizontal line in each of the boxes shows the third quartile. The red dotted circles indicate the outliers in each of the FS methods.

All the classifiers with the features obtained from the RLFS method achieved the best accuracies in comparison to other FS methods, as seen in Figure 3. The combination of RLFS with SVM showed the second-best performance by attaining an accuracy of 0.8582, as seen in Table 1. The ENET method showed the best performance among all the regularized regression models with all the FS methods, and the best accuracy was obtained with the proposed RLFS method.

The proposed combination of RLFS-ERRM has better performance than the other existing combinations of the FS and classifier without the proposed FS method RLFS and classifier ERRM itself. For example, the existing FS methods IG, Chi2, and MRMR with the eight existing individual classifiers' performances are lower than the proposed RLFS-ERRM combination, as shown in Table 1.

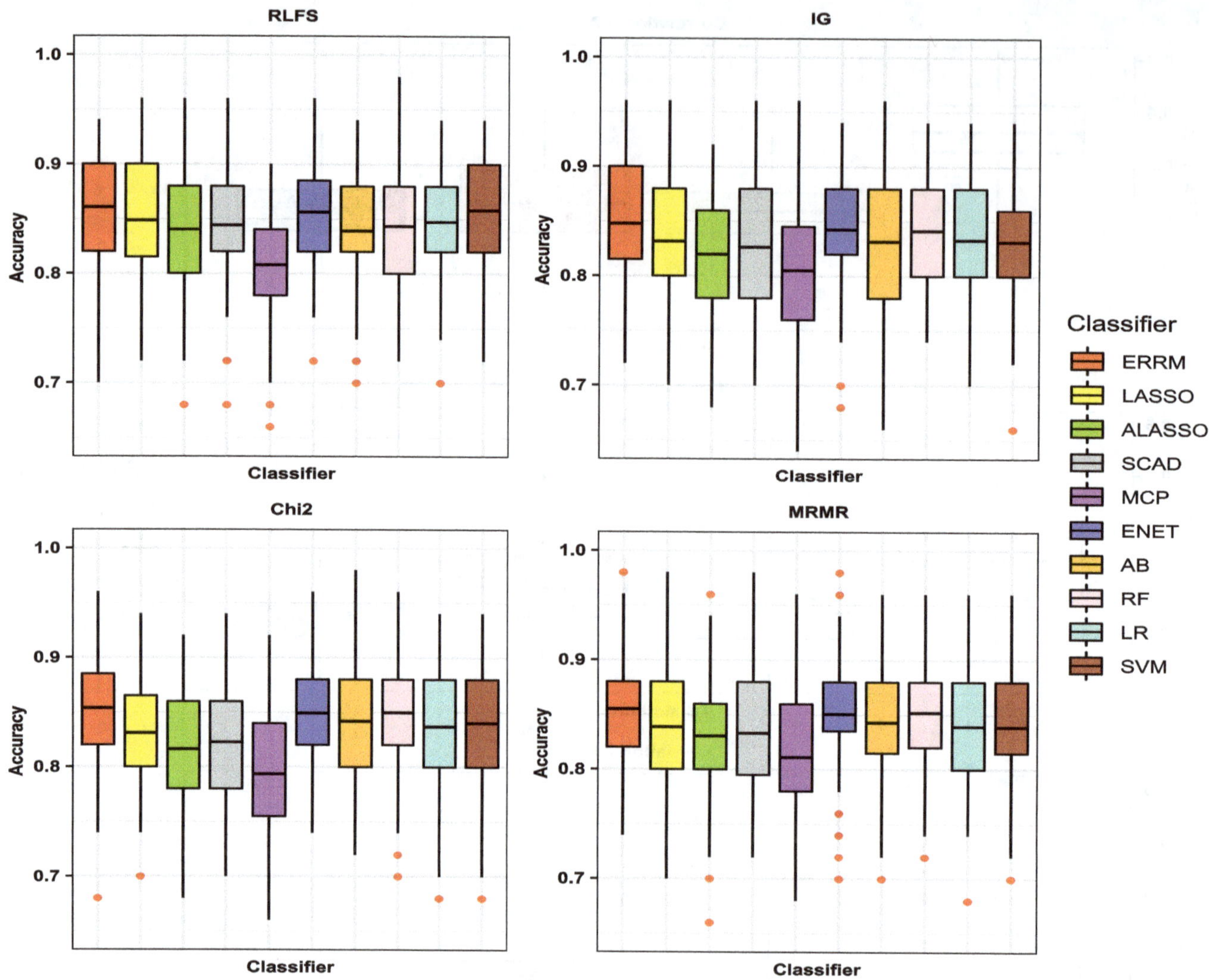

Figure 3. Comparison of accuracies of proposed combination of ensemble of regularized regression models (ERRM) with resampling-based lasso feature selection (RLFS) and other classifiers with feature selection methods when correlation = 0.2.

3.1.2. Simulation Scenario (S2): Intermediate Correlation 0.5

The predictor variables were generated using a medium correlation structure with $\rho = 0.5$. The proposed combination of the RLFS method and ERRM classifier, with the accuracy and Gmean, each of which is 0.9256 and 0.9266, respectively, attained relatively better performance compared to other combinations of the FS method and classifier such as RLFS-LASSO, RLFS-ALASSO, RLFS-ENET and the other remaining combinations. The results are shown in Table 1. From Figure 4, we see that the proposed ensemble classifier ERRM with other FS methods such as IG, Chi2, and MRMR performs best compared to the other nine individual classifiers.

The SVM and ENET classifiers with the RLFS method attained accuracies that are almost similar to the proposed combination of ERRM-RLFS. However, when Gmean is considered, the ERRM-RLFS outperforms the SVM combinations. The average SD of the proposed combination of the ERRM-RLFS is smaller than other combinations of the FS method and classifier. The accuracies of SVM and ENET classifiers with the IG method were 0.9128 and 0.9150 lower compared to the ERRM classifier with the IG method which had an accuracy of 0.9184. Similarly, the ERRM with the Chi2 method showed relatively better performance than the competitive classifiers ENET and SVM. Further, the ERRM classifier with the MRMR method having an accuracy of 0.9174 showed better performance than ENET, SVM, and other top-performing individual classifiers.

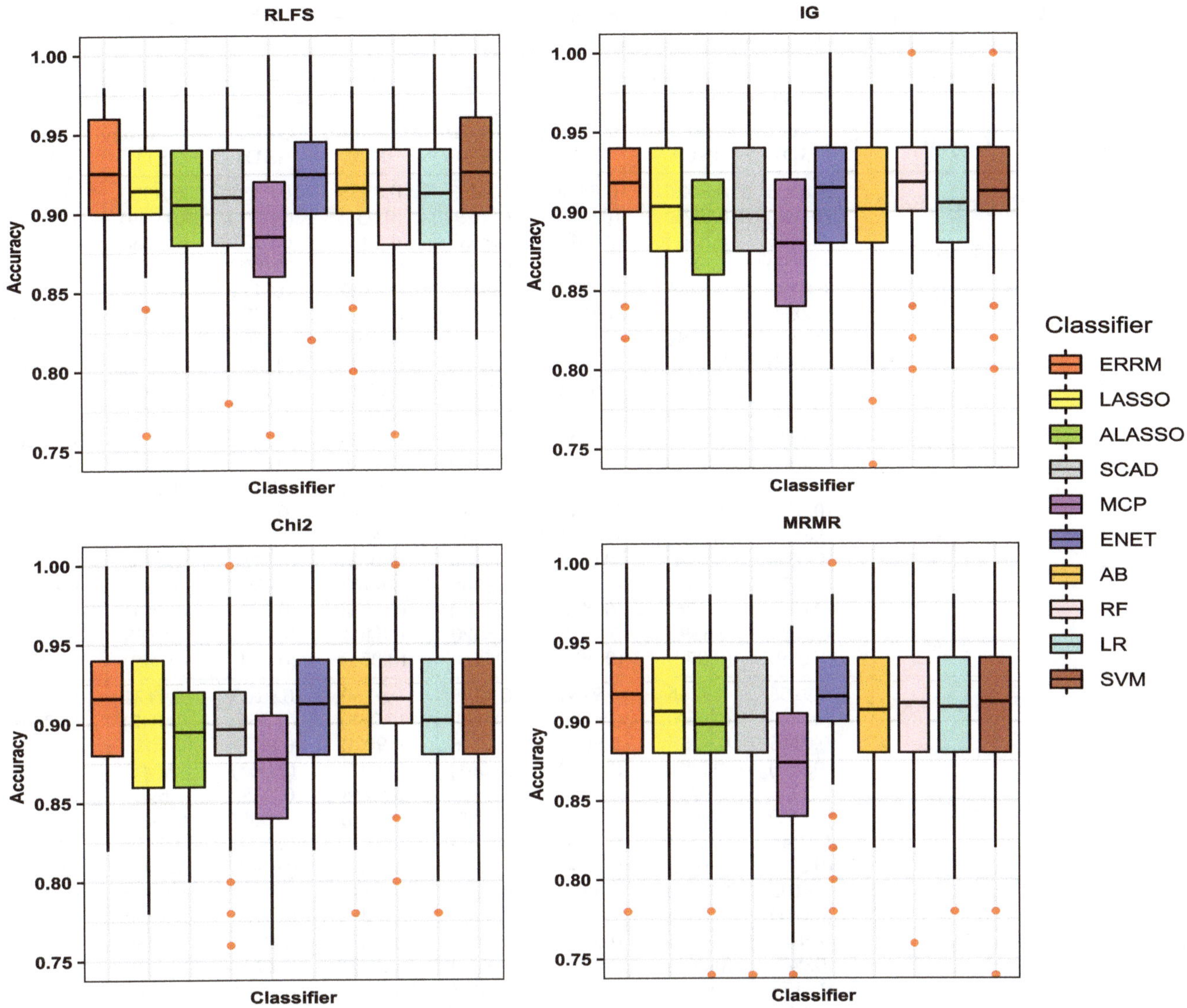

Figure 4. Boxplot showing the accuracies of classifiers with FS methods when correlation = 0.5.

While the SVM and ENET classifiers showed promising performance on the RLFS that had a good number of important features, they failed to show the same consistency on the other FS methods. On the other hand, the ensemble ERRM showed robust behavior, with being able to withstand the noise that helps in attaining better prediction accuracies and Gmean, not only with the RLFS method but also with other FS methods, such as IG, Chi2, and MRMR, as seen in Table 1.

Similar results are also found in the Simulation Scenario (S3): which has the highly correlated data with ρ set to 0.8. The results for this scenario are described in the Appendix A.

3.2. Experimental Results

Figure 5 shows the box plot of average accuracies taken over 100 iterations for all the combinations of FS and classifiers in experimental data. Each of the sub-figures in the figure shows the classifiers with the corresponding FS methods. As seen in Table 2, the performance of all the individual classifiers when applied on the RLFS method—the accuracy and Gmean—are relatively much better than the accuracies of the individual classifiers when applied on the IG, Chi2, and MRMR methods.

Table 1. Classification performance of proposed RLFS with ERRM compared to other combinations of feature selection methods with classifiers over 100 iterations.

Correlation	Classifier	Proposed RLFS		IG		Chi2		MRMR	
		Acc (SD)	Gmean (SD)	Accuracy (SD)	Gmean (SD)	Accuracy (SD)	Gmean (SD)	Accuracy (SD)	Gmean (SD)
0.2	Proposed ERRM	0.8606 (0.049)	0.8626 (0.073)	0.8476 (0.052)	0.8483 (0.079)	0.8538 (0.053)	0.8551 (0.071)	0.8550 (0.049)	0.8552 (0.075)
	LASSO	0.8486 (0.052)	0.8504 (0.075)	0.8316 (.054)	0.8335 (0.083)	0.8310 (0.052)	0.8323 (0.071)	0.8388 (0.051)	0.8393 (0.077)
	ALASSO	0.8402 (0.054)	0.8416 (0.077)	0.8198 (0.051)	0.8217 (0.079)	0.8160 (0.053)	0.8171 (0.075)	0.8304 (0.051)	0.8313 (0.079)
	ENET	0.8564 (0.048)	0.8584 (0.072)	0.8424 (0.054)	0.8441 (0.081)	0.8494 (0.046)	0.8509 (0.067)	0.8508 (0.052)	0.8508 (0.077)
	SCAD	0.8440 (0.054)	0.8457 (0.080)	0.8264 (0.057)	0.8283 (0.086)	0.8226 (0.061)	0.8239 (0.077)	0.8330 (0.056)	0.8336 (0.081)
	MCP	0.8078 (0.049)	0.8095 (0.081)	0.8050 (0.062)	0.8074 (0.088)	0.7936 (0.060)	0.7952 (0.085)	0.8110 (0.060)	0.8126 (0.082)
	AB	0.8390 (0.051)	0.8224 (0.077)	0.8314 (0.060)	0.8328 (0.080)	0.8422 (0.054)	0.8435 (0.075)	0.8432 (0.054)	0.8437 (0.075)
	RF	0.8432 (0.057)	0.8467 (0.084)	0.8414 (0.052)	0.8435 (0.078)	0.8498 (0.053)	0.8520 (0.075)	0.8522 (0.051)	0.8534 (0.077)
	LR	0.8474 (0.050)	0.8489 (0.076)	0.8330 (0.053)	0.8346 (0.080)	0.8370 (0.054)	0.8380 (0.073)	0.8394 (0.051)	0.8394 (0.080)
	SVM	0.8582 (0.049)	0.8595 (0.070)	0.8312 (0.052)	0.8320 (0.083)	0.8404 (0.054)	0.8416 (0.074)	0.8388 (0.049)	0.8378 (0.084)
0.5	Proposed ERRM	0.9256 (0.037)	0.9266 (0.053)	0.9184 (0.039)	0.9195 (0.059)	0.9160 (0.038)	0.9165 (0.056)	0.9174 (0.042)	0.9176 (0.056)
	LASSO	0.9146 (0.037)	0.9155 (0.053)	0.9034 (0.045)	0.9046 (0.061)	0.9020 (0.043)	0.9029 (0.063)	0.9066 (0.045)	0.9065 (0.062)
	ALASSO	0.9056 (0.039)	0.9062 (0.056)	0.8956 (0.044)	0.8966 (0.065)	0.8948 (0.046)	0.8954 (0.065)	0.8984 (0.046)	0.8982 (0.062)
	ENET	0.9244 (0.038)	0.9253 (0.052)	0.9150 (0.044)	0.9163 (0.061)	0.9122 (0.039)	0.9130 (0.060)	0.9158 (0.043)	0.9155 (0.058)
	SCAD	0.9102 (0.041)	0.9110 (0.060)	0.8974 (0.046)	0.8986 (0.063)	0.8964 (0.045)	0.8972 (0.065)	0.9090 (0.045)	0.9090 (0.059)
	MCP	0.8850 (0.047)	0.8855 (0.066)	0.8798 (0.050)	0.8813 (0.068)	0.8772 (0.045)	0.8782 (0.065)	0.8738 (0.049)	0.8738 (0.070)
	AB	0.9158 (0.035)	0.9166 (0.050)	0.9014 (0.046)	0.9027 (0.065)	0.9102 (0.040)	0.9112 (0.060)	0.9072 (0.047)	0.9075 (0.062)
	RF	0.9148 (0.039)	0.9166 (0.055)	0.9186 (0.041)	0.9199 (0.059)	0.9154 (0.042)	0.9167 (0.060)	0.9116 (0.043)	0.9127 (0.060)
	LR	0.9124 (0.037)	0.9127 (0.054)	0.9054 (0.043)	0.9063 (0.061)	0.9018 (0.045)	0.9024 (0.063)	0.9092 (0.043)	0.9084 (0.060)
	SVM	0.9256 (0.038)	0.9261 (0.054)	0.9128 (0.038)	0.9135 (0.056)	0.9080 (0.043)	0.9099 (0.061)	0.9126 (0.045)	0.9120 (0.062)

When we look at the performances of all the classifiers with the IG method in comparison to other FS methods, there is much variation in the accuracies, as seen in Figure 5. The SVM classifier, which attained the accuracy of 0.7026 with the RLFS method, dropped to 0.6422 with the IG method.

The proposed combination of the ERRM classifier with the RLFS method achieved the highest average accuracy of 0.7161, and the Gmean of 0.7127 outperformed the rest of the combinations of classifier with the FS method. The RLFS method is also a top-performing FS method on all individual classifiers. However, among the other FS methods, the MRMR method, when applied to all the individual classifiers, showed relatively much better performance than the application of IG and Chi2 methods to the individual classifiers. The second best-performing method is the ENET-RLFS combination, which had an accuracy of 0.7138. The SVM-IG combination showed the lowest performance with an accuracy of 0.6422 among all the combinations of the classifier with FS methods, as shown in Table 2.

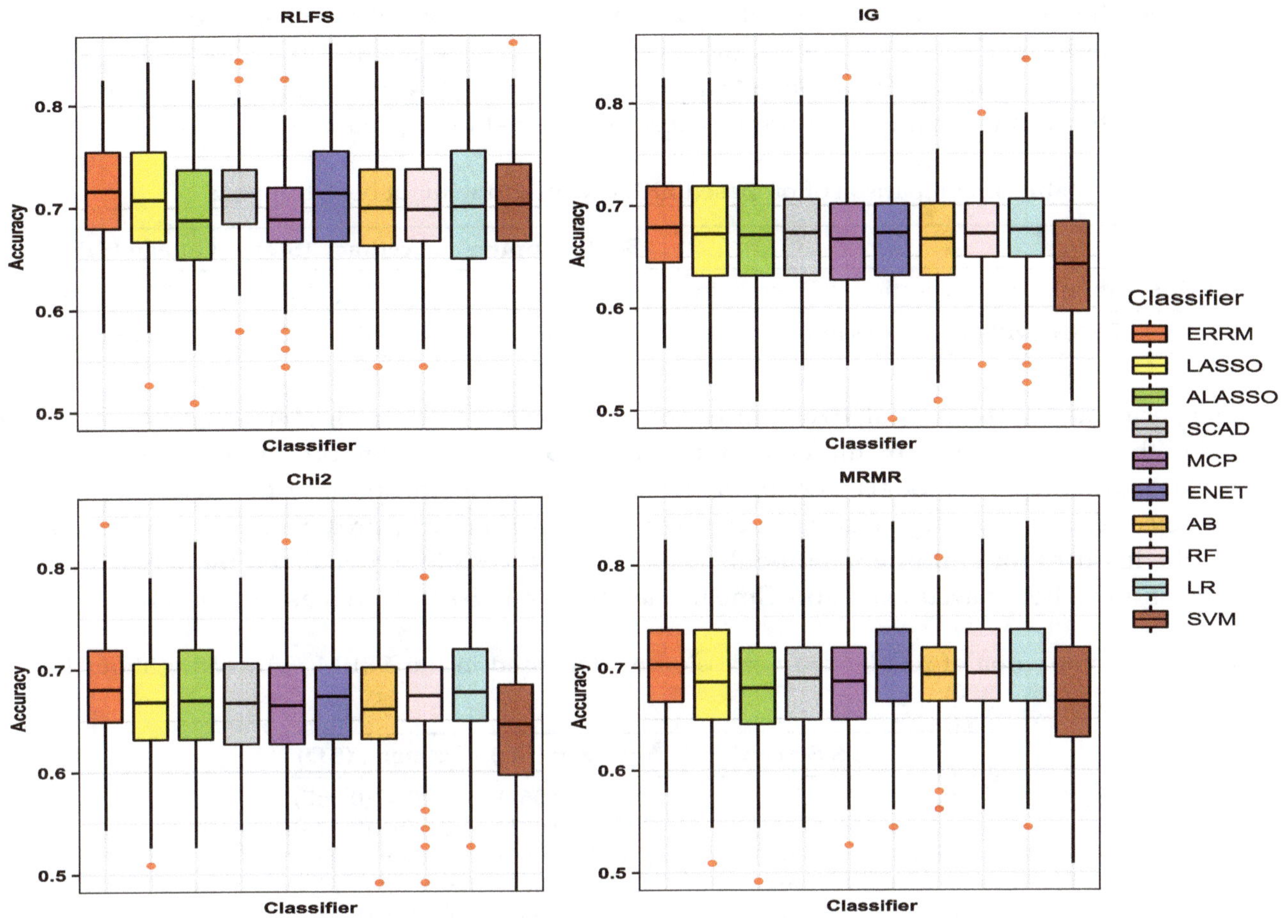

Figure 5. Boxplot showing the accuracies of classifiers with FS methods in experimental data SMK-CAN-187.

Table 2. Average values taken over 100 iterations in experimental data: SMK-CAN-187.

Classifier	Proposed RLFS		IG		Chi2		MRMR	
	Accuracy (SD)	Gmean (SD)	Accuracy (SD)	Gmean (SD)	Accuracy (SD)	Gmean (SD)	Accuracy (SD)	Gmean (SD)
Proposed ERRM	0.7161 (0.053)	0.7127 (0.082)	0.6789 (0.056)	0.6791 (0.091)	0.6807 (0.056)	0.6808 (0.091)	0.7035 (0.056)	0.7024 (0.087)
LASSO	0.7073 (0.064)	0.7058 (0.087)	0.6726 (.060)	0.6725 (0.095)	0.6680 (0.057)	0.6680 (0.090)	06859 (0.061)	0.6871 (0.097)
ALASSO	0.6878 (0.065)	0.6869 (0.091)	0.6715 (0.060)	0.6714 (0.094)	0.6696 (0.064)	0.6698 (0.092)	0.6800 (0.059)	0.6803 (0.092)
ENET	0.7138 (0.061)	0.7116 (0.085)	0.6733 (0.057)	0.6722 (0.093)	0.6733 (0.052)	0.6726 (0.090)	0.6998 (0.061)	0.6992 (0.095)
SCAD	0.7114 (0.054)	0.7098 (0.083)	0.6735 (0.056)	0.6732 (0.090)	0.6670 (0.058)	0.6669 (0.091)	0.6894 (0.059)	0.6901 (0.091)
MCP	0.6880 (0.010)	0.6870 (0.082)	0.6673 (0.057)	0.6663 (0.089)	0.6647 (0.059)	0.6639 (0.092)	0.6866 (0.057)	0.6874 (0.089)
AB	0.6991 (0.064)	0.6958 (0.087)	0.6673 (0.054)	0.6634 (0.086)	0.6605 (0.058)	0.6583 (0.094)	0.6929 (0.050)	0.6897 (0.083)
RF	0.6975 (0.056)	0.6933 (0.089)	0.6729 (0.045)	0.6691 (0.078)	0.6738 (0.054)	0.6703 (0.090)	0.6942 (0.055)	0.6902 (0.088)
LR	0.7001 (0.065)	0.6987 (0.089)	0.6761 (0.058)	0.6662 (0.097)	0.6770 (0.059)	0.6769 (0.094)	0.7008 (0.058)	0.7000 (0.086)
SVM	0.7026 (0.058)	0.7014 (0.086)	0.6422 (0.059)	0.6430 (0.099)	0.6459 (0.066)	0.6477 (0.105)	0.6668 (0.058)	0.6658 (0.092)

For assessing the importance of bootstrapping and FS screening of the proposed framework, we measured the performance of ERRM without FS screening. The results in Table 3 shows the results of ensembles method with and without bootstrapping procedure. We see that having the bootstrapping approach which is random sampling with replacement is a better approach in the ensembles.

Table 3. Comparison of proposed ERRM with and without bootstrapping.

	Bootstrapping	Accuracy (SD)	Gmean (SD)
ERRM without feature selection (FS) screening	Yes	0.7129 (0.053)	0.7093 (0.091)
ERRM without FS screening	No	0.6947 (0.057)	0.6944 (0.089)

The performance of the regularized regression models used in the proposed ensembles algorithm is tested with the FS screening method and without the FS screening method. In the former approach, the regularized regression models were built and tested using the proposed RLFS screening method with the selected amount of significant features, whereas in the latter approach, the regularized models used all the features for building the model. The performances of the penalized models with the FS screening showed better accuracies and Gmean than without FS screening, as reported in Table 4.

Table 4. Comparison of regularized regression models used in the ERRM with and without FS screening.

	FS Screening	Accuracy (SD)	Gmean (SD)
LASSO	Yes	0.7073 (0.064)	0.7058 (0.087)
	No	0.6740 (0.061)	0.6752 (0.125)
ALASSO	Yes	0.6878 (0.065)	0.6869 (0.091)
	No	0.6740 (0.061)	0.6752 (0.125)
ENET	Yes	0.7138 (0.061)	0.7116 (0.085)
	No	0.6740 (0.061)	0.6752 (0.125)
SCAD	Yes	0.7114 (0.054)	0.7098 (0.083)
	No	0.6740 (0.061)	0.6752 (0.125)
MCP	Yes	0.6880 (0.010)	0.6870 (0.082)
	No	0.6740 (0.061)	0.6752 (0.125)

4. Discussion

We investigated the performance of the proposed combination of ERRM with the RLFS method using simulation studies and a real data application. The RLFS method ranks the features by employing the lasso method with a resampling approach and the *b*-SIS criteria to set the threshold for selecting the optimal number of features, and these features are applied on the ERRM classifier, which uses bootstrapping and rank aggregation to select the best performing model across the bootstrapped samples and helps in attaining the best prediction accuracy in a high dimensional setting. The ensemble framework ERRM was built using five different regularized regression models. The regularized regression models are known for having the best performances in terms of variable selection and prediction accuracy on gene expression data.

To show the performance of our proposed framework, we used three different simulation scenarios with low, medium, and high correlation structures that matched the gene expression data. To further illustrate our point, we also used SMK-CAN-187 data. Figure 2 shows the boxplots of the average number of true important features, where the RLFS shows higher detection power than the other FS methods such as IG, Chi2, and MRMR. From the results of both simulation studies and experimental data, we showed that all the individual classifiers with the RLFS method performed much better compared to the IG, Chi2, and MRMR. We also observed that all the individual classifiers showed much instability with the other three FS methods. This means that the individual classifiers do not work well with more noise and less true important variables in the model. The SVM and ENET classifiers with all the FS methods performed a little better among all the classifiers. However, the performance was relatively still low in comparison to the proposed ERRM classifier with every FS method. The tree-based ensemble methods RF and AB with RLFS also attained good accuracies but were not the best compared to the ERRM classifier.

The proposed ERRM method was assessed with the FS screening and without the FS screening step along with the bootstrapping option. The ERRM with FS screening and bootstrapping approach works better than ERRM without the FS screening and bootstrapping technique. Also, the results from Table 3 show that the ensemble with bootstrapping is a better approach to both the filtered and unfiltered data. On comparing the performance of the individual regularized regression models used in the ensembles, the individual models with the proposed RLFS screening step showed comparatively better accuracy in comparison to the individual regularized regression models without the FS screening. This means that using the reduced number of significant features with RLFS is a better approach instead of using all the features from the data.

The importance of FS method was not addressed in any of the ensemble approaches [37–39], and the classification accuracies achieved by the corresponding proposed methods were much closer to the accuracies attained by existing approaches. In this paper, we compared the various combination of FS methods with different classifiers. The ERRM showed better overall performance not only with the RLFS but also with the other FS methods compared in this study. This means that the ERRM is robust and works much better on the highly correlated gene expression data. The rule of thumb fpr attaining the best prediction accuracy is that more the true important variables, better the prediction accuracy. Henceforth, from the results of simulation and experimental data, we see that the proposed combination of RLFS-ERRM is better compared to the other existing combinations of FS and classifiers, as seen in the Tables 1 and 2. The proposed ERRM classifier showed the best performance across all the FS methods, with the highest performance achieved with the RLFS method. The proposed RLFS method attained a higher number of significant features compared to other FS methods. However, the drawback is that with the increase in the correlation structure, there is a decreasing performance in selecting the significant features, as shown in Figure 2. The ensembles algorithms are known to be computationally expensive [39] because of the tree-based nature. However, in our proposed framework, before the ensembles of ERRM, we apply FS methods to remove the irrelevant features and keep significant features. This filtering step not only helps with improving prediction accuracy but also with overcoming the drawback of computational time required, as the number of features processed becomes lower.

5. Conclusions

In this paper, we proposed a combination of the ensembles of regularized regression models (ERRM) with resampling-based lasso feature selection (RLFS) for attaining better prediction accuracies in high dimensional data. We conducted extensive simulation studies where we showed the better performance of RLFS in detecting the significant features than other competitive FS methods. The ensemble classifier ERRM also showed better average prediction accuracy with the RLFS, IG, Chi2, and MRMR compared to other classifiers with these FS methods. We also saw an improved performance in the ensemble method when used with bootstrapping. On comparing the performances of individual regularized regression models, all the models showed an increase in their accuracies with the FS screening approach. In both the simulation study and the experimental data SMK-CAN-187, the better performance was achieved by the proposed combination of RLFS and ERRM compared to all other combinations of FS and classifiers. The minor drawback in the proposed framework is that, in the case of highly correlated data, there is smaller number of significant features selected with all the FS methods. As future work, we plan to focus on improving the detecting power of true important genes with the new FS method.

Author Contributions: Conceptualization, S.K. and A.R.P.; formal analysis and investigation, A.R.P.; Writing—Original draft preparation, A.R.P.; supervision and reviewing, S.K. All authors have read and agreed to the published version of the manuscript.

Acknowledgments: We would like to thank the editors and three anonymous reviewers for their insightful comments and helpful suggestions.

Abbreviations

The following abbreviations are used in this manuscript:

FS	feature Selection
RLFS	resampling-based lasso feature selection
ERRM	ensemble regularized regression models
IG	information gain
Chi2	chi-square
MRMR	minimum redundancy maximum relevance
ALASSO	adaptive lasso
AB	adaptive boosting
RF	random forests
LR	logistic regression
SVM	support vector machines
SD	standard deviation

Appendix A

The data are generated based on a high correlation data structure with $\rho = 0.8$. The performance of the proposed combination of RLFS-ERRM is relatively better than the other combinations of the FS methods and classifiers. The results for simulation scenario S3 are shown in Figure A1. The average accuracies and Gmeans for all the FS and classifiers are noted in Table A1. The SVM and ENET classifiers with all the FS methods showed a little better performance among all individual classifiers. However, the accuracies and Gmeans attained by the proposed ensemble classifier ERRM with the FS methods RLFS, IG, and Chi2 were relatively better compared to the individual classifiers with FS methods. The best performance was achieved by the proposed RLFS-ERRM combination with an accuracy of 0.9586 and Gmean of 0.9596. The second-best performing combination was MRMR-SVM. The lowest performance in terms of accuracy and the Gmean was shown by Chi2-MCP. The MCP classifier has the lowest accuracy with all the FS methods. This explains why the MCP does not perform well when the predictor variables are highly correlated.

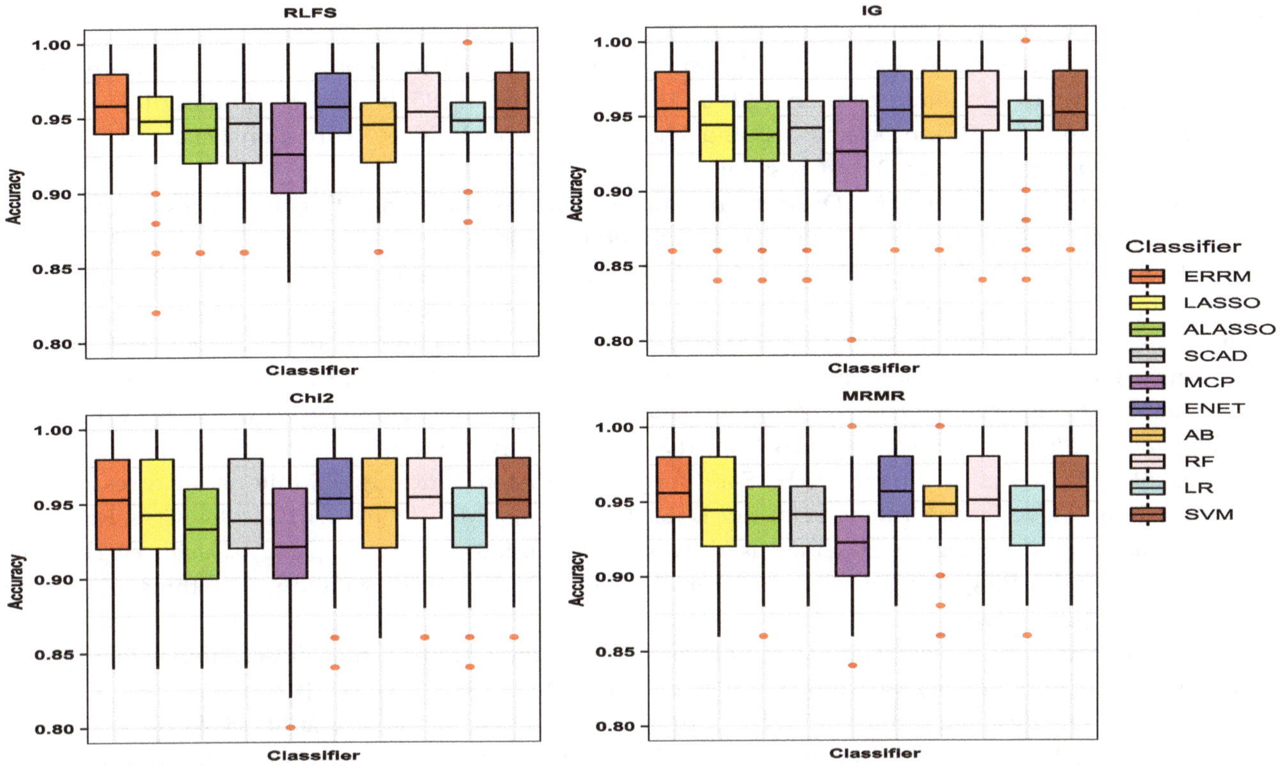

Figure A1. Boxplot showing the accuracies of Classifiers with FS methods in simulation scenario: S3 (Correlation = 0.8).

Table A1. Average values taken over 100 iterations in simulation scenario: S3 (High correlation: 0.8).

Classifier	Proposed RLFS		IG		Chi2		MRMR	
	Accuracy (SD)	Gmean (SD)	Accuracy (SD)	Gmean (SD)	Accuracy (SD)	Gmean (SD)	Accuracy (SD)	Gmean (SD)
Proposed ERRM	0.9586 (0.025)	0.9596 (0.039)	0.9556 (0.027)	0.9565 (0.041)	0.9530 (0.034)	0.9544 (0.045)	0.9560 (0.024)	0.9558 (0.037)
LASSO	0.9482 (0.033)	0.9493 (0.050)	0.9442 (.030)	0.9194 (0.045)	0.9428 (0.037)	0.9447 (0.051)	0.9444 (0.032)	0.9442 (0.042)
ALASSO	0.9420 (0.031)	0.9425 (0.051)	0.9376 (0.030)	0.9379 (0.045)	0.9328 (0.041)	0.942 (0.056)	0.9388 (0.033)	0.9389 (0.047)
ENET	0.9576 (0.025)	0.9587 (0.039)	0.9538 (0.029)	0.9546 (0.042)	0.9532 (0.034)	0.9546 (0.045)	0.9566 (0.024)	0.9562 (0.036)
SCAD	0.9464 (0.031)	0.9475 (0.049)	0.9422 (0.030)	0.9428 (0.045)	0.9386 (0.043)	0.9401 (0.055)	0.9414 (0.031)	0.9408 (0.043)
MCP	0.9256 (0.040)	0.9270 (0.062)	0.9262 (0.038)	0.9269 (0.055)	0.9210 (0.041)	0.9221 (0.058)	0.9224 (0.034)	0.9223 (0.048)
AB	0.9454 (0.032)	0.9469 (0.047)	0.9494 (0.030)	0.9501 (0.044)	0.9470 (0.034)	0.9482 (0.046)	0.9480 (0.029)	0.9481 (0.040)
RF	0.9540 (0.030)	0.9557 (0.043)	0.9560 (0.029)	0.9565 (0.043)	0.9542 (0.032)	0.9556 (0.044)	0.9508 (0.027)	0.9510 (0.039)
LR	0.9478 (0.029)	0.9482 (0.045)	0.9462 (0.030)	0.9469 (0.044)	0.9418 (0.038)	0.9432 (0.050)	0.9438 (0.028)	0.9437 (0.041)
SVM	0.9560 (0.027)	0.9568 (0.041)	0.9522 (0.030)	0.9527 (0.043)	0.9520 (0.031)	0.9526 (0.042)	0.9594 (0.026)	0.9587 (0.037)

References

1. Tariq, H.; Eldridge, E.; Welch, I. An efficient approach for feature construction of high-dimensional microarray data by random projections. *PLoS ONE* **2018**, *13*, e0196385. [CrossRef] [PubMed]
2. Bhola, A.; Singh, S. Gene Selection Using High Dimensional Gene Expression Data: An Appraisal. *Curr. Bioinform.* **2018**, *13*, 225–233. [CrossRef]
3. Dai, J.J.; Lieu, L.H.; Rocke, D.M. Dimension reduction for classification with gene expression microarray data. *Stat. Appl. Genet. Mol. Biol.* **2006**, *5*, 6. [CrossRef] [PubMed]
4. Lu, J.; Kerns, R.T.; Peddada, S.D.; Bushel, P.R. Principal component analysis-based filtering improves detection for Affymetrix gene expression arrays. *Nucleic Acids Res.* **2011**, *39*, e86. [CrossRef]
5. Bourgon, R.; Gentleman, R.; Huber, W. Reply to Talloen et al.: Independent filtering is a generic approach that needs domain specific adaptation. *Proc. Natl. Acad. Sci. USA* **2010**, *107*, E175–E175. [CrossRef]
6. Bourgon, R.; Gentleman, R.; Huber, W. Independent filtering increases detection power for high-throughput experiments. *Proc. Natl. Acad. Sci. USA* **2010**, *107*, 9546–9551. [CrossRef]
7. Ramsköld, D.; Wang, E.T.; Burge, C.B.; Sandberg, R. An Abundance of Ubiquitously Expressed Genes Revealed by Tissue Transcriptome Sequence Data. *PLoS Comput. Biol.* **2009**, *5*, e1000598. [CrossRef]
8. Li, L.; Kabesch, M.; Bouzigon, E.; Demenais, F.; Farrall, M.; Moffatt, M.; Lin, X.; Liang, L. Using eQTL weights to improve power for genome-wide association studies: A genetic study of childhood asthma. *Front. Genet.* **2013**, *4*, 103. [CrossRef]
9. Calle, M.L.; Urrea, V.; Vellalta, G.; Malats, N.; Steen, K.V. Improving strategies for detecting genetic patterns of disease susceptibility in association studies. *Stat. Med.* **2008**, *27*, 6532–6546. [CrossRef]
10. Bock, C. Analysing and interpreting DNA methylation data. *Nat. Rev. Genet.* **2012**, *13*, 705–719. [CrossRef]
11. Sun, H.; Wang, S. Penalized logistic regression for high-dimensional DNA methylation data with case-control studies. *Bioinformatics* **2012**, *28*, 1368–1375. [CrossRef]
12. Kim, S.; Halabi, S. High Dimensional Variable Selection with Error Control. *BioMed Res. Int.* **2016**, *2016*, 8209453. [CrossRef] [PubMed]
13. Kim, S.; Kim, J.M. Two-Stage Classification with SIS Using a New Filter Ranking Method in High Throughput Data. *Mathematics* **2019**, *7*, 493. [CrossRef]
14. Fan, J.; Lv, J. Sure independence screening for ultrahigh dimensional feature space. *J. R. Stat. Soc. Ser. B* **2008**, *70*, 849–911. [CrossRef] [PubMed]
15. Quinlan, J.R. *C4.5: Programs for Machine Learning*; Morgan Kaufmann Publishers Inc.: San Francisco, CA, USA, 1993.
16. Okeh, U.; Oyeka, I. Estimating the Fisher's Scoring Matrix Formula from Logistic Model. *Am. J. Theor. Appl. Stat.* **2013**, *2013*, 221–227.
17. Guyon, I.; Elisseeff, A. An Introduction to Variable and Feature Selection. *J. Mach. Learn. Res.* **2003**, *3*, 1157–1182.
18. Peng, H.; Long, F.; Ding, C. Feature Selection Based on Mutual Information: Criteria of Max-Dependency, Max-Relevance, and Min-Redundancy. *IEEE Trans. Pattern Anal. Mach. Intell.* **2005**, *27*, 1226–1238. [CrossRef]
19. Ditzler, G.; Morrison, J.C.; Lan, Y.; Rosen, G.L. Fizzy: Feature subset selection for metagenomics. *BMC Bioinform.* **2015**, *16*, 358. [CrossRef]
20. Su, C.T.; Yang, C.H. Feature selection for the SVM: An application to hypertension diagnosis. *Expert Syst. Appl.* **2008**, *34*, 754–763. [CrossRef]
21. Kursa, M.B.; Rudnicki, W.R. Feature Selection with the Boruta Package. 2010.
22. Urbanowicz, R.J.; Meeker, M.; Cava, W.L.; Olson, R.S.; Moore, J.H. Relief-based feature selection: Introduction and review. *J. Biomed. Inform.* **2017**, *85*, 189–203. [CrossRef] [PubMed]
23. Breiman, L. Random Forests. *Mach. Learn.* **2001**, *45*, 5–32. [CrossRef]
24. Freund, Y. An Adaptive Version of the Boost by Majority Algorithm. *Mach. Learn.* **2001**, *43*, 293–318. [CrossRef]
25. Hearst, M.A.; Dumais, S.T.; Osuna, E.; Platt, J.; Scholkopf, B. Support vector machines. *IEEE Intell. Syst. Appl.* **1998**, *13*, 18–28. [CrossRef]
26. Statnikov, A.R.; Wang, L.; Aliferis, C.F. A comprehensive comparison of random forests and support vector machines for microarray-based cancer classification. *BMC Bioinform.* **2008**, *9*, 319–319. [CrossRef] [PubMed]
27. Tibshirani, R. Regression Shrinkage and Selection via the Lasso. *J. R. Stat. Soc. Ser. B (Methodol.)* **1996**, *58*, 267–288. [CrossRef]
28. Marquardt, D.W.; Snee, R.D. Ridge Regression in Practice. *Am. Stat.* **1975**, *29*, 3–20.

29. Yang, X.G.; Lu, Y. Informative Gene Selection for Microarray Classification via Adaptive Elastic Net with Conditional Mutual Information. *arXiv* **2018**, arXiv:1806.01466.
30. Fan, J.; Li, R. Variable Selection via Nonconcave Penalized Likelihood and its Oracle Properties. *J. Am. Stat. Assoc.* **2001**, *96*, 1348–1360. [CrossRef]
31. Zhang, C.H. Nearly unbiased variable selection under minimax concave penalty. *Ann. Stat.* **2010**, *38*, 894–942. [CrossRef]
32. Hastie, T.; Tibshirani, R.; Friedman, J. *The Elements of Statistical Learning: Data Mining, Inference And Prediction*, 2nd ed.; Springer: Berlin/Heidelberg, Germany, 2009.
33. Dietterich, T.G. Ensemble Methods in Machine Learning. In *International Workshop on Multiple Classifier Systems*; Springer: London, UK, 2000; pp. 1–15.
34. Maclin, R.; Opitz, D.W. Popular Ensemble Methods: An Empirical Study. *arXiv* **2011**, arXiv:1106.0257.
35. Breiman, L. Bagging Predictors. *Mach. Learn.* **1996**, *24*, 123–140.:1018054314350. [CrossRef]
36. Freund, Y.; Schapire, R.E. A Decision-Theoretic Generalization of On-Line Learning and an Application to Boosting. *J. Comput. Syst. Sci.* **1997**, *55*, 119–139. [CrossRef]
37. Datta, S.; Pihur, V.; Datta, S. An adaptive optimal ensemble classifier via bagging and rank aggregation with applications to high dimensional data. *BMC Bioinform.* **2010**, *11*, 427. [CrossRef] [PubMed]
38. Ahn, H.; Moon, H.; Fazzari, M.J.; Lim, N.; Chen, J.J.; Kodell, R.L. Classification by ensembles from random partitions of high-dimensional data. *Comput. Stat. Data Anal.* **2007**, *51*, 6166–6179. [CrossRef]
39. Lim, N.; Ahn, H.; Moon, H.; Chen, J.J. Classification of high-dimensional data with ensemble of logistic regression models. *J. Biopharm. Stat.* **2009**, *20*, 160–171. [CrossRef]
40. R Development Core Team. *R: A Language and Environment for Statistical Computing*; R Foundation for Statistical Computing: Vienna, Austria, 2008; ISBN 3-900051-07-0.
41. Kursa, M.B. *Praznik: Collection of Information-Based Feature Selection Filters*; R Package Version 5.0.0; R Foundation for Statistical Computing: Vienna, Austria, 2018.
42. Natalia Novoselova, Junxi Wang, F.P.F.K. *Biocomb: Feature Selection and Classification with the Embedded Validation Procedures for Biomedical Data Analysis*; R package version 0.4; R Foundation for Statistical Computing: Vienna, Austria, 2018.
43. Friedman, J.; Hastie, T.; Tibshirani, R. Regularization Paths for Generalized Linear Models via Coordinate Descent. *J. Stat. Softw.* **2010**, *33*, 1–22. [CrossRef]
44. Breheny, P.; Huang, J. Coordinate descent algorithms for nonconvex penalized regression, with applications to biological feature selection. *Ann. Appl. Stat.* **2011**, *5*, 232–253. [CrossRef]
45. Liaw, A.; Wiener, M. Classification and Regression by randomForest. *R News* **2002**, *2*, 18–22.
46. Meyer, D.; Dimitriadou, E.; Hornik, K.; Weingessel, A.; Leisch, F. *e1071: Misc Functions of the Department of Statistics, Probability Theory Group (Formerly: E1071), TU Wien*; R Package Version 1.7-1; R Foundation for Statistical Computing: Vienna, Austria, 2019.
47. Pihur, V.; Datta, S.; Datta, S. *RankAggreg: Weighted Rank Aggregation*; R package version 0.6.5; R Foundation for Statistical Computing: Vienna, Austria, 2018.
48. The RLFS-ERRM Resources 2019. Available online: https://sites.google.com/site/abhijeetrpatil01/file-cabinet/blfs-errm-manuscript-files-2019 (accessed on 25 December 2019).
49. Feature Selection Datasets. Available online: http://featureselection.asu.edu/old/datasets.php (accessed on 25 December 2019).
50. Bolón-Canedo, V.; Sánchez-Maroño, N.; Alonso-Betanzos, A.; Benítez, J.M.; Herrera, F. A review of microarray datasets and applied feature selection methods. *Inf. Sci.* **2014**, *282*, 111–135. [CrossRef]
51. Wang, M.; Barbu, A. Are screening methods useful in feature selection? An empirical study. *PloS ONE* **2018**, *14*, e0220842. [CrossRef]
52. Tsai, C.; Chen, J.J. Multivariate analysis of variance test for gene set analysis. *Bioinformatics* **2009**, *25*, 897–903. [CrossRef] [PubMed]
53. Li, J.; Cheng, K.; Wang, S.; Morstatter, F.; Trevino, R.P.; Tang, J.; Liu, H. Feature Selection: A Data Perspective. *ACM Comput. Surv.* **2017**, *50*, 94:1–94:45. [CrossRef]
54. Chen, X.D.; Ishwaran, H. Random forests for genomic data analysis. *Genomics* **2012**, *99*, 323–329. [CrossRef] [PubMed]
55. Bielza, C.; Robles, V.; Larrañaga, P. Regularized logistic regression without a penalty term: An application to cancer classification with microarray data. *Expert Syst. Appl.* **2011**, *38*, 5110–5118. [CrossRef]

56. Liao, J.G.; Chin, K.V. Logistic regression for disease classification using microarray data: model selection in a large p and small n case. *Bioinformatics* **2007**, *23*, 1945–1951. [CrossRef] [PubMed]
57. Zou, H.; Hastie, T. Regularization and Variable Selection via the Elastic Net. *J. R. Stat. Soc. Ser. B (Stat. Methodol.)* **2005**, *67*, 301–320. [CrossRef]
58. Li, Y.; Zhang, Y.; Zhao, S. Gender Classification with Support Vector Machines Based on Non-tensor Pre-wavelets. In Proceedings of the 2010 Second International Conference on Computer Research and Development, Kuala Lumpur, Malaysia, 7–10 May 2010; pp. 770–774.

8

Outliers Detection Models in Shewhart Control Charts; An Application in Photolithography: A Semiconductor Manufacturing Industry

Ishaq Adeyanju Raji [1], Muhammad Hisyam Lee [1,*], Muhammad Riaz [2], Mu'azu Ramat Abujiya [3] and Nasir Abbas [2]

1 Department of Mathematical Sciences, Universiti Teknologi Malaysia, Skudai 81310, Malaysia; arishaq2@graduate.utm.my

2 Department of Mathematics and Statistics, King Fahd University of Petroleum and Minerals, Dhahran 31261, Saudi Arabia; riazm@kfupm.edu.sa (M.R.); nasirabbas@kfupm.edu.sa (N.A.)

3 Preparatory Year Mathematics Program, King Fahd University of Petroleum and Minerals, Dhahran 31261, Saudi Arabia; abujiya@kfupm.edu.sa

* Correspondence: mhl@utm.my.

Abstract: Shewhart control charts with estimated control limits are widely used in practice. However, the estimated control limits are often affected by phase-I estimation errors. These estimation errors arise due to variation in the practitioner's choice of sample size as well as the presence of outlying errors in phase-I. The unnecessary variation, due to outlying errors, disturbs the control limits implying a less efficient control chart in phase-II. In this study, we propose models based on Tukey and median absolute deviation outlier detectors for detecting the errors in phase-I. These two outlier detection models are as efficient and robust as they are distribution free. Using the Monte-Carlo simulation method, we study the estimation effect via the proposed outlier detection models on the Shewhart chart in the normal as well as non-normal environments. The performance evaluation is done through studying the run length properties namely average run length and standard deviation run length. The findings of the study show that the proposed design structures are more stable in the presence of outlier detectors and require less phase-I observation to stabilize the run-length properties. Finally, we implement the findings of the current study in the semiconductor manufacturing industry, where a real dataset is extracted from a photolithography process.

Keywords: average run length; control chart; median absolute deviation; outlier; photolithography; Shewhart; Tukey

1. Introduction

The two salient tools of statistical process control (SPC) are memory and memory-less control charts. The memory-less control charts are most suitable for large shift, while the memory-control charts are used to monitor moderate and small shifts. The prominent form of memory-less control chart for location monitoring is the Shewhart $\overline{X}$ control chart. In general, control charts-irrespective of the magnitude they measure-operate in two phases: phase-I, the prospective stage from which the control limits are obtained; phase-II, where we monitor the process and correct the unnatural causes of variation whenever they occur (cf. [1]). In phase-I we estimate the control limits using the parameters of the process under study which, in reality, are seldom known. The amount of data employed in phase-I for estimating process parameters varies from one practitioner to the other. As a result, this variability affects the chart performance in the monitoring stage i.e., phase-II. (see for example [2–6]).

Furthermore, the amount of data employed in estimating the process parameters does affect the accuracy of the chart, as well as its limits. As we all know, the larger the sample size, the closer we are to the parameter. Therefore, increasing the sample size used for estimating the parameters should be the remedy to this shortcoming, but there is a limit to which we can increase sample sizes in real-life situations. As a result, the Shewhart chart, like any other chart, loses its performance and credibility. The depth of the loss depends on the efficacy of the parameter estimation and sample size employed in phase-I.

The presence of outlying/extreme values in the phase-I dataset can affect the performance of the control chart. The insufficiency of the phase-I estimates could be a result of extreme sample points in the sample, and not necessarily the size of the sample (see [5,7]). The easiest remedy for the extreme values is to drop such a sample and pick another one, but this is not appropriate for small sample data. Therefore, there is a need to screen the extreme values to improve the overall performance of the control chart.

Over the years, researchers have studied different types of robust outlier detection models in a series of control charts to enhance their performance. Examples include [8–12]. These outlier detectors require the data to be from normal distribution such as the Student-type and Grubbs-type detectors. However, for a non-normal dataset, the Tukey's and median absolute deviation (MAD) outlier detection models are more accurate and robust since they are independent of mean and standard deviation. (see [13–19]). SPC is widely applied and implemented in various sectors; health, industrial, manufacturing and every service-rendering sector. Control charts, however, are most applied in manufacturing industry, with semiconductors as a case study. Semiconductor manufacturing processes are prone to high chances of assignable cause of variations, due to machine breakdown, multiple products, re-entrant flows, batching processes etc. [20]. Researchers have employed SPC in solving these recurring challenges in this industry (see [21–24]). The proposed charts in this study are applied in photolithography, a semiconductor manufacturing process.

In this article, we study the effects of parameter estimation on the Shewhart $\overline{X}$ chart for normal and non-normal environments. We also study the effect of outliers on the reliability of the control charts and the process parameters are estimated. Furthermore, we propose non-parametric outlier detectors, namely: the robust Tukey and MAD outlier detection models in designing the basic control chart structure. A fair comparison between the two-outlier detection models is also made. We achieve all of these using average run length (ARL) and standard deviation run length (SDRL) as the performance measures.

The remainder of this article is as follows: the next Section entails the methodologies employed for the study; briefing the overview of the Shewhart $\overline{X}$ control chart when the parameters are known and unknown, alongside the performance measure properties adopted in this study; the variability in Shewhart chart performance due to phase-I estimation; a scenario for the presence of outliers in the design structure of Shewhart chart, and its effect; incorporating the Turkey and MAD outlier detection models in the design structure of the Shewhart chart as remedies for rectifying the presence of outliers; Section 3 gives a concise and precise description of the simulation results. In Section 4, a detailed comparison of the results is presented; while Section 5 provides an illustrative example with a real life dataset; finally a concluding remark and future recommendations are given in Section 6.

2. Methodology

In this section, we give details of the Shewhart control chart for normal and non-normal environments. The known and unknown parameter scenarios, the practitioner–practitioner variation in the estimation stage, the presence of outliers/extreme values in the estimation sample, and incorporating some outlier detection models in the Shewhart chart are all discussed in the following subsections.

2.1. Overview of the Shewhart Control Chart

Let Y_{ij} $i = 1, 2, \ldots, n$ and $j = 1, 2, \ldots$ represent a ith observation from jth sample of an ongoing (continuous) process. Further Y_{ij} follows a normal distribution with mean $\mu_0 + \delta\sigma_0$ and variance ${\sigma_0}^2$ i.e., $Y_{ij} \sim N\left(\mu_0 + \delta\sigma_0, {\sigma_0}^2\right)$. The process is said to be in the in-control (IC) state if $\delta = 0$, and out-of-control (OoC) otherwise. A default Shewhart set-up monitors a process by plotting the sample mean ($\overline{Y}_i = 1/n \sum_{j=1}^{n} Y_{ij}$) of Y_{ij} against the following control chart limits.

$$\text{UCL} = \mu_0 + L\frac{\sigma_0}{\sqrt{n}}, \quad \text{LCL} = \mu_0 - L\frac{\sigma_0}{\sqrt{n}} \tag{1}$$

where UCL and LCL denote the upper and lower control limits, respectively. Limits in (1) are useful when the parameters (μ_0 and $\sigma_0{}^2$) of the process are known. However, when they are unknown, their respective unbiased estimators from the phase-I are used, and the resulting control chart structures will be in estimated form.

For phase-I, let Y_{il} represents ith observation from lth random sample $\forall\ i = 1, 2, 3, \ldots, n$ and $l = 1, 2, 3, \ldots, m$, regarded to be under statistically IC state. It is good to mention here that the choice of m and n varies from one practitioner-to another. Therefore, it affects the accuracy of the control limits implying an influenced ARL in phase-II. The unbiased estimators for the parameters μ and σ of an IC process are defined as:

$$\begin{aligned} \hat{\mu}_0 &= (1/m) \textstyle\sum_{l=1}^{m} \overline{Y}_l \\ \hat{\sigma}_0 &= (1/mC_4) \textstyle\sum_{l=1}^{m} S_l \end{aligned} \tag{2}$$

where $\overline{Y}_l = \frac{\sum_{l=1}^{n} Y_{il}}{n}$, $S_l = \sqrt{\frac{\sum_{i=1}^{n}(Y_{il}-\overline{Y}_l)^2}{n-1}}$ and $c_4 = \frac{\sqrt{2/(n-1)}\Gamma(n/2)}{\Gamma[(n-1)/2]}$ is the bias correction constant. Subsequently, the resulting control limits in (1) are modified to the following:

$$\widehat{\text{UCL}} = \frac{\sum_{l=1}^{m} \overline{Y}_l}{m} + \hat{L}\frac{\sum_{l=1}^{m} S_l}{mC_4\sqrt{n}}, \quad \widehat{\text{LCL}} = \frac{\sum_{l=1}^{m} \overline{Y}_l}{m} - \hat{L}\frac{\sum_{l=1}^{m} S_l}{mC_4\sqrt{n}} \tag{3}$$

In phase-II, $\overline{Y}_l$s are plotted against the control limits in (3) and the chart is said to have given an OoC signal if any value of $\overline{Y}_l$ is plotted outside the limits. Here, the sample number at which the statistic is plotted outside the limits is recorded as run length (RL). RL is an important variable in measuring the performance of control charts in general, and the Shewhart is not an exception. The most widely used property of RL is ARL, which is the average number of samples observed before the chart sends an OoC signal. Mathematically, ARL $= \sum_{k=1}^{s} \text{RL}_k/s$ where s is the number of RLs recorded. In addition to ARL, standard deviation of the RL (SDRL) gives more information about the behavior of the RL variable in evaluating the performance of a control chart. Furthermore, the ARL is of two types i.e., the IC ARL, denoted as ARL_0 and the OoC ARL, referred to as ARL_1. ARL_0 is expected to be sufficiently large enough to avoid false alarms. On the other hand, ARL_1 is anticipated to be sufficiently small to enable the process to send a signal as soon as there is a shift in the process parameter(s).

2.2. *Variability in the Shewhart Chart Performance*

In this section, we explain the effect of the practitioner to practitioner variability on the Shewhart chart, both in normal and non-normal distribution, by using the Monte Carlo simulation approach. See ([25–29]) for more information about the effect of sample size and practitioners' variability. To achieve this aim, we develop an algorithm in R programing language to simulate the Shewhart chart environment, using the standard Shewhart chart as our benchmark and reference point. The $\overline{X}$ chart has a control limits width determinant L that influences RL properties. We use the standard $L = 3$, that corresponds to the $\text{ARL}_0 = 370$ (see [1] for more details). Without any loss of generality, we generate random samples from a standard normal distribution $N(\mu = 0, \sigma = 1)$, each of sample size $n = 5$, assuming the process parameters are known. While for the non-normal distribution, we considered the t-distribution with degrees of freedom $v = 5$, 25, and 100. Since all the three categories of v exhibit the same pattern, we report only the results for $v = 100$. In both environments, normal and t-distributions, we set up the chart limits as given in Equation (1) and plot the sample means against the UCL and LCL. As soon as a value of $\overline{Y}_j$ is plotted outside the limits, RL is recorded and saved. The process is iterated 10^5 times to get ARL and SDRL.

For the unknown parameters, we estimate the parameter from phase-I. The number of samples employed for the estimation differs from on practitioner to another and so does the accuracy of the charts in phase-II. To depict that, we estimated both μ_0 and σ_0 from different number of in-control phase-I samples i.e., $m = 25, 50, 100, 250, 500$ and 1000 each of sample size $n = 5$. The estimated parameters $\hat{\mu}_0$ and $\hat{\sigma}_0$ from the phase-I IC stage are, therefore, used in the same algorithm instead of μ_0 and σ_0 respectively. Subsequently the parameter L, changes as the amount of phase-I samples changes. The corresponding L's for the different m's are $L = 2.962$, 2.983, 2.9925, 2.997, 2.999, and 3 respectively for the normal distribution, and $L = 2.974$, 2.995, 3.005, 3.010, 3.012, and 3.012 respectively for the t-distribution of $v = 100$. These L's are determined through simulations to obtain $ARL_0 = 370$. We carry out the simulation with different level of shifts δ ranging from 0 to 5 i.e., $\delta \in (0, 0.5, 5)$, as shown in Tables 1 and 2.

Table 1. Average run length (ARL) of the Shewhart chart with estimated parameters for standard normal and t ($v = 100$) distributions.

ARL	Standard Normal Distribution						T-Distribution $v = 100$					
δ/m	25	50	100	250	500	1000	25	50	100	250	500	1000
0	370.93	369.66	370.43	369.09	371.10	370.98	370.67	370.00	369.58	370.89	369.63	370.00
0.5	190.76	173.92	165.35	159.25	157.00	156.54	194.06	177.76	167.85	161.75	160.48	159.04
1	53.91	48.74	46.37	44.73	44.41	44.21	55.14	50.13	47.75	46.26	46.02	45.32
1.5	17.19	16.05	15.52	15.20	15.09	15.04	17.70	16.53	16.06	15.65	15.65	15.52
2	6.87	6.59	6.46	6.35	6.33	6.33	7.05	6.80	6.65	6.57	6.54	6.57
2.5	3.40	3.33	3.29	3.26	3.25	3.25	3.52	3.44	3.40	3.36	3.35	3.35
3	2.05	2.03	2.02	2.01	2.00	2.00	2.10	2.08	2.06	2.06	2.05	2.05
3.5	1.47	1.46	1.45	1.45	1.45	1.45	1.50	1.49	1.48	1.48	1.48	1.47
4	1.20	1.19	1.19	1.19	1.19	1.19	1.21	1.21	1.21	1.21	1.20	1.20
4.5	1.08	1.08	1.07	1.07	1.07	1.07	1.09	1.08	1.08	1.08	1.08	1.08
5	1.03	1.02	1.02	1.02	1.02	1.02	1.03	1.03	1.03	1.03	1.03	1.03
L	2.962	2.983	2.9925	2.997	2.999	3	2.974	2.995	3.005	3.010	3.012	3.012

Table 2. Standard deviation of the run length (SDRL) of the Shewhart chart with estimated parameters for standard normal and t ($v = 100$) distributions.

SDRL	Standard Normal Distribution						T-Distribution $v = 100$					
δ/m	25	50	100	250	500	1000	25	50	100	250	500	1000
0	601.54	473.20	420.75	387.76	379.99	375.44	601.69	468.55	416.69	389.69	376.91	373.30
0.5	333.88	232.92	193.19	169.15	161.43	158.10	340.19	241.98	197.21	172.49	164.11	160.46
1	88.35	61.71	52.00	46.39	45.00	44.24	91.10	64.04	53.93	47.96	46.61	45.29
1.5	24.05	18.52	16.40	15.20	14.85	14.67	25.00	19.21	16.92	15.71	15.33	15.20
2	8.24	6.88	6.32	5.97	5.90	5.85	8.73	7.10	6.48	6.22	6.11	6.11
2.5	3.40	3.04	2.86	2.76	2.72	2.71	3.60	3.16	2.99	2.86	2.84	2.83
3	1.66	1.54	1.48	1.44	1.43	1.42	1.70	1.59	1.54	1.49	1.48	1.47
3.5	0.90	0.85	0.83	0.81	0.81	0.81	0.94	0.89	0.86	0.84	0.84	0.84
4	0.51	0.49	0.48	0.48	0.48	0.48	0.54	0.51	0.51	0.50	0.50	0.50
4.5	0.30	0.29	0.28	0.28	0.28	0.28	0.32	0.31	0.30	0.29	0.29	0.29
5	0.17	0.16	0.16	0.16	0.16	0.16	0.18	0.17	0.17	0.17	0.17	0.17
L	2.962	2.983	2.9925	2.997	2.999	3	2.974	2.995	3.005	3.010	3.012	3.012

2.3. *Presence of Outliers in the Shewhart Chart with Estimated Parameters*

Although the estimation of the unknown parameters in phase-I samples plays its role on the efficiency of the control chart in phase-II. The drop in the efficacy of the chart performance is not limited to this fact alone, rather it extends to presence of outlying/extreme values in the phase-I samples.

In this Section, we study the effect of outliers in the phase-I samples on the performance and accuracy of the Shewhart chart. Here, through Monte Carlo simulation, we generate the m phase-I samples from a mixture distribution i.e., $(1-\alpha)100\%$ from assumed (normal or t-distribution) and the remaining $\alpha 100\%$ from a chi-square distribution with n degrees of freedom denoted by $\chi^2_{(n)}$. Subsequently, the estimated parameters emerging from the m samples have an extreme values effect on the control chart in phase-II. That is, each observation of the phase-I sample is generated from the following expression:

$$(1-\alpha)N\left(\mu,\sigma^2\right)+\alpha\left[N\left(\mu,\sigma^2\right)+w\,\chi^2_{(n)}\right] \text{ or } \\ (1-\alpha)t(v)+\alpha\left[t(v)+w\,\chi^2_{(n)}\right] \tag{4}$$

where $\alpha > 0$, is the probability of having a multiple of $\chi^2_{(n)}$ added to the assumed distribution, serving as the outliers in the samples. In addition, $w \geq 1$ is the magnitude of the outlier. We develop an algorithm from the R language, similar to that in Section 2.2, but the samples are from the environment described in (4). We set $\mu = 0$, $\sigma^2 = 1$, $v = 100, w = 3$, and $\alpha \in [0, 0.01]$. We design the Shewhart chart using the same parameters L and m as in Section 2.2.

In general, the pattern exhibited by the RL properties implies the following:

- Increasing the m phase-I samples in the presence of outliers, gets the ARL_0's closer to the theoretical values.
- Reducing the value of α, the percentage of outliers present in the m samples also brings the ARL_0's closer to the theoretical values.

Unfortunately, neither of the two suggested remedies is practicable in real life. Thus, we propose outliers detecting structures through the robust Turkey and MAD detection models.

2.4. *Shewhart Chart with Outlier Detection Models*

In the section, we propose two outlier-detecting models as remedy to the issues raised in Sections 2.2 and 2.3. The Tukey and the MAD model-based Shewhart charts. Their procedures applied in parallel to the Shewhart chart are described in the sub sections below:

2.4.1. The Tukey Shewhart Control Chart

For the phase-I samples, $\widetilde{Y}$ be the median of all $m \times n$ observations. For any observation y_o if $\left|y_o - \widetilde{Y}\right| > p \times IQR$, then y_o is declared an outlier. Here $IQR = Q_3 - Q_1$ is the inter-quartile range of the sample. Q_3 and Q_1 are the third and first quartiles, respectively, of all $m \times n$ phase-I observations. The constant p on the other hand is the confidence factor of the Tukey's detector, commonly chosen between 1.5 and 3.0. The confidence factor should be carefully chosen, and not too small, to avoid over detection. Also it should not be too large, to prevent under detection [18]. In this study, we choose $p = 2.2$. Applying the same algorithm, parameters and limits employed in Section 2.2, we incorporate the Tukey outlier-detector model on the phase-I samples to screen out the extreme values present there in. Then we compute the IC ARL and SDRL values for the Shewhart chart based on the Tukey model in phase-II, when the parameters are estimated.

2.4.2. The Median Absolute Deviation (MAD) Shewhart Control Chart

We define median absolute deviation (MAD) as the deviation of the dataset about the median as $\text{MAD} = median\left(\left|Y_{il} - \widetilde{Y}\right|\right)/0.6574$. Then it follows, that any observation y_o from the sample that falls outside the expression $\widetilde{Y} \pm b * \text{MAD}$, is declared an outlier. Here b is the outlier detecting constant and chosen 3.642 so that the percentage of screening by MAD is the same as Tukey. This has been done to keep the comparison between two outlier detectors valid [19].

Furthermore, it is worth distinguishing between outlying and OoC sample points. The former emerges from $m$phase-I samples, which are used to construct the control limits for the monitoring stage; phase-II; while the latter are the sample points that fall beyond the control limits in phase-II. Therefore, the presence of outlying sample points in phase-I leads to wider control limits, rendering the control charts less effective. A flowchart summarizing the procedure is depicted in Figure 1.

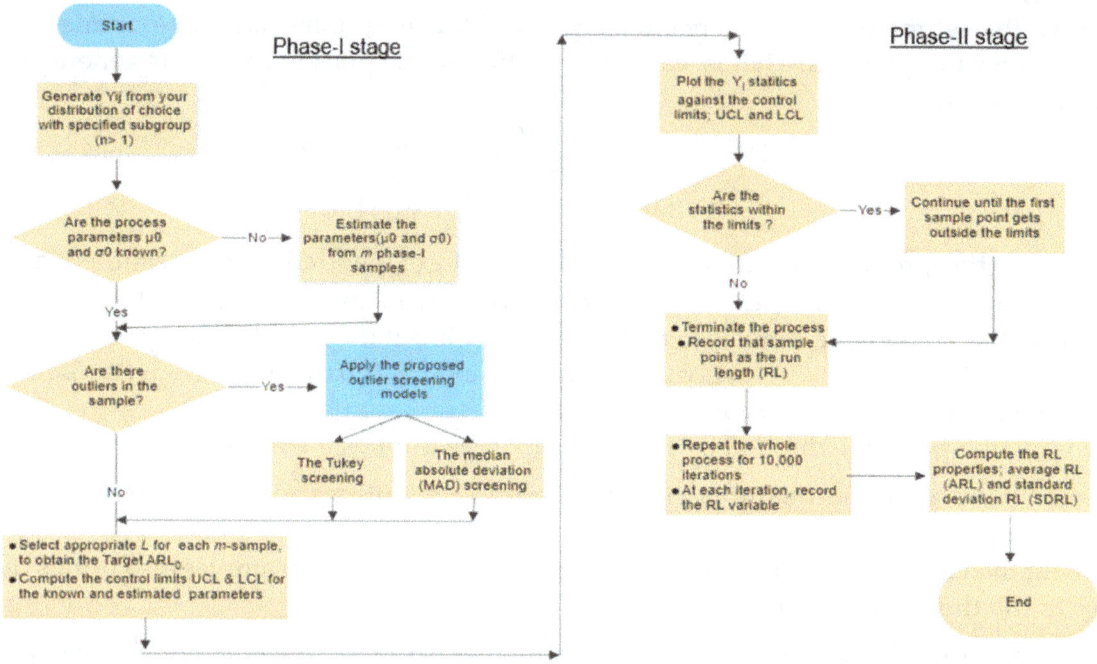

Figure 1. Flowchart of the procedures of proposed Shewhart control chart.

3. Results

In this section, we provide the results of the methodologies discussed in Section 2. These results are presented in three folds, so is the discussion in the next section.

3.1. Practitioners' Estimation Variability

Here, through the simulation results of the algorithm explained in Section 2.2, we observe the variability that appears in the Shewhart control chart due to different choices of sample size m, amongst practitioners. Tables 1 and 2 depict the Shewhart chart whose parameters, both mean and variance, are estimated from m phase-I samples for both normal and non-normal distributions. It is evident from the result, the effect of parameter estimation on the performance of the chart. The ARL_0s when $\delta = 0$, are clustering around the target 370 with their respective L's. However, when $\delta \neq 0$, we observe that the smaller m becomes, the less effective the Shewhart chart performance. The ARL_1's are expected to be sufficiently small in order to detect any drift in the ongoing process, but as m gets smaller, ARL_1's get bigger. Which implies the chart is less sensitive in identifying the presence of shifts in the ongoing process early enough. Another noticeable effect of the parameter estimation on the Shewhart chart is the decrement in the limits L, as m reduces. This should be recorded as an edge if the corresponding phase-II charts detects shift earlier than when the parameters are known.

3.2. Effect of Outliers on the Shewhart Control Charts

In Tables 3 and 4, we present the simulation results of environment (4) discussed in Section 2.3. From these results, the gross impact of outliers in the phase-I samples on the performance of the Shewhart chart cannot be over emphasized. Having seen the pattern of the IC and OoC RL properties in Tables 1 and 2, in order to save space, we restrict the performance evaluation to the IC RL properties. That is, considering the case when $\delta = 0$ only. From Tables 3 and 4, when $\alpha = 0$, in the absence of outlier, the ARL_0's are clustering around its target 370, irrespective of the amount of phase-I sample m. However, when $\alpha > 0$, the ARL_0's deviate from the target, vigorously. As the amount of phase-I samples m reduces, and the percentage of outliers present in the samples α increases, the more the ARL_0's deviate from the target. Similarly the pattern of the SDRL, even more.

Table 3. ARL of the Shewhart chart in the presence of outliers with estimated parameters for standard normal and t ($v = 100$) distributions.

ARL	Standard Normal Distribution						T-Distribution $v = 100$					
α/m	25	50	100	250	500	1000	25	50	100	250	500	1000
0	371.20	369.51	370.29	369.59	370.72	370.85	370.65	369.57	369.28	370.85	369.11	370.87
0.001	424.76	401.49	392.98	389.92	388.74	389.13	418.21	394.60	386.19	385.76	385.04	381.04
0.002	484.38	435.14	417.35	410.86	409.61	408.88	478.03	424.90	408.83	405.04	403.56	398.94
0.003	540.09	469.66	440.42	431.91	429.47	428.70	534.47	458.93	435.77	424.67	424.75	422.44
0.004	602.97	507.51	469.33	452.76	450.30	449.72	587.12	500.66	457.98	443.64	444.15	437.50
0.005	664.58	545.85	494.67	477.03	471.98	471.09	650.79	528.62	489.31	469.04	461.71	461.22
0.006	731.55	588.72	525.81	500.69	495.46	493.34	718.10	569.85	512.77	491.40	486.42	481.51
0.007	788.48	629.88	554.41	525.13	519.74	516.88	758.54	604.64	539.10	514.95	509.78	504.97
0.008	861.51	673.43	586.11	551.83	544.06	541.56	837.89	652.55	578.41	541.64	531.29	528.75
0.009	931.14	726.12	618.88	579.38	569.55	565.85	916.99	693.94	606.71	568.85	560.07	552.75
0.01	996.35	773.78	654.18	607.70	596.56	592.55	953.61	745.34	634.00	593.84	586.09	575.08
L	2.962	2.983	2.9925	2.997	2.999	3	2.974	2.995	3.005	3.010	3.012	3.012

Table 4. SDRL of the Shewhart chart in the presence of outliers with estimated parameters for standard normal and t ($v = 100$) distributions.

SDRL	Standard Normal Distribution						T-Distribution $v = 100$					
α/m	25	50	100	250	500	1000	25	50	100	250	500	1000
0	607.93	473.75	419.35	388.27	379.67	375.44	595.89	468.00	418.16	382.43	375.42	369.72
0.001	1238.32	664.83	467.76	414.51	400.57	394.71	1150.79	635.34	463.74	411.63	395.24	385.68
0.002	1704.83	838.46	515.41	441.77	423.19	415.18	1802.06	760.16	508.22	433.04	416.23	406.03
0.003	2041.64	991.95	568.33	468.17	446.51	436.26	2096.30	913.63	558.95	460.42	441.81	431.04
0.004	2410.92	1176.26	630.64	497.30	469.60	459.66	2354.27	1184.84	606.26	486.65	462.63	447.94
0.005	2689.45	1294.24	687.44	528.00	494.90	481.60	2626.16	1192.76	683.26	517.24	485.89	473.05
0.006	2994.88	1502.07	756.39	559.42	521.35	505.08	2871.17	1449.52	699.03	546.74	512.57	494.30
0.007	3193.80	1637.48	822.18	592.02	549.29	530.76	2956.37	1514.95	757.01	573.97	540.59	519.76
0.008	3480.50	1799.40	888.29	627.68	578.43	557.35	3370.66	1664.42	912.65	618.19	564.09	544.14
0.009	3772.38	1994.86	959.76	664.49	607.34	582.18	3703.48	1766.87	955.78	655.56	594.72	571.46
0.01	4012.72	2139.41	1028.74	702.52	638.62	612.00	3823.24	2009.14	980.24	685.94	628.22	591.55
L	2.962	2.983	2.9925	2.997	2.999	3	2.974	2.995	3.005	3.010	3.012	3.012

3.3. *Improvement of Tukey and MAD Outlier Detection Models on Shewhart Chart Performance*

While incorporating the procedures in Sections 2.4.1 and 2.4.2, the simulation results are presented in Tables 5–8 respectively. Tables 5 and 7 represents the ARL result for Tukey and MAD outlier detection models respectively, as Tables 6 and 8 are the corresponding SDRL results. The effect of these detection models are noticed as ARLs and SDRLs are closer to when there is an absence of outliers or even better.

Table 5. ARL of the Shewhart chart with Tukey outlier detection for standard normal and t ($v = 100$) distributions.

ARL	Standard Normal Distribution						T-Distribution $v = 100$					
α/m	25	50	100	250	500	1000	25	50	100	250	500	1000
0	363.24	363.44	364.17	364.36	365.34	366.80	358.67	358.71	357.97	358.10	359.64	357.52
0.001	365.71	366.33	366.48	366.83	367.56	368.22	361.59	360.26	357.98	362.04	359.85	360.39
0.002	367.81	367.30	368.51	368.71	369.89	366.67	364.59	363.25	359.73	359.63	363.42	361.61
0.003	370.30	370.04	370.91	371.14	370.90	371.03	365.85	365.75	364.12	361.29	362.85	362.73
0.004	372.35	371.48	372.44	372.76	372.72	374.13	368.84	365.44	365.73	365.64	365.87	366.25
0.005	376.13	374.05	374.49	373.95	375.73	377.23	370.27	368.13	366.95	366.59	366.89	367.05
0.006	377.97	376.06	376.96	376.61	376.88	379.44	372.27	368.44	368.28	371.20	371.40	370.77
0.007	380.13	378.38	379.03	378.98	379.15	379.44	374.78	372.44	371.54	370.82	370.45	371.76
0.008	383.48	380.17	380.51	380.63	380.56	381.29	378.04	372.69	372.06	373.31	376.35	374.45
0.009	386.35	383.02	383.50	382.73	383.29	382.84	377.58	374.74	371.66	375.88	375.09	372.29
0.01	388.82	384.62	385.57	383.80	384.95	386.39	381.84	377.80	378.17	377.29	378.48	374.09
L	2.962	2.983	2.9925	2.997	2.999	3	2.974	2.995	3.005	3.010	3.012	3.012

Table 6. SDRL of the Shewhart chart with Tukey outlier detection for standard normal and t ($v = 100$) distributions.

SDRL	Standard Normal Distribution						T-Distribution $v = 100$					
α/m	25	50	100	250	500	1000	25	50	100	250	500	1000
0	596.32	468.02	414.04	383.16	374.68	368.66	607.33	467.10	411.17	377.81	369.81	360.34
0.001	600.30	472.43	417.55	385.96	377.25	372.35	612.79	467.20	406.53	380.03	366.57	366.34
0.002	607.28	476.91	419.54	388.33	379.39	371.64	627.84	471.38	410.10	376.44	372.50	365.67
0.003	615.99	477.95	422.41	389.95	380.10	375.45	617.88	474.03	415.42	378.88	374.34	366.23
0.004	615.39	481.58	423.16	392.96	382.24	377.77	636.02	476.02	416.09	385.14	377.04	371.24
0.005	625.95	486.00	427.29	393.54	385.71	380.37	650.31	480.25	420.69	387.36	377.24	370.45
0.006	636.60	488.82	430.22	396.23	386.61	383.77	637.39	483.89	423.95	390.68	383.17	376.49
0.007	646.05	490.89	433.11	399.86	389.30	384.98	630.84	488.73	429.10	390.52	381.42	377.88
0.008	656.27	496.12	435.44	400.84	391.44	385.54	654.64	489.53	430.19	391.87	386.60	378.02
0.009	654.98	500.43	439.24	403.24	393.19	387.01	675.27	499.78	423.47	395.23	385.19	380.11
0.01	669.83	502.93	441.89	404.61	395.16	390.51	672.71	493.77	436.38	399.23	390.17	380.26
L	2.962	2.983	2.9925	2.997	2.999	3	2.974	2.995	3.005	3.010	3.012	3.012

Table 7. ARL of the Shewhart chart with median absolute deviation (MAD) outlier detection for standard normal and t ($v = 100$) distributions.

ARL	Standard Normal Distribution						T-Distribution $v = 100$					
α/m	25	50	100	250	500	1000	25	50	100	250	500	1000
0	363.24	363.44	364.17	364.36	365.34	366.80	358.67	358.71	357.97	358.10	359.64	357.52
0.001	365.71	366.33	366.48	366.83	367.56	368.22	361.59	360.26	357.98	362.04	359.85	360.39
0.002	367.81	367.30	368.51	368.71	369.89	366.67	364.59	363.25	359.73	359.63	363.42	361.61
0.003	370.30	370.04	370.91	371.14	370.90	371.03	365.85	365.75	364.12	361.29	362.85	362.73
0.004	372.35	371.48	372.44	372.76	372.72	374.13	368.84	365.44	365.73	365.64	365.87	366.25
0.005	376.13	374.05	374.49	373.95	375.73	377.23	370.27	368.13	366.95	366.59	366.89	367.05
0.006	377.97	376.06	376.96	376.61	376.88	379.44	372.27	368.44	368.28	371.20	371.40	370.77
0.007	380.13	378.38	379.03	378.98	379.15	379.44	374.78	372.44	371.54	370.82	370.45	371.76
0.008	383.48	380.17	380.51	380.63	380.56	381.29	378.04	372.69	372.06	373.31	376.35	374.45
0.009	386.35	383.02	383.50	382.73	383.29	382.84	377.58	374.74	371.66	375.88	375.09	372.29
0.01	388.82	384.62	385.57	383.80	384.95	386.39	381.84	377.80	378.17	377.29	378.48	374.09
L	2.962	2.983	2.9925	2.997	2.999	3	2.974	2.995	3.005	3.010	3.012	3.012

Table 8. SDRL of the Shewhart chart with MAD outlier detection for standard normal and t (v = 100) distributions.

SDRL	Standard Normal Distribution						T-Distribution v = 100					
α/m	25	50	100	250	500	1000	25	50	100	250	500	1000
0	596.32	468.02	414.04	383.16	374.68	368.66	607.33	467.10	411.17	377.81	369.81	360.34
0.001	600.30	472.43	417.55	385.96	377.25	372.35	612.79	467.20	406.53	380.03	366.57	366.34
0.002	607.28	476.91	419.54	388.33	379.39	371.64	627.84	471.38	410.10	376.44	372.50	365.67
0.003	615.99	477.95	422.41	389.95	380.10	375.45	617.88	474.03	415.42	378.88	374.34	366.23
0.004	615.39	481.58	423.16	392.96	382.24	377.77	636.02	476.02	416.09	385.14	377.04	371.24
0.005	625.95	486.00	427.29	393.54	385.71	380.37	650.31	480.25	420.69	387.36	377.24	370.45
0.006	636.60	488.82	430.22	396.23	386.61	383.77	637.39	483.89	423.95	390.68	383.17	376.49
0.007	646.05	490.89	433.11	399.86	389.30	384.98	630.84	488.73	429.10	390.52	381.42	377.88
0.008	656.27	496.12	435.44	400.84	391.44	385.54	654.64	489.53	430.19	391.87	386.60	378.02
0.009	654.98	500.43	439.24	403.24	393.19	387.01	675.27	499.78	423.47	395.23	385.19	380.11
0.01	669.83	502.93	441.89	404.61	395.16	390.51	672.71	493.77	436.38	399.23	390.17	380.26
L	2.962	2.983	2.9925	2.997	2.999	3	2.974	2.995	3.005	3.010	3.012	3.012

For better visuals of the results, we depict the ARL results (Tables 3, 5 and 7) in Figures 2 and 3 and the SDRL results (Tables 4, 6 and 8) in Figures 4 and 5.

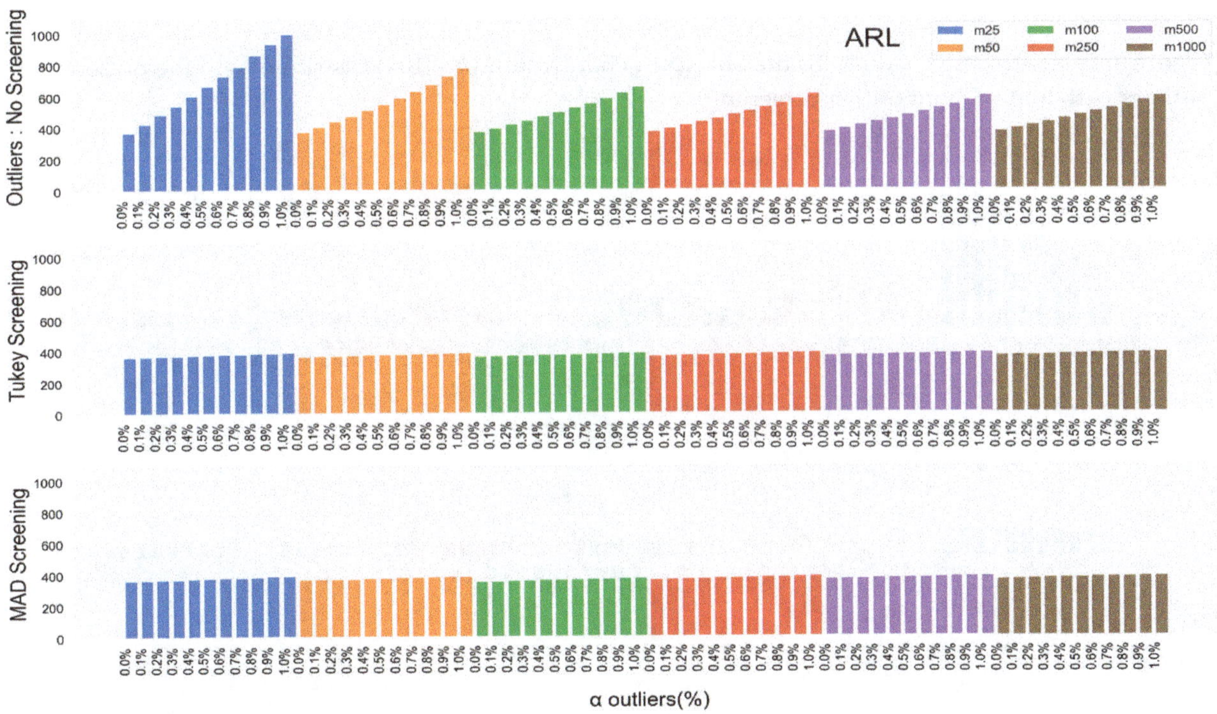

Figure 2. In-control ARL values for the Shewhart chart from standard normal distribution in the presence of outliers with and without outlier screening.

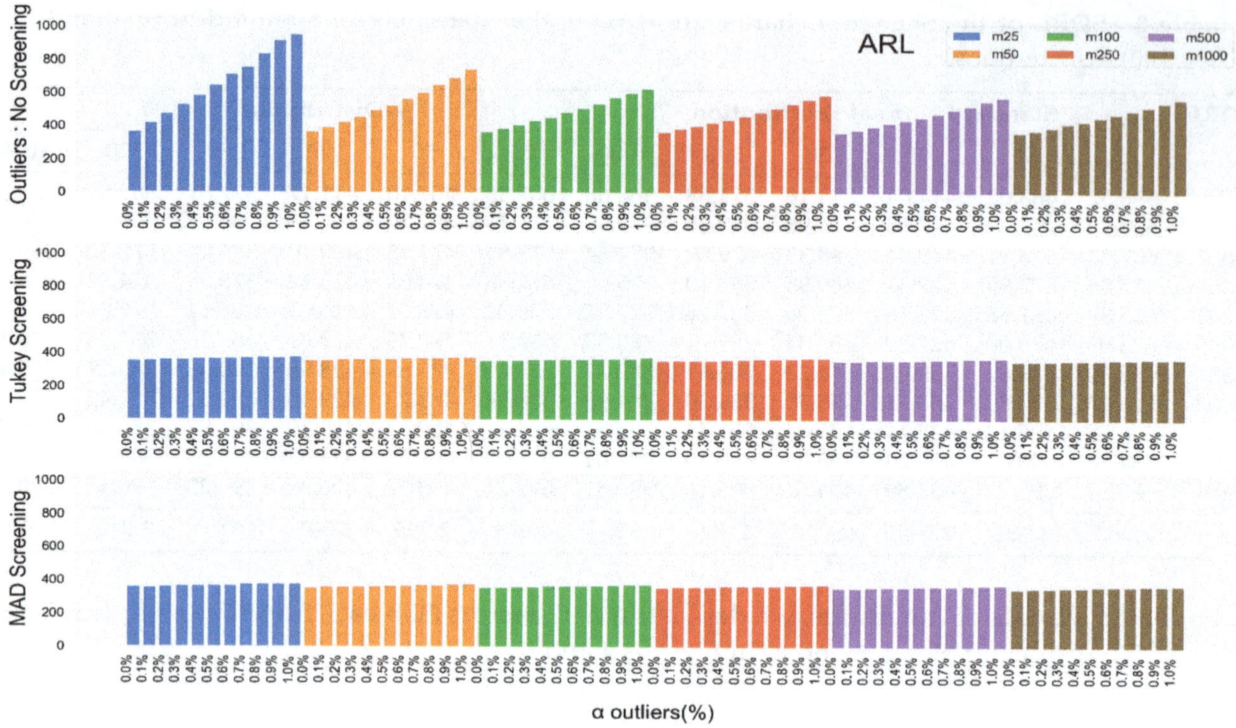

Figure 3. In-control ARL values for the Shewhart chart from t-(v = 100) distribution in the presence of outliers with and without outlier screening.

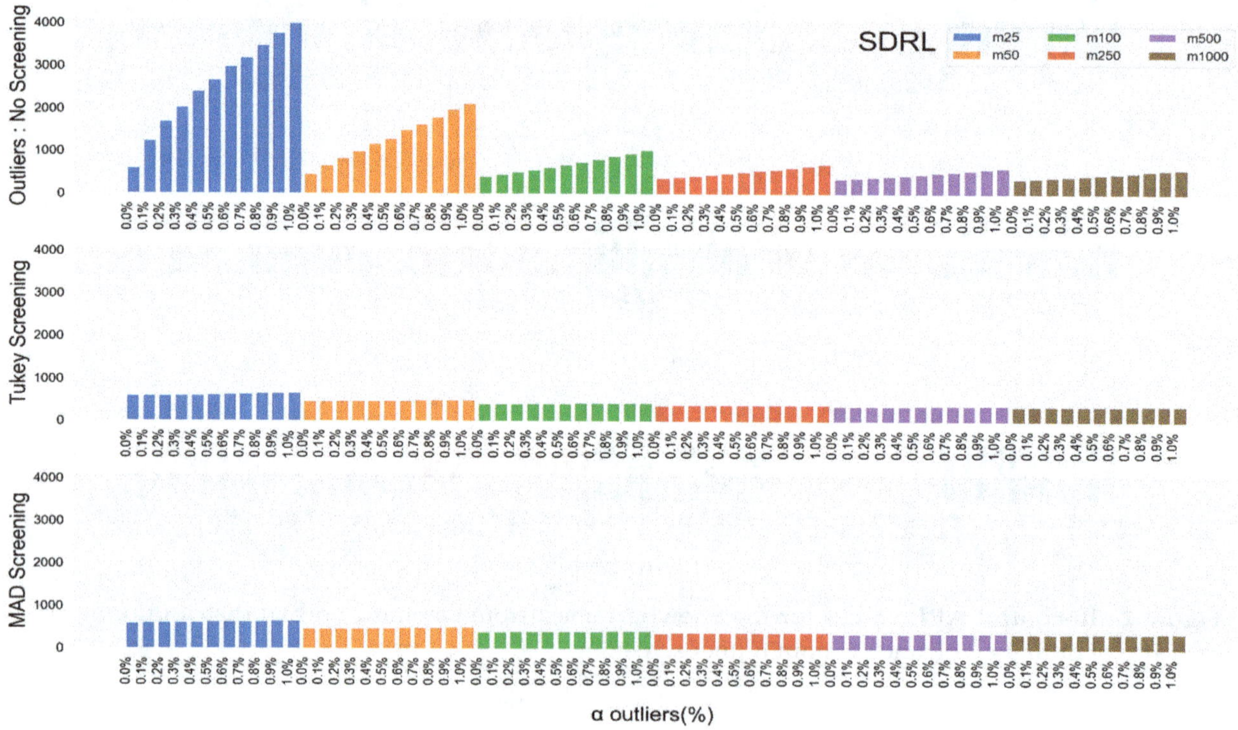

Figure 4. In-control SDRL values for the Shewhart chart from standard normal distribution in the presence of outliers with and without outlier screening.

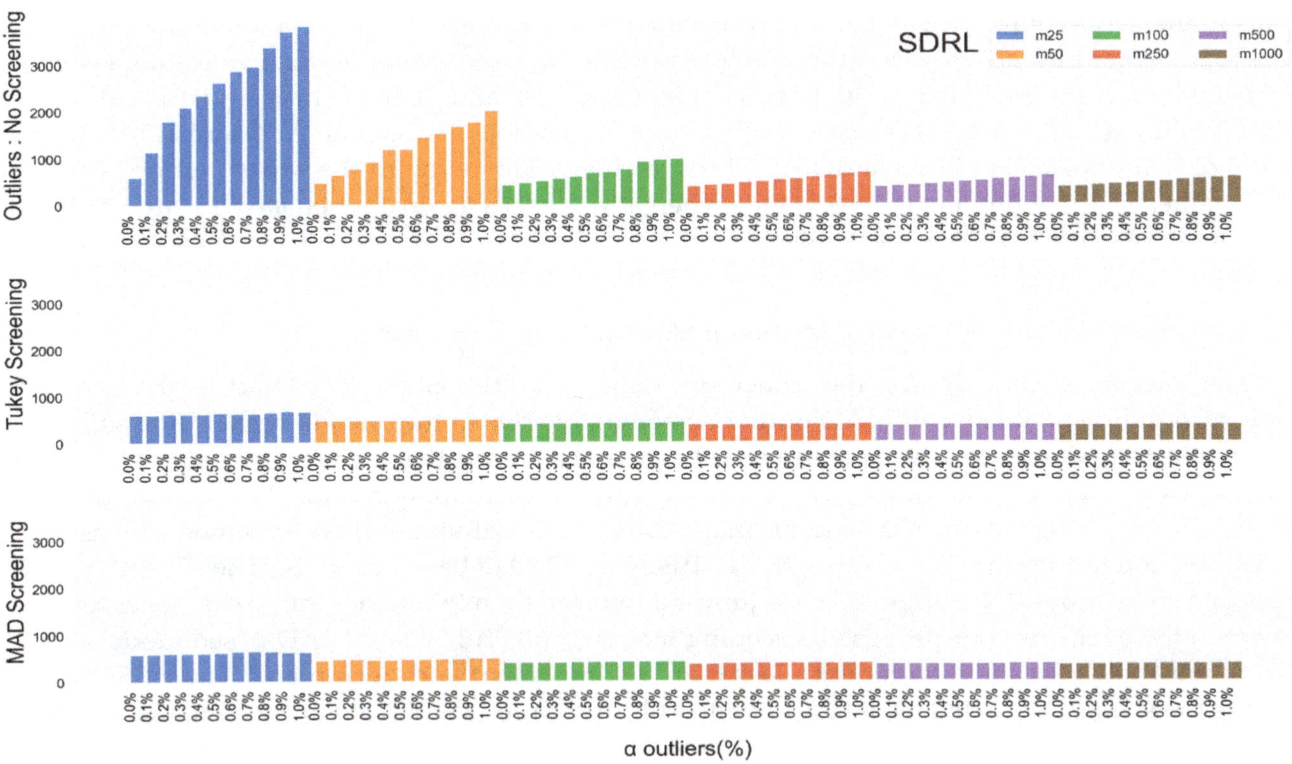

Figure 5. In-control SDRL values for the Shewhart chart from t-(v = 100) distribution in the presence of outliers with and without outlier screening.

4. Discussion

We summarize the findings of the study under the following subsections: (a) parameter estimation effect on the Shewhart control chart, (b) effect of outliers on Shewhart $\overline{X}$ chart performance, and (c) improvement of outliers screening models on the Shewhart $\overline{X}$ chart performance. Through the discussion, we use the run length properties as a yardstick for measuring the performance of the charts.

4.1. Parameter Estimation Effect on the Shewhart $\overline{X}$ Control Chart

Theoretically, when the Shewhart charts parameters are known, the limit L corresponding to the IC $ARL_0 = 370$ is $L = 3$. When the parameters are estimated from phase-I samples, the first effect of the estimation is the change in L. The control limit L deviates from its theoretical value as much as the sample size m reduces. That implies, the smaller the sample size m, the farther the control limit from the theoretical value. This is noticeable in Tables 1 and 2, as L changes as the sample size does. We compute Ls based on 100,000 iterations of simulation. Secondly, in the introduction of shifts, which makes the process OC, the RL properties values of the estimated parameters are bigger than the theoretical values. This indicates that the chart with estimated parameters are slower in detecting shifts in the process as compared to the chart with known parameters. For instance, (cf. Tables 1 and 2), with $m = 1000$, $\delta = 0.5$ the resulting ARL_1 and $SDRL_1$ are 156.42 and 158.84 for normal distribution and 150.92 and 160.46 for t-distribution respectively. However, with $m = 25$, $\delta = 0.5$ ARL_1 and $SDRL_1$ are 190.12 and 333.88 for normal distribution and 194.06 and 340.19 for t-distribution respectively.

4.2. Effect of Outliers on Shewhart $\overline{X}$ Control Chart performance

Haven noticed the effect of parameter estimation on Shewhart chart performance; one major cause could be the presence of outliers in the dataset. The results in Tables 3 and 4 prove that extreme values in the sample causes great havoc to the performance of the process. As discussed earlier in Section 4, $\alpha = 0$ indicates absence of outliers, and the presence of outliers if otherwise. We observe jumps in the values of IC ARL and SDRL from Tables 3 and 4. With different combinations of α and m, we say the bigger the value of α and the smaller the value of m, the gross the effect of the outliers on the chart. Take for instance, in the normal environment, the ARL and SDRL values of just 1% of outliers ($\alpha = 0.01$) for when $m = 1000$ as against when $m = 25$. It shocks to see the ARL and SDRL jumped from 592.55 and 612.00 to 996.3 and 4012.72 respectively. However, in the t-distribution, ARL and SDRL values of 1% of outliers ($\alpha = 0.01$) for when $m = 1000$ as against when $m = 25$, are 575.08 and 591.55 to 953.61 and 3823.24 respectively.

4.3. *Improvement of Outliers Screening Models on Shewhart Chart Performance*

The proposed remedy for the effect of outliers on the Shewhart chart works perfectly. The incorporation of Tukey and MAD outlier-screening models in the Shewhart chart normalizes the outlier effects and restores the performance even much better than it was. To access the effect of these two screening methods, we present Figures 2–5, displaying the IC ARL values with $m = 25$, 50, 100, 250, 500 and 1000, and the magnitude $w = 3$, without outliers screening, alongside the IC ARL whose outliers are screened with the Tukey and MAD-based models. The IC ARL that are supposed to be around the target 370 has jumped to more than 250% increment due to the effect of outliers. However, with our proposed screening models; both Tukey and MAD-based models; the IC ARL is returned back to its target with less than 5% increment and decrement. The IC SDRL also exhibits the same pattern; in fact, its improvement is more appreciable as compared to the ARL's.

5. Illustrative Example

In the manufacturing industry, semiconductor lithography (photolithography) refers to the formation of three-dimensional images on the substrate for subsequent transfer of the pattern to the substrate. A keynote aspect of this process is the bake process, both the pre (soft)-bake and post (hard)-bake. In this section, we implement the Shewhart chart with the proposed outlier detection models on the flow width measurement of a hard bake process. In the subsequent subsections, we give a brief overview of the hard-bake process and then application of the Shewhart chart on the dataset extracted from such a process (the Basics of Microlithography n.d.).

5.1. *The Post (Hard) Bake Process*

A typical photolithography process consist of the following sequence of operation: substrate preparation, photoresist spin coat, pre-bake, exposure, post-exposure bake, development and finally the post-bake. The hard-bake process, as the name implies, is used to harden the final resist image so that it will withstand the harsh environments of etching. This post-bake ensures complete removal of solvent, improving adhesion in wet etch processes and resistance to plasma etches. Practitioners use different temperatures depending on the material under study. However, the temperature should be carefully chosen and not more than 200 °C. A major characteristic of this process is the wafer. Recall that the word lithography is a combination of two Greek words: lithos meaning stones and graphia, meaning to write. Our stones in this case are silicon wafers and the patterns are written with photoresist, which are sensitive polymers. Figures 6 and 7 depict a typical photolithography flowchart and the hard-bake process.

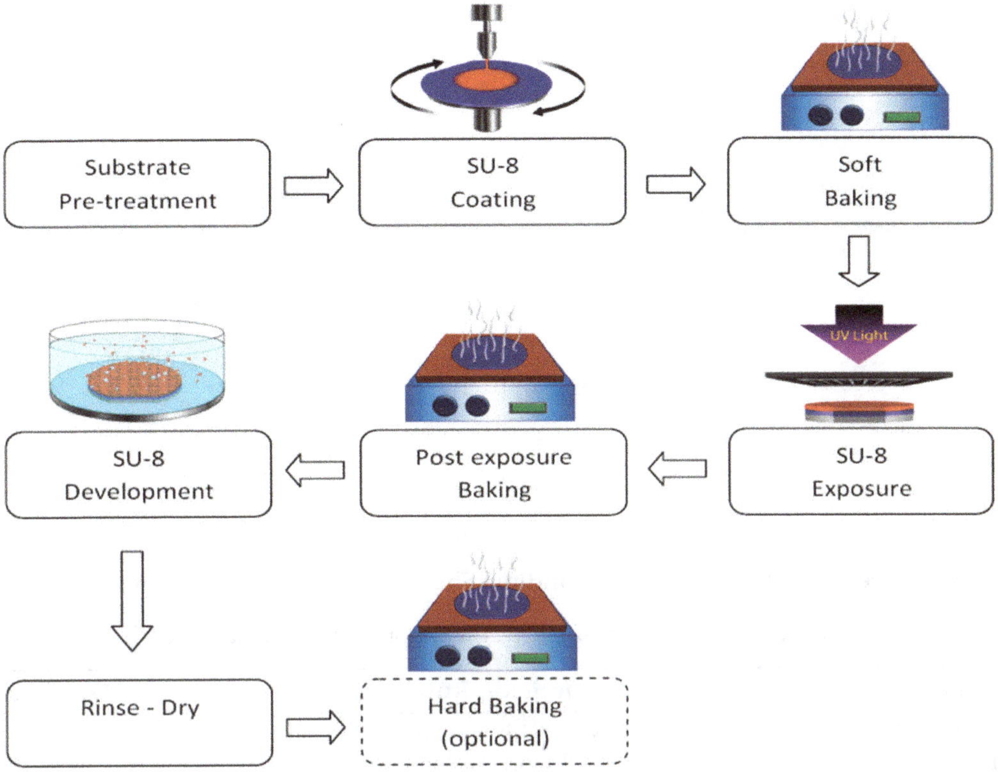

Figure 6. A flowchart of a photolithography process of semiconductor manufacturing industry.

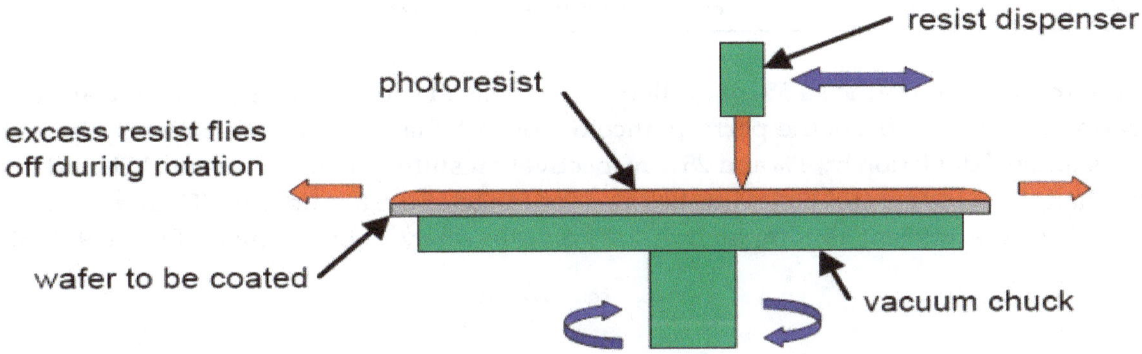

Figure 7. Illustration of hard-bake process.

5.2. Application of Shewhart Control Charts with Outlier

In this section, we implement the findings of this study on a set of data generated from a semiconductor manufacturing of a hard-bake process, which monitors the flow width measurement of wafers [1]. The variable of interest is the flow width measurement (in microns) for the hard-brake process. The data consist of 25 IC phase-I samples and 10 phase-II samples each of sample size 5. The process mean and standard deviation of the phase-I samples are 16.7163 and 3.5167, respectively. Therefore, we use these estimates to setup Shewhart chart control limits for monitoring phase-II samples. Figure 8 shows all phase-I sample points staying within the limits and 3 of the phase-II sample points stretching beyond the LCL making them OoC due to some assignable cause of variation.

Prior to setting the limits, we test the data for possible autocorrelation. The data is autocorrelation-free as the Durbin–Watson (DW) test result proves. The value of the DW test statistics is DW $= 1.7564$ and the critical values at 1% level of significance are $d_L = 1.19,\ and\ d_U = 1.31$. By the interpretation explained in Table 9, we fail to reject the null hypothesis and conclude that there is no evidence of autocorrelation in the data.

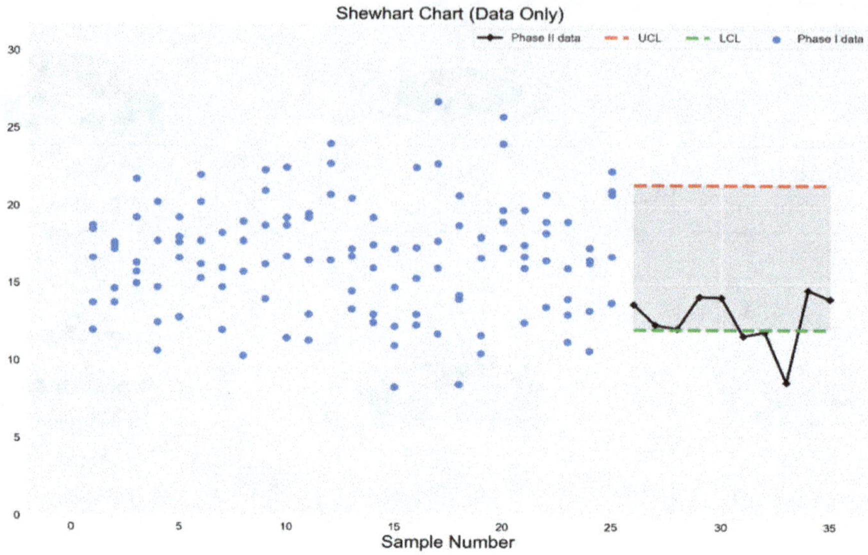

Figure 8. Scatter plot of phase-I sample and the Shewhart chart with estimated parameters.

Table 9. Interpretation of Durbin–Watson autocorrelation test.

Categories	Decision Rules	Decisions
$0 \leftrightarrow d_L = 1.19$	Reject H_0 : positive autocorrelation	
$d_L = 1.19 \leftrightarrow d_U = 1.31$	Inconclusive	
$d_U = 1.31\ 4 \leftrightarrow d_L = 2.69$	Do not reject H_0 : no autocorrelation	$1.31 < \text{DW} = 1.7564 < 2.69$
$4 - d_L = 2.69 \leftrightarrow 4 - d_U = 2.81$	Inconclusive	
$4 - d_U = 2.81 \leftrightarrow 4$	Reject H_0 : negative autocorrelation	

Furthermore, we introduce a 5% of outliers to the phase-I samples, to illustrate the argument that the presence of outliers affects the performance of control charts. This subsequently increased the mean and standard deviation by 4% and 25% respectively resulting to an increased UCL and decreased LCL. The changes in the control limits implies a wider range of the boundaries. Therefore the resulting control charts is less efficient as compared to the previous one without outliers. Figure 9 depicts this.

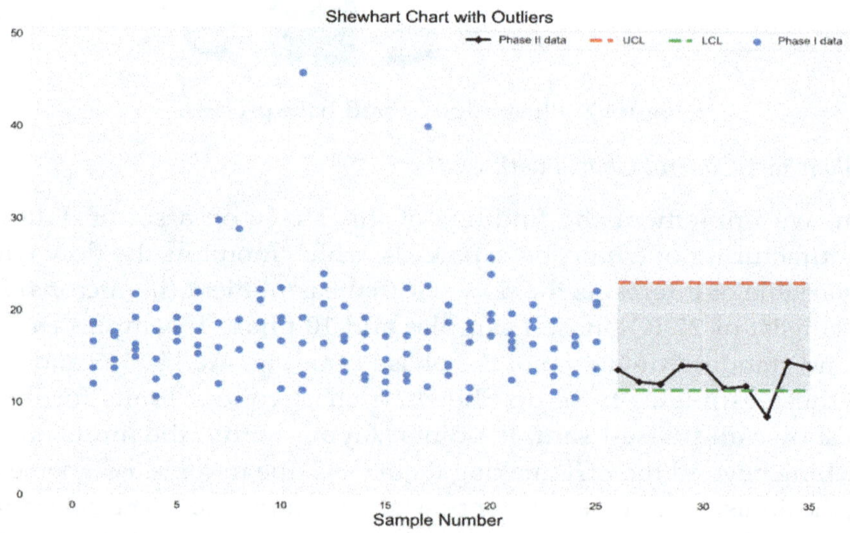

Figure 9. Scatter plot of phase-I sample and the resulting Shewhart chart with estimated parameters and 5% of outliers with magnitude 3.

5.3. *Application of Shewhart Outlier Detection Model*

Having established the deficiency of the Shewhart chart with outliers on the dataset; we employ our proposed outlier detection model with the Shewhart chart explained in Section 2.4 to rectify this shortcoming. Figure 10 shows the application of the Shewhart Tukey-based model. It is evident there in that the chart was not only able to restore the efficiency of the chart as there were no outliers, detecting 3 OoC sample points, but also to identify the outliers in the phase-I sample points. Similarly, Figure 11 portrays the scenario when the Shewhart MAD-based model is applied on the monitoring stage. Despite the presence of outlier in the dataset, the chart is able to detect the OC sample points as much as it does when there were no outliers.

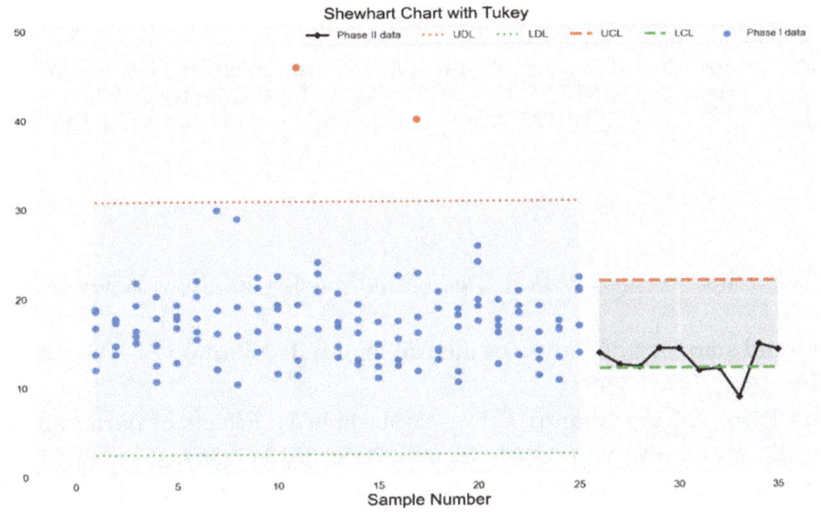

Figure 10. Scatter plot of phase-I sample and the resulting Shewhart chart with Tukey outlier detection screening.

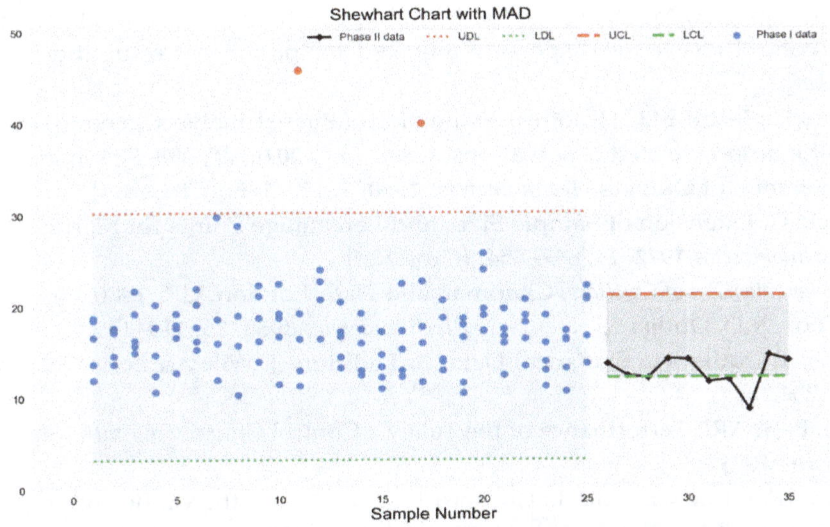

Figure 11. Scatter plot of phase-I sample and the resulting Shewhart chart with MAD-outlier detection screening.

6. Conclusions

In this article, we evaluate the performance of the Shewhart control chart for location monitoring with estimated parameters. The study substantiates the effect of estimation error and the variability in the practitioners' choice of phase-I samples on the chart, especially when the samples are prone to outliers. Increasing the phase-I sample size (although not practicably) will to some extent reduce the gross impact on the Shewhart chart. The results of this study further prove that incorporation of the non-parametric outlier screening models, Tukey and MAD, in the design of the Shewhart chart is more practicable as it requires less phase-I samples and yields better results. Another advantage of this study lies in the simplicity of its design and ease of usage. The study rounds up with an illustrative example with a photolithography real data. A comparison of the two detection models, Tukey and MAD, reveals that duo relatively efficient. The study is limited to operate within the univariate setup, while focusing on multivariate setup will be a great advantage and we plan a future study for that. Also, proposed charts are memory-less, which implies they are suitable for monitoring large shift. However, the idea of the study is not only applicable in Shewhart multivariate setup, but also extendable to other control charts, like exponentially weighted moving average and cumulative sum charts both univariate and multivariate setups.

Author Contributions: Conceptualization, I.A.R. and N.A.; Data curation, I.A.R. and M.R.A.; Investigation, M.H.L.; Methodology, N.A.; Resources, M.H.L.; Supervision, M.R. and M.R.A.; Writing—original draft, I.A.R.; Writing—review & editing, M.H.L., M.R. and N.A. All authors have read and agreed to the published version of the manuscript.

References

1. Montgomery, D.C. *Introduction to Statistical Quality Control*, 6th ed.; John Wiley & Sons Inc.: Hoboken, NJ, USA, 2009.
2. Chen, G. The mean and standard deviation of the run length distribution of X charts when control limits are estimated. *Stat. Sin.* **1997**, *7*, 789–798.
3. Jensen, W.A.; Jones-Farmer, L.A.; Champ, C.W.; Woodall, W.H. Effects of parameter estimation on control chart properties: A literature review. *J. Qual. Technol.* **2006**, *38*, 349–364. [CrossRef]
4. Psarakis, S.; Vyniou, A.K.; Castagliola, P. Some Recent Developments on the Effects of Parameter Estimation on Control Charts. *Qual. Reliab. Eng. Int.* **2014**, *30*, 1113–1129. [CrossRef]
5. Saleh, N.A.; Mahmoud, M.A.; Keefe, M.J.; Woodall, W.H. The difficulty in designing shewhart X and X control charts with estimated parameters. *J. Qual. Technol.* **2015**, *47*, 127–138. [CrossRef]
6. Goedhart, R.; Da Silva, M.M.; Schoonhoven, M.; Epprecht, E.K.; Chakraborti, S.; DoesA, R.J.M.M. Veiga Shewhart control charts for dispersion adjusted for parameter estimation. *IISE Trans.* **2017**, *49*, 838–848. [CrossRef]
7. Dasdemir, E.; Weiß, C.; Testik, M.C.; Knoth, S. Evaluation of Phase I analysis scenarios on Phase II performance of control charts for autocorrelated observations. *Qual. Eng.* **2016**, *28*, 293–304. [CrossRef]
8. Anscombe, F.J. Rejection of Outliers. *Technometrics* **1960**, *2*, 123–146. [CrossRef]
9. Grubbs, F.E.; Beck, G. Extension of Sample Sizes and Percentage Points for Significance Tests of Outlying Observations. *Technometrics* **1972**, *14*, 847–854. [CrossRef]
10. Hawkins, D.M. *Identification of Outliers*; Chapman and Hall: London, UK, 1980.
11. Beckman, R.J.; Cook, R.D. Outlier ... . s. *Technometrics* **1983**, *25*, 119–149.
12. Barnett, V.; Lewis, T. Outliers in Statistical Data. 3rd edition. J. Wiley & Sons 1994, XVII. 582 pp., £49.95. *Biom. J.* **1995**, *37*, 256.
13. Torng, C.-C.; Lee, P.-H. ARL Performance of the Tukey's Control Chart. *Commun. Stat. Simul. Comput.* **2008**, *37*, 1904–1913. [CrossRef]
14. Tercero-Gomez, V.; Ramirez-Galindo, J.; Cordero-Franco, A.; Smith, M.; Beruvides, M. Modified Tukey's Control Chart. *Commun. Stat. Simul. Comput.* **2012**, *41*, 1566–1579. [CrossRef]
15. Fan, S.-K.S.; Huang, H.-K.; Chang, Y.-J. Robust Multivariate Control Chart for Outlier Detection Using Hierarchical Cluster Tree in SW2. *Qual. Reliab. Eng. Int.* **2013**, *29*, 971–985. [CrossRef]
16. Khaliq, Q.-U.-A.; Riaz, M. Robust Tukey-CUSUM Control Chart for Process Monitoring. *Qual. Reliab. Eng. Int.* **2016**, *32*, 933–948. [CrossRef]
17. Fitriyah, H.; Budi, A.S. Outlier Detection in Object Counting based on Hue and Distance Transform using Median Absolute Deviation (MAD). In Proceedings of the 2019 4th International Conference on Sustainable Information Engineering and Technology, Lombok, West Nusa Tenggara, Indonesia, 28–30 September 2019; pp. 217–222.

18. Abbas, N. A robust S^2 control chart with Tukey's and MAD outlier detectors. *Qual. Reliab. Eng. Int.* **2020**, *36*, 403–413. [CrossRef]
19. Abbas, N.; Abujiya, M.R.; Riaz, M.; Mahmood, T. Cumulative Sum Chart Modeled under the Presence of Outliers. *Mathematics* **2020**, *8*, 269. [CrossRef]
20. Dequeant, K.; Vialletelle, P.; Lemaire, P.; Espinouse, M.L. A literature review on variability in semiconductor manufacturing: The next forward leap to Industry 4.0. In Proceedings of the Winter Simulation, Arlington, VA, USA, 11–14 December 2016; pp. 2598–2609.
21. Spanos, C.J. Statistical Process Control in Semiconductor Manufacturing. *Proc. IEEE* **1992**, *80*, 819–830. [CrossRef]
22. Huh, I. Multivariate EWMA Control Chart and Application to a Semiconductor Manufacturing Process. Ph.D. Thesis, Department of Statistics, McMaster University, Hamilton, ON, Canada, 2010.
23. Baud-Lavigne, B.; Bassetto, S.; Penz, B. A broader view of the economic design of the X-bar chart in the semiconductor industry. *Int. J. Prod. Res.* **2010**, *48*, 5843–5857. [CrossRef]
24. Higashide, M.; Nishina, K.; Kawamura, H.; Ishii, N. Statistical process control for semiconductor manufacturing processes. *Front. Stat. Qual. Control* **2010**, *9*, 71–84.
25. Quesenberry, C.P. The Effect of Sample Size on Estimated Limits for $\overline{X}$ and X Control Charts. *J. Qual. Technol.* **1993**, *25*, 237–247. [CrossRef]
26. Albers, W.; Kallenberg, W.C.M. Estimation in Shewhart control charts: Effects and corrections. *Metrika* **2004**, *59*, 207–234. [CrossRef]
27. Dovoedo, Y.H.; Chakraborti, S. Effects of Parameter Estimation on the Multivariate Distribution-free Phase II Sign EWMA Chart. *Qual. Reliab. Eng. Int.* **2017**, *33*, 431–449. [CrossRef]
28. Maleki, M.R.; Castagliola, P.; Amiri, A.; Khoo, M.B.C. The effect of parameter estimation on phase II monitoring of poisson regression profiles. *Commun. Stat. Simul. Comput.* **2019**, *48*, 1964–1978. [CrossRef]
29. Jardim, F.S.; Chakraborti, S.; Epprecht, E.K. Two perspectives for designing a phase II control chart with estimated parameters: The case of the Shewhart $\overline{X}$ Chart. *J. Qual. Technol.* **2020**, *52*, 198–217. [CrossRef]

9

Comparing Groups of Decision-Making Units in Efficiency based on Semiparametric Regression

Hohsuk Noh [1] and Seong J. Yang [2,*]

1 Department of Statistics, Sookmyung Women's University, Seoul 04310, Korea; hsnoh@sookmyung.ac.kr
2 Department of Statistics (Institute of Applied Statistics), Jeonbuk National University, Jeollabuk-do 54896, Korea
* Correspondence: sjyang@jbnu.ac.kr

Abstract: We consider a stochastic frontier model in which a deviation of output from the production frontier consists of two components, a one-sided technical inefficiency and a two-sided random noise. In such a situation, we develop a semiparametric regression-based test and compare the technical efficiencies of the different decision-making unit groups, assuming that the production frontier function is the same for all the groups. Our test performs better than the previously proposed ones for the same purpose in numerical studies, and also has the theoretical advantage of working under more general assumptions. To illustrate our method, we apply the proposed test to Program for International Student Assessment (PISA) 2015 data and investigate whether an efficiency difference exists between male and female student groups at a specific age in terms of learning time and achievement in mathematics.

Keywords: data envelopment analysis; stochastic frontier model; semiparametric regression; group efficiency comparison

---

1. Introduction

Efficiency comparison between groups is currently used in various fields such as banking, insurance, sports, and R&D investment evaluation. Numerous empirical studies frequently analyze group efficiency using so-called Data Envelopment Analysis (DEA). DEA is a body of techniques for measuring relative efficiency by comparing it with the possible frontiers of decision-making units (DMUs) with multiple inputs and outputs. Here, the term DMU is used to collectively refer to all the units in which the production activity takes place. In the DEA framework, the DMU efficiency scores of each group can be obtained after specifying some assumptions appropriate to the situation, and then the comparison of the efficiency distributions of the groups is made on the basis of their obtained scores. For example, Golany and Storberg [1] and Lee et al. [2] applied non-parametric tests, such as the Mann–Whitney (MW) and Kruskal–Wallis tests, to the efficiency scores. Cummins et al. [3] introduced a dummy variable to indicate the groups, and then regressed the efficiency scores on the dummy variable. Simar and Zelenyuk [4] adapted the test developed in Li [5] to the DEA context and applied it to the obtained scores, to test the equality of efficiency distributions. O'Donnell et al. [6] used the concept of a meta-frontier to compare the technical efficiencies of firms that may be classified into different groups.

However, this stream of research under the DEA framework has a limitation in that it does not consider the noise factor in the production process. DEA typically assumes that the inefficiency of the DMU is the only cause of its production not reaching its maximum output, but obviously there are many uncontrolled factors which need to be considered as the cause. From this recognition,

Aigner and Chu [7] and Meeusen and van den Broeck [8] first proposed the stochastic frontier model (SFM), which allows for both unobserved variation in output: the technical inefficiency of the production unit and the noise which represents the effect of innumerable uncontrollable factors. For illustrative comparison between DEA and SFM frameworks, see Figure 1.

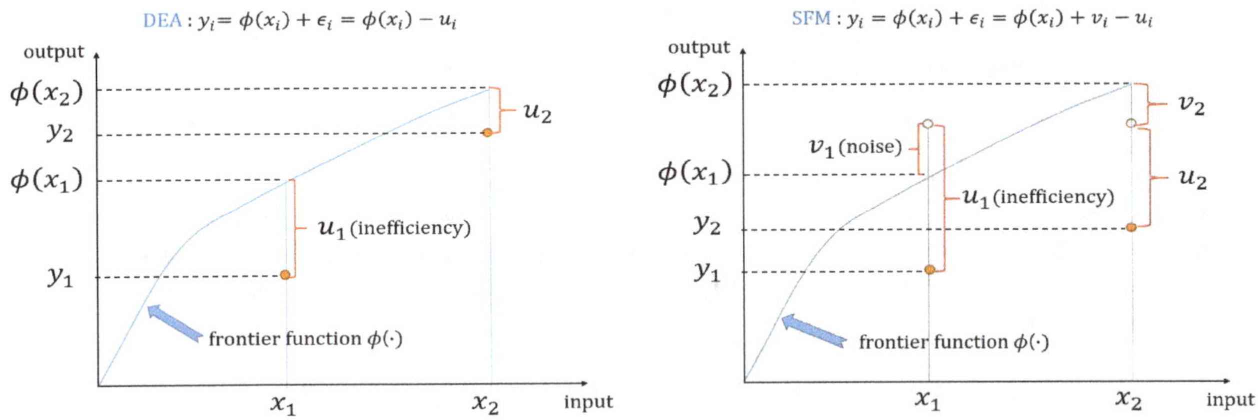

Figure 1. Comparison between Data Environment Analysis (DEA) **(left panel)** and SFM **(right panel)** frameworks (y_i: output, x_i: input, $\phi(x_i)$: the maximum output which can be obtained from the input x_i, ϵ_i: deviation from the production frontier function $\phi(x_i)$, u_i: technical inefficiency, v_i: noise). Note that the technical inefficiency u_i is a nonnegative random variable with unknown distribution.

Nowadays, the stochastic frontier model is used in a large literature of studies of production. Hence, we feel the need to develop a method and compare the efficiency difference between groups under SFM framework. One pioneering work in this direction is Banker et al. [9]. They developed five DEA-based hypothesis tests to compare the efficiency of groups under SFM. Although the paper referred above is an important development toward group efficiency comparison under SFM, their tests need to improve further.

First, their rather strong assumptions might limit the applicability of the proposed methods. For their parametric tests, they assumed the equality of both noise variance and inefficiency variance across groups. Second, their theoretical justification of the proposed methods needs to be checked. As regards their ordinary least squares (OLS) test of the mean difference in inefficiency, they provided its asymptotic normality as theoretical basis, but to our knowledge, such asymptotic normality is difficult to obtain because of the slow convergence rate of the DEA estimator when the number of input variables is greater than or equal to 2. The same comment is given in Section 3.2 of Simar and Wilson [10] on a similar type of asymptotic normality result as proof of Proposition 1 in Banker and Natarajan [11]. Finally, because they used the DEA methods for SFM, the tests they developed were based not directly on inefficiency itself, but on the inefficiency contaminated by positive measurement error due to noise. This indirect approach can lower the performance of their tests.

This observation has motivated us to develop a theoretically sound tool for comparison of group inefficiencies in the presence of noise. We develop such a methodology using a semi-parametric regression technique instead of DEA methods. The newly developed test performs better than the tests of Banker et al. [9] in numerical studies. It also has the theoretical advantage of working under more general assumptions compared to Banker et al. [9].

The rest of this paper is organized as follows. Section 2 describes our proposed test for group inefficiency comparison. We then perform some simulation studies and compare our test with the tests proposed by Banker et al. [9] in Section 3. We illustrate our method by applying the proposed test to Program for International Student Assessment (PISA) 2015 data and investigate whether an efficiency difference exists between male and female student groups at a specific age in terms of learning time and achievement in mathematics in Section 4. Section 5 provides some discussion and future research topics.

2. Group Efficiency Comparison under SFM

Assume that we have observations on n DMUs, where each observation consists of a vector of p inputs $\mathbf{X}_i = (X_{1,i}, \ldots, X_{p,i})^\top$ and the corresponding output Y_i. We consider the case where n DMUs can be divided into two distinct groups with n_l observations ($n = n_1 + n_2$). We assume the following stochastic frontier model for two groups of DMUs:

$$\begin{aligned} \text{(The first group)} \quad Y_i &= \phi(\mathbf{X}_i) + \varepsilon_{1,i},\ i \in \{1, \ldots, n_1\}; \\ \text{(The second group)} \quad Y_i &= \phi(\mathbf{X}_i) + \varepsilon_{2,i},\ i \in \{n_1 + 1, \ldots, n_1 + n_2\}, \end{aligned} \tag{1}$$

where $\varepsilon_{l,i} = V_{l,i} - U_{l,i}$, $V_{l,i}$ is a random noise term of the lth group with $E(V_{l,i}|\mathbf{X}_i) = 0$, and $U_{l,i}$ is an inefficiency term of the lth group with $U_{l,i} \geq 0$ for $l = 1, 2$. We assume that the same production technology is applied to both DMU groups. Hence, the production frontier function $\phi(\cdot)$ is the same throughout the groups, as in Banker et al. [9]. Under this model, we need to estimate the difference $E(U_1) - E(U_2)$ and test the hypothesis

$$H_0 : E(U_1) - E(U_2) = 0 \text{ vs. } H_1 : E(U_1) - E(U_2) > 0\,(< 0) \tag{2}$$

to know which DMU group is more efficient. A novelty of our approach in developing the test is to implement the test without imposing any parametric assumption on the frontier function $\phi(\cdot)$, and with minimal assumptions on inefficiency and random noise. Banker et al. [9] also implemented the test without any parametric assumption on $\phi(\cdot)$, but with additional restrictive parametric assumptions on noise and inefficiency. In the following sections, we first review the work of Banker et al. [9] and then explain the development of our semiparametric regression-based test.

2.1. The Previous Work

To apply the DEA methods to SFM, Banker et al. [9] assumed that the random noise variables $V_{1,i}$, $V_{2,i}$ are bounded above by V^{max}, that is, $V_{1,i}$, $V_{2,i} \leq V^{max}$. Under this assumption, they transformed model (1) as

$$Y_i = (\phi(\mathbf{X}_i) + V^{max}) - (V^{max} - V_{l,i} + U_{l,i}) \equiv \tilde{\phi}(\mathbf{X}_i) - \tilde{U}_{l,i},\ l = 1, 2. \tag{3}$$

Since $\tilde{U}_{l,i} = (V^{max} - V_{1,i}) + U_{l,i} \geq 0$, they considered the translated production function $\tilde{\phi}(\cdot) = \phi(\cdot) + V^{max}$ as a new production function, and $\tilde{U}_{l,i}$ as the inefficiency of the DEA framework. The new inefficiency $\tilde{U}_i^l$ can be estimated as $\hat{\tilde{\phi}}(\mathbf{X}_i) - Y_i$ after $\tilde{\phi}(\cdot)$ is estimated using the conventional DEA methods. After estimating $\tilde{U}_{l,i}$ using DEA methods, they used it for group efficiency comparison. This approach is advantageous in that we use the strength of the existing well-developed DEA techniques. However, the approach has one disadvantage in that the tests developed are based not on inefficiency ($U_{l,i}$) itself, but on the inefficiency contaminated by the positive measurement error ($V^{max} - V_{l,i}$) due to random noise. Additionally, the distributional property of the inefficiency estimated using DEA methods is generally hard to derive or quite complicated, making it very difficult to develop a statistical test theory based on estimated inefficiency (estimate of $\tilde{U}_{l,i}$). Hence, we are motivated to develop a test for (2) directly based on inefficiency $U_{l,i}$. We will explain this in the following section.

2.2. The Proposed Test

This section introduces our approach to testing the hypothesis in (2). Unlike Banker et al. [9], we do not require that neither the noise variance nor inefficiency should be equal across groups. Moreover, we allow for distributional difference in the composite error ε and input vector $\mathbf{X}_i$ from the production environmental factors of each group. Specifically, the variance of V_l and mean of inefficiency U_l can differ by the group as well as conditional distribution of $\mathbf{X}_i$, given group l.

First, model (1) can be written as two nonparametric mean regression models as follows:

$$\begin{aligned} Y_i &= [\phi(\mathbf{X}_i) - E(U_1)] + [V_{1,i} - (U_{1,i} - E(U_1))] \\ &\equiv \phi^*(\mathbf{X}_i) + \varepsilon^*_{1,i},\ i \in \{1,\dots,n_1\}; \quad (4) \\ Y_i &= [E(U_1) - E(U_2)] + [\phi(\mathbf{X}_i) - E(U_1)] + [V_{2,i} - (U_{2,i} - E(U_2))] \\ &\equiv \beta_0 + \phi^*(\mathbf{X}_i) + \varepsilon^*_{2,i},\ i \in \{n_1+1,\dots,n_1+n_2\}, \quad (5) \end{aligned}$$

where $E(\varepsilon^*_{1,i}) = E(\varepsilon^*_{2,i}) = 0$, and $\beta_0 = E(U_1) - E(U_2)$. If a dummy variable is defined for groups letting $T_i = 0$ for $i \in \{1,\dots,n_1\}$ and $T_i = 1$ for $i \in \{n_1+1,\dots,n_1+n_2\}$, the two models (4) and (5) can be integrated into a single partial linear semiparametric regression model as follows:

$$Y_i = \beta_0 T_i + \phi^*(\mathbf{X}_i) + \varepsilon^*_i,\ i \in \{1,\dots,n\}, \quad (6)$$

where $\varepsilon^*_i = (1 - I(T_i = 1))\varepsilon^*_{1,i} + I(T_i = 1)\varepsilon^*_{2,i}$ and $E(\varepsilon^*_i|T_i,\mathbf{X}_i) = 0$. Using this model (6), we can test hypothesis (2) by testing hypothesis

$$H_0 : \beta_0 = 0 \text{ vs. } H_1 : \beta_0 > 0(< 0). \quad (7)$$

Note that $Var(\varepsilon^*_i|T_i,\mathbf{X}_i) = (1 - I(T_i = 1))Var(\varepsilon^*_{1,i}) + I(T_i = 1)Var(\varepsilon^*_{2,i})$. Thus, model (6) is a heteroscedastic partial linear model. Liang [12] and Ma et al. [13] studied model (6) when $\mathbf{X}_i$ is univariate. By extending the theory from there to the case where $\mathbf{X}_i$ is multivariate, we can test hypothesis (7). In Appendix A, we prove the asymptotic normality of the kernel-based profile estimator of β_0 based on a local linear model smoother when $\mathbf{X}_i$ is multivariate, and provide the necessary assumptions for it. As with the estimator in Liang [12], the kernel-based profile estimator of β_0 when $\mathbf{X}_i$ is multivariate is given as

$$\hat{\beta}_0 = (\mathbf{T}^\top(\mathbf{I}-\mathbf{S})^\top(\mathbf{I}-\mathbf{S})\mathbf{T})^{-1}\mathbf{T}^\top(\mathbf{I}-\mathbf{S})^\top(\mathbf{I}-\mathbf{S})\mathbf{Y} \equiv \mathbf{H}\mathbf{Y}, \quad (8)$$

where $\mathbf{T} = (T_1,\dots,T_n)^\top$, $\mathbf{Y} = (Y_1,\dots,Y_n)^\top$, and $\mathbf{S}$ the smoother matrix for estimating the vector $(E(\cdot|\mathbf{X}_1),\dots,E(\cdot|\mathbf{X}_n))^\top$. If we choose local linear regression as the smoothing method, the smoothing matrix $\mathbf{S} = [\mathbf{s}_{\mathbf{X}_1} \cdots \mathbf{s}_{\mathbf{X}_n}]^\top$ will be a collection of row vectors, each of which is the smoother vector

$$\mathbf{s}_\mathbf{x}^\top = \mathbf{e}_1^\top(\mathbf{X}_\mathbf{x}^\top\mathbf{W}_\mathbf{x}\mathbf{X}_\mathbf{x})^{-1}\mathbf{X}_\mathbf{x}^\top\mathbf{W}_\mathbf{x}, \quad (9)$$

where $\mathbf{e}_1^\top = (1,0,\dots,0)$ is a $(p+1)\times 1$ vector; $\mathbf{W}_\mathbf{x} = \text{diag}\{K_\mathbf{h}(\mathbf{X}_1-\mathbf{x}),\cdots,K_\mathbf{h}(\mathbf{X}_n-\mathbf{x})\}$ for some kernel function K and bandwidth vector $\mathbf{h} = (h_1,\dots,h_p)^\top$; and

$$\mathbf{X}_\mathbf{x} = \begin{bmatrix} 1 & (\mathbf{X}_1-\mathbf{x})^\top \\ \vdots & \vdots \\ 1 & (\mathbf{X}_n-\mathbf{x})^\top \end{bmatrix}. \quad (10)$$

Here, $K_\mathbf{h}(\mathbf{X}_i-\mathbf{x}) = \prod_{j=1}^p h_j^{-1}K((X_{j,i}-x_j)/h_j)$. From the theorem in Appendix A, under some regularity conditions, $\sqrt{n}(\hat{\beta}_0 - \beta_0)$ is asymptotically normal with mean zero and variance $\sigma^2 = E(\tilde{T}^2)^{-2}E(\varepsilon^*\tilde{T})^2$, where $\tilde{T} = T - E(T|\mathbf{X})$. Using a consistent estimator of σ^2, we can test (7) with significance level α by rejecting H_0 if $Z = \hat{\beta}_0/(\hat{\sigma}/\sqrt{n}) \geq z_\alpha$ (or $\leq -z_\alpha$), where z_α is the $(1-\alpha)$-quantile of the standard normal distribution.

As regards the estimation of σ^2, we can first directly estimate variance σ^2 using the estimates $\hat{\varepsilon}^*_i$ and $\hat{E}(T|\mathbf{X}_i)$, where $\hat{\varepsilon}^*_i = Y_i - \hat{\beta}_0 T_i - \widehat{\phi^*}(\mathbf{X}_i)$ and $\widehat{\phi^*}(\cdot)$ is the local linear estimator of $\phi^*(\cdot)$ based on $Y_i - \hat{\beta}_0 T_i$, $i = 1,\dots n$. We can also estimate it using the sandwich covariance estimate based on (8),

$$\widehat{Var}(\hat{\beta}_0|\mathbf{T},\mathbf{X}_1,\dots,\mathbf{X}_n) = \mathbf{H}\,\widehat{Var}(\mathbf{Y}|\mathbf{T},\mathbf{X}_1,\dots,\mathbf{X}_n)\,\mathbf{H}^\top. \quad (11)$$

Matrix $\widehat{Var}(\mathbf{Y}|\mathbf{T}, \mathbf{X}_1, \ldots, \mathbf{X}_n)$ is diagonal, with the ith diagonal element equal to

$$\widehat{E}(\widehat{\varepsilon^{*2}}|T_i, \mathbf{X}_i) = (1 - I(T_i = 1))\widehat{Var}(\varepsilon_1^*) + I(T_i = 1)\widehat{Var}(\varepsilon_2^*), \tag{12}$$

where

$$\widehat{Var}(\varepsilon_1^*) = n_1^{-1} \sum_{i:\, T_i=0} (\hat{\varepsilon}_i^*)^2 - \left(n_1^{-1} \sum_{i:\, T_i=0} \hat{\varepsilon}_i^*\right)^2, \tag{13}$$

$$\widehat{Var}(\varepsilon_2^*) = n_2^{-1} \sum_{i:\, T_i=1} (\hat{\varepsilon}_i^*)^2 - \left(n_2^{-1} \sum_{i:\, T_i=1} \hat{\varepsilon}_i^*\right)^2. \tag{14}$$

Since the frontier function is generally (coordinatewise) non-decreasing with respect to the input variables, one might consider it necessary to impose such a monotonicity on $\phi^*(\cdot)$. However, from Theorem 2.1 in Huang [14], such imposition will not decrease the asymptotic variance of $\hat{\beta}_0$; that is, it shows no theoretical improvement in performance. We therefore choose to develop the test without the monotonicity assumption for simplicity.

Note that our test directly estimates the mean difference in inefficiency $E(U_1) - E(U_2)$ using the semiparametric regression technique. Thus, the proposed test can work under assumptions that are more general than those in Banker et al. [9]. Additionally, we do not have to assume that noise has a finite upper support bound (V_{max}). However, the tests in Banker et al. [9] need such assumptions because they estimate $\tilde{U}_{l,i} = (V_{max} - V_{l,i}) + U_{l,i}$ and use it as a surrogate estimate of $U_{l,i}$. However, $V_{max} - V_{l,i}$ may hamper the tests and degrade their performance.

3. Numerical Studies

In this section, we compare the performance of our test with those of Banker et al. [9]. We consider single and multiple input cases and use sandwich formulas to estimate the variance in estimators.

3.1. Single Input Case

We first consider a single input case using the following model:

$$\begin{aligned} Y_i &= \phi(X_i) + V_{1,i} - U_{1,i},\ i \in \{1, \ldots, n_1\} \\ Y_i &= \phi(X_i) + V_{2,i} - U_{2,i},\ i \in \{n_1 + 1, \ldots, n_1 + n_2\}, \end{aligned}$$

where $\phi(x) = 30x - 9x^2$, $X \sim U(0,1)$, $U_{l,i} \sim N_+(0, \sigma_{l,u}^2)$, and $V_{l,i}$ follow the truncated normal distribution with mean 0 and variance $\sigma_{l,v}^2$, which lies within $(-6\sigma_{l,v}, 6\sigma_{l,v})$, $l = 1, 2$. Here, N_+ stands for a normal distribution limited to the domain $[0, \infty)$. As for $V_{l,i}$, we try two cases to reflect both the equal and unequal error variances between groups. We set $\sigma_{1,v} = \sigma_{2,v} = 1$ for the equal error variance case and $\sigma_{1,v} = \sqrt{2}$, $\sigma_{2,v} = 1$ for the unequal variance case. To evaluate the type I error rate and power, we again consider two cases based on whether a mean difference (β_0) exists or does not exist between group inefficiencies: $\sigma_{1,u} = \sigma_{2,u} = 1$ ($\beta_0 = 0$) and $\sigma_{1,u} = \sqrt{2}$, $\sigma_{2,u} = 1$ ($\beta_0 = 0.3305$). Here, the type I error rate implies the rate of supporting group difference in mean inefficiency when there is no difference and the power means the rate of supporting group difference in mean inefficiency when there are really inefficiency differences between groups.

We consider three sample sizes, n = 100, 200, and 400; the proportion of each group is approximately 50% and number of replications is 1000. For a comparison, we report the type I error and power of the following five tests with significance level $\alpha = 0.05$: our proposed test (PT), the OLS test, the T-test, the Mann-Whitney (MW) test, the Kolmogorov-Smirnov (KS) test, and the F-test. The last five tests are from Banker et al. [9]. We used a plug-in principle (see Ruppert et al. [15]) to find the bandwidth for our PT.

The test results are depicted in Table 1. Four of these tests, that is, except the KS test and the F-test, seem to respect the significance level in both the equal and unequal variance cases. However, the KS test obviously shows a larger type I error rate than expected for unequal error variances and the F-test seems to be a conservative test, which gives much smaller type I error probabilities than expected. As regards the power, our PT performs best, with the largest power among all the tests. In unequal variance cases where $n = 200, 400$, the KS test has larger power than our PT. However, the KS test is not reliable since it tends to reject the null hypothesis too easily in those cases. Finally, all tests tend to show higher power with larger sample sizes.

Table 1. Type I error and power of the single input case with equal and unequal error variances.

		Type I Error (Rejection Rate When β_0 = 0)						Power (Rejection Rate When β_0 = 0.3305)					
Variances	n	**PT**	**OLS**	**T**	**MW**	**KS**	**F**	**PT**	**OLS**	**T**	**MW**	**KS**	**F**
Equal ($\sigma_{1,v} = \sigma_{2,v}$)	100	0.052	0.050	0.050	0.050	0.037	0.013	0.377	0.336	0.320	0.283	0.200	0.152
	200	0.062	0.058	0.056	0.062	0.047	0.006	0.595	0.555	0.547	0.483	0.397	0.300
	400	0.063	0.064	0.062	0.061	0.052	0.003	0.848	0.822	0.818	0.761	0.665	0.533
Unequal ($\sigma_{1,v} \neq \sigma_{2,v}$)	100	0.047	0.065	0.060	0.050	0.062	0.037	0.337	0.337	0.328	0.276	0.277	0.279
	200	0.053	0.064	0.063	0.052	0.099	0.029	0.505	0.468	0.463	0.388	0.521	0.401
	400	0.044	0.051	0.050	0.048	0.179	0.026	0.738	0.722	0.712	0.645	0.836	0.650

3.2. *Multiple Input Case*

We next consider a multiple input case with $p = 3$. All the components in the simulation, except for the frontier function ϕ, are the same as in the single input case. As regards the frontier function, we consider two scenarios: the production function has an additive form, and the production function does not have an additive form. The additive assumption on the production function is used in Ferrara and Vidoli [16], but it may not be satisfied in some cases. However, in case of multiple covariates, the practical applicability of our proposed method may become worse since it requires multivariate smoothing and therefore suffers from the well-known "curse of dimensionality" problem as the dimension of the covariates becomes higher. In this case, additive modeling can be a meaningful alternative. For this, we try to estimate the difference in group efficiencies and test whether it is zero with an alternative estimating strategy, where we employ a backfitting procedure, which is a well-known estimating approach under the additive assumption. See Appendix B for details of the alternative method. Considering these two scenarios (additive and non-additive production functions), we examine how our PT(n) and its alternative based on the additive assumption, PT(a), behave depending on the validity of the assumption. The model considered here is

$$\begin{aligned} Y_i &= \phi(\mathbf{X}_i) + V_{1,i} - U_{1,i},\ i \in \{1, \ldots, n_1\} \\ Y_i &= \phi(\mathbf{X}_i) + V_{2,i} - U_{2,i},\ i \in \{n_1 + 1, \ldots, n_1 + n_2\}, \end{aligned}$$

where $X_{j,i}$, $j = 1, 2, 3$, are generated from $U(0,1)$ independently. For the first scenario, we set $\phi(\mathbf{x}) = (30x_1 - 9x_1^2) + (5 + 2\arctan(10(x_2 - 0.5))) + (4\sqrt{x_3})$; this has an additive structure. For the second scenario, we consider $\phi(\mathbf{x}) = (4\sqrt{x_1} + 7\sqrt{x_2} + 5\sqrt{x_3} + 8\sqrt{x_1 x_2} + 10\sqrt{x_2 x_3} + 9\sqrt{x_1 x_3})^{1.1}$. Note that both these production functions are concave. For PT(n), we select bandwidths by a generalized cross validation method (see Hastie and Tibshirani [17]), and for PT(a), we adopt a plug-in principle, as in the single input case.

The performances of PT(n) and PT(a) as well as the other five tests are reported in Tables 2 and 3. From Table 2, our PT(n) outperforms the competitors overall, especially in the unequal variances case. Note that its type I error rates do not deviate much from 0.05, which means that the type I error rate is under control as desired; those of other competitors such as the OLS, T and KS tests tend to be a bit smaller than this level in case of equal variances, and considerably larger in case of unequal variances. The MW test seems to respect the level like PT(n) but PT(n) turns to be more powerful than MW. Note that PT(a) shows good results in terms of type I error rate, with comparable power to the MW test. Its power is lower than that of PT(n), but this is natural since the true production function is not additive. The F-test seems to be anticonservative leading to too high of a type I error probability, especially in the unequal variances case. However, in the equal variance case, the F-test shows shows very good performance in terms of type I error rate and power in large samples, as reported in Banker et al. [9]. From Table 3, our proposed two tests outperform their competitors when the additive assumption is true. Note that under the additive assumption, both PT(a) and PT(n) correctly specify the model. From our simulation, PT(a) slightly outperforms PT(n), since their type I error rates are close to 0.05 and the power of PT(a) is larger than that of PT(n). In case of equal variances, the type I error rates of the five competitors tend to be below 0.05, but when the variances are unequal, their power becomes much lower than our proposed tests although overall they show satisfactory type I errors. The only exception is the F-test. It shows the largest power in the unequal variances case but such merit is dimmed by considerably larger type I errors than other tests.

Table 2. Type I error and power of the multiple input case with equal and unequal error variances when the true production function is not additive.

		Type I Error (Rejection Rate When $\beta_0 = 0$)							Power (Rejection Rate When $\beta_0 = 0.3305$)						
Variances	n	PT(a)	PT(n)	OLS	T	MW	KS	F	PT(a)	PT(n)	OLS	T	MW	KS	F
Equal	100	0.066	0.053	0.044	0.043	0.054	0.040	0.141	0.220	0.294	0.234	0.235	0.170	0.133	0.492
$(\sigma_{1,v} = \sigma_{2,v})$	200	0.059	0.077	0.049	0.050	0.044	0.035	0.113	0.328	0.488	0.473	0.471	0.356	0.268	0.644
	400	0.048	0.056	0.039	0.039	0.037	0.031	0.064	0.509	0.765	0.705	0.706	0.580	0.517	0.815
Unequal	100	0.061	0.063	0.101	0.099	0.054	0.046	0.316	0.199	0.258	0.346	0.340	0.173	0.143	0.649
$(\sigma_{1,v} \neq \sigma_{2,v})$	200	0.058	0.074	0.139	0.137	0.062	0.053	0.381	0.302	0.433	0.583	0.580	0.347	0.331	0.849
	400	0.048	0.060	0.128	0.130	0.046	0.093	0.385	0.468	0.682	0.793	0.792	0.530	0.650	0.947

Table 3. Type I error and power of the multiple input case with equal and unequal error variances when the true production function is additive.

		Type I Error (Rejection Rate When $\beta_0 = 0$)							Power (Rejection Rate When $\beta_0 = 0.3305$)						
Variances	n	PT(a)	PT(n)	OLS	T	MW	KS	F	PT(a)	PT(n)	OLS	T	MW	KS	F
Equal	100	0.054	0.072	0.046	0.045	0.041	0.034	0.065	0.364	0.328	0.213	0.208	0.170	0.132	0.279
$(\sigma_{1,v} = \sigma_{2,v})$	200	0.066	0.060	0.063	0.061	0.056	0.045	0.045	0.578	0.551	0.387	0.384	0.318	0.258	0.403
	400	0.054	0.055	0.034	0.034	0.033	0.033	0.018	0.820	0.801	0.572	0.571	0.503	0.419	0.517
Unequal	100	0.057	0.070	0.066	0.062	0.044	0.048	0.144	0.301	0.295	0.236	0.228	0.153	0.121	0.381
$(\sigma_{1,v} \neq \sigma_{2,v})$	200	0.068	0.060	0.089	0.089	0.054	0.060	0.137	0.501	0.481	0.415	0.414	0.281	0.290	0.566
	400	0.057	0.054	0.058	0.058	0.032	0.056	0.096	0.719	0.695	0.588	0.586	0.446	0.535	0.720

4. Application to PISA 2015 Data

In this section, we applied our PT to PISA 2015 data and test the efficiency difference between male and female student groups at a specific age in terms of learning time (X_i) and achievement (Y_i) in mathematics. The data can be downloaded from http://www.oecd.org/pisa/data/. PISA is a worldwide study to evaluate educational systems by measuring the scholastic performance of 15-year-old school students in mathematics, science, and reading. We considered the regional averages of the students' learning time and achievement in mathematics based on test results of the 2015 version as production data. Out of the 103 regions in the data, two regions, Nova Scotia in Canada and Chile, were excluded from our analysis in view of their outlier characteristics in efficiency analysis.

Usually, international large-scale assessments data include measurement errors at the individual as well as group level. Therefore, we considered the following stochastic frontier model for such data:

$$\begin{aligned} \text{(Male Students) } Y_i &= \phi(X_i) + V_i^{male} - U_i^{male},\ i \in \{1, \dots, 101\} \\ \text{(Female Students) } Y_i &= \phi(X_i) + V_i^{female} - U_i^{female},\ i \in \{102, \dots, 202\}. \end{aligned}$$

In this model, we assumed that there would be no gender difference in learning ability from a biological point of view and use the same production frontier for both gender group. It means that all the socio-economic characteristics of differentiation between the gender groups were in the random error terms and not introduced in the frontier function.

Table 4 shows summary statistics of each student group data. We applied the six tests in Section 3 to the data and calculated the p-values for the following hypothesis testing:

$$H_0 : E(U^{male}) = E(U^{female}) \text{ vs. } H_1 : E(U^{male}) < E(U^{female})$$

Table 4. Summary statistics of our PISA 2015 data.

		min	**Q_1**	**median**	**mean**	**Q_3**	**max**
male	X_i	27.89	39.52	41.72	43.10	47.81	56.70
	Y_i	338.5	470.6	499.7	483.9	513.8	565.6
female	X_i	25.23	38.99	41.49	41.98	45.26	56.67
	Y_i	339.0	456.9	487.9	474.8	501.9	565.0

From Samuelsson and Samuelsson [18], it is known that male students are often more involved in mathematics classes than female students. Additionally, since women are more involved in domestic chores than men and for men time is often made free by their families and relatives for the learning activity, male students are likely to be in an environment where they can focus more on studying than female students. Hence, we expected the results of the test to indicate that the effectiveness of male students was greater than that of female students in average.

Table 5 gives the test results. At a significance level of around $\alpha = 0.05$, our PT, the MW test, and the KS test (with p-value slightly higher that $\alpha = 0.05$) supported the hypothesis that on average male students are more efficient in mathematics than female students. However, the OLS test, the T-test and the F-test reported no significant difference in learning efficiency between the two groups. The reason for this could be the somewhat restrictive assumptions for test validity. Thus, the three tests seem to face the risk of unreasonable results if the assumptions are not satisfied in practice, but our PT does not seem to suffer from this problem.

Table 5. P-values of the six tests to detect efficiency difference in groups of male and female students in terms of learning time and achievement in mathematics.

test	PT	OLS	T	MW	KS	F
p-value	0.049	0.150	0.150	0.044	0.057	0.322

5. Discussion and Conclusions

In this study, we developed tests with sound statistical theory for group efficiency comparison under SFM with considerably better performance than the previous tests proposed in numerical simulations. However, there is still room for improvement in our methods.

First, since we perform full nonparametric modeling for the frontier function $\phi(\cdot)$, which can be multivariate, our test might suffer from the "curse of dimensionality" and require high-order kernels for implementation with four or more input variables. In such a situation, we can consider an alternative test with spirit as in our test when the frontier function $\phi(\mathbf{X})$ has an additive structure, that is, $\phi(\mathbf{X}) = \sum_{j=1}^{p} \phi_j(X_j)$, or could be well-approximated by it.

Second, we only deal with one output case, which limits practical applications. Our methods should be extended to cover the case of multi-output production frontiers, which DEA methods cover.

Third, we assume the same production frontier for both group, which is a clear limitation in practice since such situation is not frequently observed. If it is important to assume separate production frontier functions for different groups, one can use the meta-frontier approach. O'Donnell et al. [6] proposed a meta-frontier approach to compare the group technical efficiencies under stochastic frontier framework. The proposed method has the advantage that it can be used without assuming a common frontier function. However, the use of their method sometimes can be restricted by their assumption that the frontier production function is log-linear.

Finally, if one is interested in estimating the mean inefficiency of each group, we refer to Noh and Van Keiligom [19], which is a recent work along that direction.

Author Contributions: Conceptualization, H.N.; methodology, H.N. and S.J.Y.; software, H.N. and S.J.Y.; formal analysis, H.N. and S.J.Y.; investigation, H.N. and S.J.Y.; writing–original draft preparation, H.N. and S.J.Y.; writing–review and editing, H.N. and S.J.Y.. All authors have read and agree to the published version of the manuscript.

Appendix A

In this appendix, we provide details of the asymptotic normality of the proposed estimator $\hat{\beta}_0$ in Section 2.2. For this, we first list the relevant assumptions.

Assumption

1. The kernel function K is symmetric, and Lipschitz continuous in $[-1,1]$.
2. ϕ is twice partially continuously differentiable.
3. The density functions of X_j ($j \in \{1,\ldots,p\}$) are continuous, and bounded away from zero and infinity on their supports $\mathcal{C}_j$, which are bounded.
4. V_1, V_2, U_1 and U_2 have finite second moments.
5. For $j \in \{1,\ldots,p\}$, h_j are asymptotic to n^{-a} for $a > 0$ such that $n(\prod_{j=1}^{p} h_j)^2/\log n \to \infty$ and $nh_j^8 \to 0$ as n goes to infinity.

Theorem A1. *Under the above assumptions,*

$$\sqrt{n}(\hat{\beta}_0 - \beta_0) \longrightarrow N(0, \sigma^2)$$

where

$$\begin{aligned} \sigma^2 &= E(\tilde{T}^2)^{-2} E(\varepsilon^{*2}\tilde{T}^2) \\ \tilde{T} &= T - E(T|\mathbf{X}) \end{aligned}$$

Proof. We write

$$\begin{aligned} \sqrt{n}(\hat{\beta}_0 - \beta_0) &= \left[n^{-1}\mathbf{T}^\top(\mathbf{I}-\mathbf{S})^\top(\mathbf{I}-\mathbf{S})\mathbf{T}\right]^{-1} \\ &\quad \times n^{-1/2}\left[\mathbf{T}^\top(\mathbf{I}-\mathbf{S})^\top(\boldsymbol{\varepsilon}^* + \boldsymbol{\phi}^* - \mathbf{S}(\mathbf{Y}-\beta_0\mathbf{T}))\right] \end{aligned}$$

where $\boldsymbol{\phi}^* = (\phi^*(X_1), \ldots, \phi^*(X_n))^\top$ and $\boldsymbol{\varepsilon}^* = (\varepsilon_1^*, \ldots, \varepsilon_n^*)^\top$. It suffices to show that

$$n^{-1}\mathbf{T}^\top(\mathbf{I}-\mathbf{S})^\top(\mathbf{I}-\mathbf{S})\mathbf{T} \xrightarrow{p} E(\tilde{T}^2), \text{ and} \tag{A1}$$

$$n^{-1/2}\left[\mathbf{T}^\top(\mathbf{I}-\mathbf{S})^\top(\varepsilon + \boldsymbol{\phi}^* - \mathbf{S}(\mathbf{Y}-\beta_0\mathbf{T}))\right] \xrightarrow{d} N(0, E(\epsilon^*)\tilde{T}^2). \tag{A2}$$

To prove these, we first give the following fact.

$$\sup_{\mathbf{x}\in C_1\times\cdots\times C_p} |\hat{\xi}(\mathbf{x}) - \xi(\mathbf{x})| = O_p(n^{-2a} + n^{-(1-ap)/2}\log n), \tag{A3}$$

where $\xi(\mathbf{x}) = E(R|\mathbf{X} = \mathbf{x})$ and $\hat{\xi}(\mathbf{x})$ is its local linear estimator. That is, $\hat{\xi}(\mathbf{x}) = \mathbf{s}_\mathbf{x}^\top \mathbf{R}$ with $\mathbf{R} = (R_1, \ldots, R_n)^\top$ when R_j is a response variable. (A3) can be shown from the standard theory of kernel smoothing. Note that

$$(\mathbf{I}-\mathbf{S})\mathbf{T} = (\mathbf{T} - E(\mathbf{T}|\mathbf{X})) + (E(\mathbf{T}|\mathbf{X}) - \mathbf{ST}).$$

The second term of the right-hand side of the above equation is $o_p(1)$ from (A3). This proves (A1). Next, we write

$$\begin{aligned} & n^{-1/2}\left[\mathbf{T}^\top(\mathbf{I}-\mathbf{S})^\top(\boldsymbol{\varepsilon}^* + \boldsymbol{\phi}^* - \mathbf{S}(\mathbf{Y}-\beta_0\mathbf{T}))\right] \\ = \; & n^{-1/2}(\mathbf{T} - E(\mathbf{T}|\mathbf{X}))^\top\boldsymbol{\varepsilon}^* + n^{-1/2}(\mathbf{T} - E(\mathbf{T}|\mathbf{X}))^\top(\boldsymbol{\phi}^* - \mathbf{S}(\mathbf{Y}-\beta_0\mathbf{T})) \\ & + n^{-1/2}(E(\mathbf{T}|\mathbf{X}) - \mathbf{ST})^\top\boldsymbol{\varepsilon}^* + n^{-1/2}(E(\mathbf{T}|\mathbf{X}) - \mathbf{ST})^\top(\boldsymbol{\phi}^* - \mathbf{S}(\mathbf{Y}-\beta_0\mathbf{T})) \\ \equiv \; & n^{-1/2}(\mathbf{T} - E(\mathbf{T}|\mathbf{X}))^\top\boldsymbol{\varepsilon}^* + \mathbf{A}_{1,n} + \mathbf{A}_{2,n} + \mathbf{A}_{3,n}. \end{aligned}$$

Since $n^{-1/2}(\mathbf{T} - E(\mathbf{T}|\mathbf{X}))^\top\boldsymbol{\varepsilon}^* = N(0, E(\epsilon^*)\tilde{T}^2) + o_p(1)$, it is enough to show that $\mathbf{A}_{j,n} = o_p(1)$, $j = 1, 2, 3$, to claim (A2).

To treat $A_{1,n}$, we note that

$$\sup_{x_j\in C_j}\left|\frac{\partial}{\partial x_j}\hat{\xi}(\mathbf{x}) - \frac{\partial}{\partial x_j}\xi(\mathbf{x})\right| = O_p(n^{-2a} + n^{-(1-ap-2a)/2}\log n),\; j \in \{1, \ldots, p\} \tag{A4}$$

and denote $\hat{\boldsymbol{\phi}}^* = \mathbf{S}(\mathbf{Y} - \beta_0\mathbf{T})$. Then,

$$\mathbf{A}_{1,n} = n^{-1/2}\sum_{i=1}^{n}(T_i - E(T_i|\mathbf{X}_i))(\phi^*(\mathbf{X}_i) - \hat{\phi}^*(\mathbf{X}_i)).$$

Let $\mathcal{G}$ denote a class of functions satisfying $|g(\mathbf{x}) - g(\mathbf{y})| \le \|\mathbf{x} - \mathbf{y}\|$ for $\mathbf{x}, \mathbf{y} \in [0,1]^p$. Then, $n^{a_0}(\hat{\phi}^*(\cdot) - \phi^*(\cdot))$ belongs to $\mathbf{G}$ with probability tending to 1 from (A3) and (A4), where $a_0 < \max\{2a, (1 - ap - 2a)/2\}$. We can show that the δ-entropy of $\mathbf{G}$ for the supremum norm satisfies

$$H_\infty(\delta, \mathcal{G}) \le K\left(\log\frac{1}{\delta} + \frac{1}{\delta^p}\right)$$

for some constant K. Here, we consider an empirical process

$$n^{-1/2}\sum_{i=1}^{n}(T_i - E(T_i|\mathbf{X}_i))g(\mathbf{X}_i),\ g \in \mathcal{G}$$

indexed by $\mathcal{G}$. Then, $E[(T_i - E(T_i|\mathbf{X}_i))g(\mathbf{X}_i)] = 0$, and by the Corollary 8.8 of van de Geer [20], we conclude that $\sup_{g\in\mathcal{G}}\left|n^{-1/2}\sum_{i=1}^{n}(T_i - E(T_i|\mathbf{X}_i))g(\mathbf{X}_i)\right| = O_p(1)$, to result in $\mathbf{A}_{1,n} = o_p(1)$. Note that the exponential tail condition, required to apply the empirical process technique, is automatically satisfied in our case since T is a binary variable.

As for $\mathbf{A}_{2,n}$, we first note that $E(\mathbf{A}_{2,n}|(T_1,\mathbf{X}_1),\ldots,(T_n,\mathbf{X}_n)) = 0$. Moreover,

$$\begin{aligned} & E(\mathbf{A}_{2,n}^2|(T_1,\mathbf{X}_1),\ldots,(T_n,\mathbf{X}_n)) \\ \le\ & n^{-1}[var(\varepsilon_1^{1,*}) + var(\varepsilon_1^{2,*})](E(\mathbf{T}|\mathbf{X}) - \mathbf{ST})^\top(E(\mathbf{T}|\mathbf{X}) - \mathbf{ST}) \\ =\ & O_p(n^{-4a} + n^{-(1-ap)}\log n) \end{aligned}$$

from (A3). This establishes $\mathbf{A}_{2,n} = o_p(1)$. Finally, $\mathbf{A}_{3,n} = O_p(n^{1/2-4a} + n^{-1/2+ap/2}\log n) = o_p(1)$ from (A3), to complete the proof. □

Appendix B

In this appendix, we describe an alternative test for (2) when the frontier function $\phi(\mathbf{X})$ has an additive structure, that is, $\phi(\mathbf{X}) = \sum_{j=1}^{p}\phi_j(X_j)$. Under the additive structure assumption, model (1) can be written as two nonparametric mean regression models:

$$\begin{aligned} Y_i &= \left[\sum_{j=1}^{p} E\phi_j(X_{j,i}) - E(U_1)\right] + \left[\sum_{j=1}^{p}(\phi_j(X_{j,i}) - E\phi_j(X_j))\right] + [V_{1,i} - (U_{1,i} - E(U_1))] \\ &\equiv \mu + \sum_{j=1}^{p}\phi_j^*(X_{j,i}) + \varepsilon_{1,i}^*,\ i \in \{1,\ldots,n_1\}; \end{aligned} \quad \text{(A5)}$$

$$\begin{aligned} Y_i &= \left[\sum_{j=1}^{p} E\phi_j(X_{j,i}) - E(U_1)\right] + [E(U_1) - E(U_2)] + \left[\sum_{j=1}^{p}(\phi_j(X_{j,i}) - E\phi_j(X_j))\right] \\ &\quad + [V_{2,i} - (U_{2,i} - E(U_2))] \\ &\equiv \mu + \beta_0 + \sum_{j=1}^{p}\phi_j^*(X_{j,i}) + \varepsilon_{2,i}^*,\ i \in \{n_1+1,\ldots,n_1+n_2\}, \end{aligned} \quad \text{(A6)}$$

where $\phi_j^*(X_j) = \phi_j^*(X_j) - E\phi_j(X_j)$ and $E(\phi^*(X_j)) = 0$ for $j \in \{1,\ldots,p\}$. If we introduce the same dummy variable T_i as in the single input case, models (A5) and (A6) can be integrated into one single semiparametric regression model, which would be a (heteroscedastic) partial linear additive model:

$$Y_i = \mu + \beta_0 T_i + \sum_{j=1}^{p}\phi_j^*(X_{j,i}) + \varepsilon_i^*,\ i \in \{1,\ldots,n\}, \quad \text{(A7)}$$

where $\varepsilon_i^* = (1 - I(T_i = 1))\varepsilon_{1,i}^* + I(T_i = 1)\varepsilon_{2,i}^*$ and $E(\varepsilon_i^*|T_i,\mathbf{X}_i) = 0$. Partial linear additive models have been studied by several authors; for example, Fan et al. [21], Fan and Li [22], Li [23], and Wei and Liu [24]. For the test, we use the profile least square estimator of β_0 in Wei and Liu [24]. However, Wei and Liu [24] only showed the asymptotic distribution of the estimator of the parametric component vector (in our case, the estimator of $\boldsymbol{\beta} = (\mu, \beta_0)^\top$) under the homoscedasticity assumption of the error (Theorem 2.1 of their paper), and so we extended their result to the heteroscedasticity case.

To introduce the profile least square estimator of β_0 using the method of Wei and Liu [24], we define some notations. Let

$$\mathbf{X}_{des} = \begin{bmatrix} 1 & X_1 \\ \vdots & \vdots \\ 1 & X_n \end{bmatrix} = [\mathbf{1}_n\ \mathbf{X}], \mathbf{S}_T = \begin{bmatrix} \mathbf{I}_n & \mathbf{S}_1^* & \cdots & \mathbf{S}_1^* \\ \mathbf{S}_2^* & \mathbf{I}_n & \cdots & \mathbf{S}_2^* \\ \vdots & \vdots & \ddots & \vdots \\ \mathbf{S}_p^* & \mathbf{S}_1^* & \cdots & \mathbf{I}_n \end{bmatrix}, \mathbf{C} = \begin{bmatrix} \mathbf{S}_1^* \\ \mathbf{S}_2^* \\ \vdots \\ \mathbf{S}_p^* \end{bmatrix}, \tag{A8}$$

where $\mathbf{S}_k$ is the smoothing matrix for local linear regression with respect to the jth ($j \in \{1, \ldots, p\}$) covariate vector $\mathbf{X}_j = (X_{j,1}, \ldots, X_{j,n})^\top$ with kernel function $K(\cdot)$ and bandwidth h_j, $\mathbf{S}_j^* = (\mathbf{I}_n - \mathbf{1}_n\mathbf{1}_n^\top/n)\mathbf{S}_j$, and $\mathbf{1}_n = (1, \ldots, 1)^\top$ with length n. Additionally, we define the additive smoother matrix $\mathbf{W}_j$ as $\mathbf{W}_j = \mathbf{E}_j\mathbf{S}_T^{-1}\mathbf{C}$, where $\mathbf{E}_j$ is a partitioned matrix of dimension $n \times np$ with $n \times n$ identity matrix as the jth "block" and zeros elsewhere. Then, the profile least squares estimator of $\boldsymbol{\beta} = (\mu, \beta_0)^\top$ is obtained as the estimator of the coefficient vector $\boldsymbol{\beta}$ of a synthetic linear regression model

$$Y_i - \tilde{Y}_i = (\mathbf{T}_{des,i} - \tilde{\mathbf{T}}_{des,i})^\top\boldsymbol{\beta} + \varepsilon_i, \tag{A9}$$

where $\mathbf{W}_M = \sum_{j=1}^p \mathbf{W}_j$, $\tilde{Y} = (\tilde{Y}_1, \ldots, \tilde{Y}_n)^\top = \mathbf{W}_M\mathbf{Y}$ and $\tilde{\mathbf{T}}_{des} = (\tilde{\mathbf{T}}_{des,1}, \ldots, \tilde{\mathbf{T}}_{des,n})^\top = \mathbf{W}_M\mathbf{T}_{des}$. Additionally, since $\mathbf{W}_M\mathbf{1}_n = (0, \ldots, 0)^\top$, we know that the linear model (A9) becomes

$$Y_i - \tilde{Y}_i = \mu + (T_i - \tilde{T}_i)\beta_0 + \varepsilon_i^*, \tag{A10}$$

where $\tilde{\mathbf{T}} = (\tilde{T}_1, \ldots, \tilde{T}_n)^\top = \mathbf{W}_M\mathbf{T}$. Hence, after a standard calculation in linear model theory, we obtain the profile least squares estimator of β_0 as

$$\hat{\beta}_0 = \left[\mathbf{T}^\top(\mathbf{I}_n - \mathbf{W_M})^\top(\mathbf{I}_n - \mathbf{J})(\mathbf{I}_n - \mathbf{W_M})\mathbf{T}\right]^{-1}\mathbf{T}^\top(\mathbf{I}_n - \mathbf{W_M})^\top(\mathbf{I}_n - \mathbf{J})(\mathbf{I}_n - \mathbf{W_M})\mathbf{Y}, \tag{A11}$$

where $\mathbf{J} = \mathbf{1}_n\mathbf{1}_n^\top/n$. Using the results to prove Theorem 2.1 in Wei and Liu [24], we show below that under some regularity conditions, estimator $\hat{\beta}_0$ is asymptotically normal with mean zero and variance

$$\sigma_{add}^2 = E(\tilde{T}^2)^{-2}E(\varepsilon^{*2}\tilde{T}^2), \tag{A12}$$

where $\tilde{T} = T - E(T) - \sum_{j=1}^p \left[E(T|X_j) - E(T)\right]$. Once we have a consistent estimate of σ_{add}^2, we can test (7) for a given significance level α. As with the case of single input variable, we can directly estimate the variance $\sigma_{\beta_0}^2$ using estimates $\hat{\varepsilon}_i^*$ and $\hat{E}(T|X_{j,i})$. Here, $\hat{E}(T|X_{j,i})$ can be obtained as the ith element of $\mathbf{S}_j(T_1, \ldots, T_n)^\top$. Alternatively, we can estimate the variance via the sandwich formula estimate based on (A11) following similar steps in Section 2.2.

Now, we can show the asymptotic property of the profile least square estimator of β_0. For this, we first list the relevant assumptions.

Assumption

1. The kernel function K is symmetric, and Lipschitz continuous in $[-1, 1]$.
2. ϕ_j ($j \in \{1, \ldots, p\}$) are twice continuously differentiable.
3. The density functions of X_j ($j \in \{1, \ldots, p\}$) are continuous, and bounded away from zero and infinity on their supports, which are bounded.
4. V_1, V_2, U_1 and U_2 have finite second moments.
5. For $j \in \{1, \ldots, p\}$, $h_j \to 0$, $nh_j/\log n \to \infty$ and $nh_j^8 \to 0$ as n goes to infinity.

Theorem A2. *Under the above assumptions,*

$$\sqrt{n}(\hat{\beta}_0 - \beta_0) \longrightarrow N(0, \sigma^2_{add})$$

where

$$\begin{aligned} \sigma^2_{add} &= E(\tilde{T}^2)^{-2}E(\varepsilon^{*2}\tilde{X}^2) \\ \tilde{T} &= T - E(T) - \sum_{j=1}^{p}\left[E(T|X_j) - E(T)\right] \end{aligned}$$

Proof. $\hat{\beta}_0$ can be expressed as follows:

$$\hat{\beta}_0 = \left[\mathbf{T}^\top(\mathbf{I}_n - \mathbf{W_M})^\top(\mathbf{I}_n - \mathbf{J})(\mathbf{I}_n - \mathbf{W_M})\mathbf{T}\right]^{-1}\mathbf{T}^\top(\mathbf{I}_n - \mathbf{W_M})^\top(\mathbf{I}_n - \mathbf{J})(\mathbf{I}_n - \mathbf{W_M})\mathbf{Y}$$

where $\mathbf{T} = (T_1, \ldots, T_n)^\top$ and $\mathbf{J} = \mathbf{1}_n\mathbf{1}_n^\top/n$. Then,

$$\begin{aligned} &\sqrt{n}(\hat{\beta}_0 - \beta_0) \\ = &\left[n^{-1}\mathbf{T}^\top(\mathbf{I}_n - \mathbf{W_M})^\top(\mathbf{I}_n - \mathbf{J})(\mathbf{I}_n - \mathbf{W_M})\mathbf{T}\right]^{-1} \\ &\times n^{-1/2}\mathbf{T}^\top(\mathbf{I}_n - \mathbf{W_M})^\top(\mathbf{I}_n - \mathbf{J})(\mathbf{I}_n - \mathbf{W_M})(\boldsymbol{\phi}^* + \boldsymbol{\varepsilon}^*), \end{aligned}$$

where $\boldsymbol{\phi}^* = (\sum_{j=1}^{p}\phi^*(X_{j,1}), \ldots, \sum_{j=1}^{p}\phi^*(X_{j,n}))^\top$ and $\boldsymbol{\varepsilon}^* = (\varepsilon_1^*, \ldots, \varepsilon_n^*)^\top$. Here, the term associated with the intercept μ vanishes because $\mathbf{W_M}\mathbf{1}_n = (0, \ldots, 0)^\top$. To prove the theorem, it suffices to show that

$$\begin{aligned} &n^{-1}\mathbf{T}^\top(\mathbf{I}_n - \mathbf{W_M})^\top(\mathbf{I}_n - \mathbf{J})(\mathbf{I}_n - \mathbf{W_M})\mathbf{T} \\ = &n^{-1}\sum_{i=1}^{n}\left[T_i - E(T_i) - \sum_{j=1}^{p}\left[E(T_i|X_{j,i}) - E(T_i)\right]\right]^2 + o_p(1) \end{aligned} \quad \text{(A13)}$$

and

$$\begin{aligned} &n^{-1/2}\mathbf{T}^\top(\mathbf{I}_n - \mathbf{W_M})^\top(\mathbf{I}_n - \mathbf{J})(\mathbf{I}_n - \mathbf{W_M})(\boldsymbol{\phi}^* + \boldsymbol{\varepsilon}^*) \\ = &n^{-1/2}\sum_{i=1}^{n}\left[T_i - E(T_i) - \sum_{j=1}^{p}\left[E(T_i|X_{j,i}) - E(T_i)\right]\right]\varepsilon_i^* + o_p(1). \end{aligned} \quad \text{(A14)}$$

Note that $(\mathbf{I}_n - \mathbf{J})(\mathbf{I}_n - \mathbf{W_M})\mathbf{X} = (\mathbf{I}_n - \mathbf{W_M})\mathbf{X} - \mathbf{1}_n\bar{X}$, where $\bar{X} = n^{-1}\sum_{i=1}^{n} X_i$ because fact $\mathbf{1}_n^\top\mathbf{W_M} = (0, \ldots, 0)$. Then, one can easily see that

$$\begin{aligned} &n^{-1}\mathbf{X}^\top(\mathbf{I}_n - \mathbf{W_M})^\top(\mathbf{I}_n - \mathbf{J})(\mathbf{I}_n - \mathbf{W_M})\mathbf{T} \\ = &n^{-1}(\mathbf{T} - \mathbf{1}_n\mu_X)^\top(\mathbf{I}_n - \mathbf{W_M})^\top(\mathbf{I}_n - \mathbf{W_M})(\mathbf{T} - \mathbf{1}_n\mu_T) + O_p(n^{-1/2}) \end{aligned}$$

for $\mu_T = E(T_1)$. Therefore, Equation (A13) can be verified as in the proof of Lemma 6.2 in Wei and Liu (2012). For Equation (A14), note that

$$\begin{aligned} &n^{-1/2}\mathbf{T}^\top(\mathbf{I}_n - \mathbf{W_M})^\top(\mathbf{I}_n - \mathbf{J})(\mathbf{I}_n - \mathbf{W_M})\boldsymbol{\phi}^* \\ = &n^{-1/2}(\mathbf{T} - \mathbf{1}_n\mu_T)^\top(\mathbf{I}_n - \mathbf{W_M})^\top(\mathbf{I}_n - \mathbf{W_M})(\boldsymbol{\phi}^* - \mathbf{1}_n\mu_{\phi^*}) + O_p(n^{-1/2}) \end{aligned} \quad \text{(A15)}$$

and

$$\begin{aligned} &n^{-1/2}\mathbf{T}^\top(\mathbf{I}_n - \mathbf{W_M})^\top(\mathbf{I}_n - \mathbf{J})(\mathbf{I}_n - \mathbf{W_M})\boldsymbol{\varepsilon}^* \\ = &n^{-1/2}(\mathbf{T} - \mathbf{1}_n\mu_X)^\top(\mathbf{I}_n - \mathbf{W_M})^\top(\mathbf{I}_n - \mathbf{W_M})\boldsymbol{\varepsilon}^* + O_p(n^{-1/2}), \end{aligned} \quad \text{(A16)}$$

where $\mu_{\phi^*} = E(\sum_{j=1}^{p} \phi^*(X_{j,1}))$. Then, we have

$$(\mathbf{I}_n - \mathbf{W_M})\boldsymbol{\varepsilon}^* = \boldsymbol{\varepsilon}^* - \sum_{j=1}^{p} \mathbf{S}_j \boldsymbol{\varepsilon}^* + O_p\left(n \sum_{j=1}^{p} h_j^4\right)$$

from Lemma B.6 in [25]. Note that this is true as long as the conditional variance of the ε^* given covariates exists. This is guaranteed by assumption 4. Wei and Liu (2012) used a similar fact under the homoscedastic error assumption. Then, with a derivation similar to that in Lemma 6.3 of Wei and Liu (2012), we can show that (A15) converges to zero in probability and (A16) can be written as:

$$n^{-1/2} \sum_{i=1}^{n} \left[T_i - E(T_i) - \sum_{j=1}^{p} \left[E(T_i | X_{j,i}) - E(T_i) \right] \right] \varepsilon_i^* + o_p(1),$$

to complete the proof. □

References

1. Golany, B.; Storberg, J. A data envelopment analysis of the operational efficiencies of bank branches. *Interfaces* **1999**, *29*, 14–26. [CrossRef]
2. Lee, H.; Park, Y.; Choi, H. Comparative evaluation of performance of national R&D programs with heterogeneous objectives: A DEA approach. *Eur. J. Oper. Res.* **2009**, *196*, 847–855.
3. Cummins, J.D.; Weiss, M.A.; Zi, H. Organizational form and efficiency: The coexistence of stock and mutual property-liability insurers. *Manag. Sci.* **1999**, *45*, 1254–1269. [CrossRef]
4. Simar, L.; Zelenyuk, V. On testing equality of distributions of technical efficiency scores. *Econom. Rev.* **2006**, *25*, 497–522. [CrossRef]
5. Li, Q. Nonparametric testing of closeness between two unknown distribution functions. *Econom. Rev.* **1996**, *15*, 261–274. [CrossRef]
6. O'Donnell, C.J.; Rao, D.S.P.; Battese, G.E. Metafrontier frameworks for the study of firm-level efficiencies and technology ratios. *Empir. Econ.* **2008**, *34*, 231–255. [CrossRef]
7. Aigner, D.; Chu, S. On estimating the industry production function. *Am. Econ. Rev.* **1968**, *58*, 826–839.
8. Meeusen, W.; van den Broeck, J. Efficiency estimation from Cobb-Douglas production functions with composed error. *Int. Econ. Rev.* **1977**, *18*, 435–444. [CrossRef]
9. Banker, R.D.; Zheng, Z.; Natarajan, R. DEA-based hypothesis tests for comparing two groups of decision making units. *Eur. J. Oper. Res.* **2010**, *206*, 231–238. [CrossRef]
10. Simar, L.; Wilson, P.W. Two-stage DEA: Caveat emptor. *J. Product. Anal.* **2011**, *36*, 205–218. [CrossRef]
11. Banker, R.D.; Natarajan, R. Evaluating contextual variables sffecting productivity using data envelopment analysis. *Oper. Res.* **2008**, *56*, 48–58. [CrossRef]
12. Liang, H. Estimation in partially linear models and numerical comparisons. *Comput. Stat. Data Anal.* **2006**, *50*, 675–687. [CrossRef] [PubMed]
13. Ma, Y.; Chiou, J.M.; Wang, N. Efficient semiparametric estimator for heteroscedastic partially linear models. *Biometrika* **2006**, *93*, 75–84. [CrossRef]
14. Huang, J. A note on estimating a partly linear model under monotonicity constraints. *J. Stat. Plan. Inference* **2002**, *107*, 343–351. [CrossRef]
15. Ruppert, D.; Sheather, S.J.; Wand, M.P. An effective bandwidth selector for local least squares regression. *J. Am. Stat. Assoc.* **1995**, *90*, 1257–1270. [CrossRef]
16. Ferrara, G.; Vidoli, F. Semiparametric stochastic frontier models: A generalized additive model approach. *Eur. J. Oper. Res.* **2017**, *258*, 761–777. [CrossRef]
17. Hastie, T.J.; Tibshirani, R.J. *Generalized Additive Models*; Chapman and Hall/CRC: New York, NY, USA, 1990.
18. Samuelsson, M.; Samuelsson, J. Gender differences in boys' and girls' perception of teaching and learning mathematics. *Open Rev. Educ. Res.* **2016**, *3*, 18–34. [CrossRef]
19. Noh, H.; Van Keiligom, I. On relaxing the distributional assumption of stochastic frontier models. *J. Korean Stat. Soc.* **2020**, in press.

20. Van de Geer, S. *Empirical Processes in M-Estimation*; Cambridge University Press: Cambridge, UK, 2000.
21. Fan, J.; Hardle, W.; Mammen, E. Direct estimation of low-dimensional components in additive models. *Ann. Stat.* **1998**, *26*, 943–971. doi:10.1214/aos/1024691083. [CrossRef]
22. Fan, Y.; Li, Q. A kernel-based method for estimating additive partially linear models. *Stat. Sin.* **2003**, *13*, 739–762.
23. Li, Q. Efficient estimation of additive partially linear models. *Int. Econ. Rev.* **2000**, *41*, 1073–1092. [CrossRef]
24. Wei, C.H.; Liu, C. Statistical inference on semi-parametric partial linear additive models. *J. Nonparametric Stat.* **2012**, *24*, 809–823. [CrossRef]
25. Fan, J.; Jiang, J. Nonparametric inferences for additive models. *J. Am. Stat. Assoc.* **2005**, *100*, 890–907. [CrossRef]

10

Functional Location-Scale Model to Forecast Bivariate Pollution Episodes

Manuel Oviedo-de La Fuente [1], Celestino Ordóñez [2,*] and Javier Roca-Pardiñas [3]

1 Department of Statistics, Mathematical Analysis and Optimization, Universidad de Santiago de Compostela, 15782 Santiago de Compostela, Spain; manuel.oviedo@usc.es
2 Department of Mining Exploitation and Propsecting, Universidad de Oviedo, Escuela Politécnica de Mieres, 33600 Mieres, Spain
3 Department of Statistics and Operation Research, Universidad de Vigo, 36310 Vigo, Spain; roca@uvigo.es
* Correspondence: ordonezcelestino@uniovi.es

Abstract: Predicting anomalous emission of pollutants into the atmosphere well in advance is crucial for industries emitting such elements, since it allows them to take corrective measures aimed to avoid such emissions and their consequences. In this work, we propose a functional location-scale model to predict in advance pollution episodes where two pollutants are involved. Functional generalized additive models (FGAMs) are used to estimate the means and variances of the model, as well as the correlation between both pollutants. The method not only forecasts the concentrations of both pollutants, it also estimates an uncertainty region where the concentrations of both pollutants should be located, given a specific level of uncertainty. The performance of the model was evaluated using real data of SO_2 and NO_x emissions from a coal-fired power station, obtaining good results.

Keywords: pollution episodes; functional data; bivariate analysis; uncertainty region; generalized additive models

1. Introduction

Forecasting air quality and concentrations of pollutants in the atmosphere by means of statistical methods is an active area of research given the transcendence of the problem and the difficulty to find optimal solutions using deterministic mathematical models. Among the different methods that can be found in the literature to tackle this problem, models for time series analysis such as the integrated autoregressive moving average—ARIMA [1–3], multivariate regression [4–7], generalized linear or additive models (GAM) [8–11] and artificial neural networks (ANN) [12–19] are the most extended. Due to the increased access to continuous data over time, functional data analysis [20,21] was also proposed for air quality forecasting and outlier detection [22–24]. Parametric [25,26] and nonparametric [27–29] functional regression methods were tested. A functional framework allows considering the inherent correlation between observations, instead of considering them as independent realizations of an underlying stochastic process. Some functional approaches add related meteorological variables to the models [30–34], which can improve the result of the predictions and help to understand the process underlying the evolution of the pollutants.

Most of the documents in the literature propose solutions to predict the concentration of each pollutant individually, being much scarcer those focused on predicting more than one pollutant at a time. Vector autoregressive moving average (VARMA) [35,36] and vector autoregressive integrated moving average (VARIMA) [37] models were applied to reach this objective. In this work, we proposed a method for the simultaneous forecasting of pollution episodes when two pollutants, i.e., SO_2 and NO_x, are involved. Apart from transport, one of the main sources of these pollutants is public electricity and heating. Their negative effects on human health are well known, and goes for mild (i.e., eyes irritated, nose or headache) to severe (i.e., lung damage or reduced oxygenation of tissues). They also have negative effects on animals and plants, as well as in other substances, such as water and soils. In addition, NO_x is a precursor of the tropospheric ozone. High levels of ozone contributes to climate change, cause adverse impacts on health and can damage vegetation.

Pollution episodes (incidents) are abnormally large emissions of one or more pollutants in short periods of time. Although the improvement of the chemical processes and particle filter systems have significantly reduced the amount and intensity of the pollution episodes, they are still of particular interest for the industries, as they may be subject to sanctions, or for other reasons, such as public health deterioration or industry discredit. Therefore, pollution industries, such as coal-fired power plants, are very interested in determining in advance when these episodes of excessive contamination might occur. Specifically, this is the purpose of our work: forecasting pollution episodes of SO_2 and NO_x early enough to allow corrective measures to be taken. Our approach uses a location-scale model [11,38,39] that treats the predictors, the concentrations of both pollutants over time, as functions, while the response is a scalar, the concentration of the pollutants some time in advance. The novelty of our approach is the combination of a biviariate location-scale model with functional additive models. This method combines the simplicity of the location and scale models with the capacity of functional data analysis to deal with data in the form of functions.

The document is structured as follows: In Section 2 we show the mathematical model proposed to solve the problem under analysis and the algorithm used to estimate a solution from the data. Section 3 is devoted to test the validity of the model using real data. Finally, a discussion of the results and the main conclusions of our work are exposed in Section 4.

2. Methodology

2.1. Mathematical Model

Let $\{\mathbf{X}_i, \mathbf{Y}_i\}_{i=1}^n$ be a set of observations of a stochastic process, $\mathbf{X} = (X^1(t), \ldots, X^p(t))$, where $X^j(t) \in L_2[0,T], j = 1, \ldots, p$, are predictor covariates and $\mathbf{Y} = (Y_1, Y_2)$, with $Y_j \in \mathbb{R}$, a response variable. In this context, the following bivariate location-scale model [40,41] is assumed

$$\begin{pmatrix} Y_1 \\ Y_2 \end{pmatrix} = \begin{pmatrix} \mu_1(\mathbf{X}) \\ \mu_2(\mathbf{X}) \end{pmatrix} + \mathbf{\Sigma}^{1/2}(\mathbf{X}) \begin{pmatrix} \varepsilon_1 \\ \varepsilon_2 \end{pmatrix} \tag{1}$$

where $\mathbf{\Sigma}^{1/2}(\mathbf{X})$ represents the Cholesky decomposition of the variance-covariance matrix $\mathbf{\Sigma}(\mathbf{X})$

$$\mathbf{\Sigma}(\mathbf{X}) = \begin{pmatrix} \sigma_1^2(\mathbf{X}) & \sigma_{12}(\mathbf{X}) \\ \sigma_{12}(\mathbf{X}) & \sigma_2^2(\mathbf{X}) \end{pmatrix} \tag{2}$$

so that $\mathrm{Var}(\mathbf{Y}|\mathbf{X}) = \mathbf{\Sigma}(\mathbf{X}) = \mathbf{\Sigma}^{1/2}(\mathbf{X}) \left(\mathbf{\Sigma}^{1/2}(\mathbf{X})\right)^{\mathbf{T}}$. To guarantee the model identification in (1), the bivariate residuals $(\varepsilon_1, \varepsilon_2)$ are assumed to be independent of the covariates, with zero mean, unit variance, and zero correlation. Despite we do not assume any distribution for the error term, within the framework of functional data analysis this work might be addressed under the assumption of other structures for error distribution: generalized Gauss-Laplace distribution that relax the constrictive assumption of the normal distribution errors [42], generalized linear mixed models (GLMMs) [38] to estimate random effects and dependent (temporal or spatial) errors, and generalized additive models for location, scale and shape (GAMLSS) [43] to model the dynamically variable distribution, considering skewness and kurtosis.

We define the unconditionally probabilistic region for the errors $(\varepsilon_1, \varepsilon_2)$ as

$$\varepsilon_\tau(k) = \{(\varepsilon_1, \varepsilon_2) \in \mathbb{R}^2 | f(\varepsilon_1, \varepsilon_2) \geq k\}$$

f being the density function of the bivariate residuals $(\varepsilon_1, \varepsilon_2)$ and k the $\tau-$quantile of $f(\varepsilon_1, \varepsilon_2)$. Then, for a given $\mathbf{X}$, we define a conditional τ^{th}- uncertainty region for (Y_1, Y_2) containing $\tau\%$ of the observations as:

$$R_\tau(\mathbf{X}) = \begin{pmatrix} \mu_1(\mathbf{X}) \\ \mu_2(\mathbf{X}) \end{pmatrix} + \boldsymbol{\Sigma}^{1/2}(\mathbf{X})\varepsilon_\tau \tag{3}$$

2.2. Estimation Algorithm

To implement an algorithm that allows applying the mathematical model exposed in the previous section, we propose using a functional additive models to estimate the means, variances and covariances in (1). Given a sample of size n, $\{\mathbf{X}_i, (Y_{i1}, Y_{i2})\}_{i=1}^n$, where $\mathbf{X}_i = \left(X_i^1(t), \ldots, X_i^p(t)\right)$, the steps of the proposed estimation algorithm are the following:

Step 1: Perform a decomposition of each covariate $X^j(t)$ in basis functions of the form $X^j(t) \approx \sum_{k=1}^K \xi_k^j \phi_k(t)$, where ϕ_k $(k = 1, \ldots, K)$ are K basis functions (i.e., B-splines, wavelets), and ξ_{il} are either the coefficients of an expansion in fixed basis or the principal component scores of the Karhunen-Loève expansion [44,45]. As a result, we obtain the transformed covariates

$$\tilde{\mathbf{X}}_i = \left(\left(\xi_{i1}^1, \ldots, \xi_{iK}^1\right); \left(\xi_{i1}^2, \ldots, \xi_{iK}^2\right); \ldots; \left(\xi_{i1}^p, \ldots, \xi_{iK}^p\right)\right) \quad i = 1, \ldots, n$$

Step 2: For $r = 1, 2$, fit an additive model to the sample $\{\tilde{\mathbf{X}}_i, Y_{i1}, Y_{i2}\}_{i=1}^n$ and obtain an estimation of the means

$$\hat{\mu}_r(\mathbf{X}_i) = \alpha_r + \sum_{j=1}^p \sum_{k=1}^K \hat{f}_{rk}^j(\xi_{ik}^j) \tag{4}$$

and then estimate $\sigma_r^2(\mathbf{X})$ from the sample $\{\tilde{\mathbf{X}}_i, (Y_{ir} - \hat{\mu}_r(\mathbf{X}_i))^2\}_{i=1}^n$ as

$$\hat{\sigma}_r^2(\mathbf{X}_i) = \exp\left(\hat{\beta}_r + \sum_{j=1}^p \sum_{k=1}^K \hat{g}_{rk}^j(\xi_{ik}^j)\right) \tag{5}$$

Then, compute the correlation $\rho(\mathbf{X})$, which is related to the covariance by $\sigma_{12}(\mathbf{X}) = \sigma_1(\mathbf{X})\sigma_2(\mathbf{X})\rho(\mathbf{X})$, using the sample $\{\mathbf{X}_i, \hat{\delta}_i\}_{i=1}^n$, as follows:

$$\hat{\rho}(\mathbf{X}_i) = \tanh\left(\hat{\gamma} + \sum_{j=1}^p \sum_{k=1}^K \hat{m}_k^j(\xi_{ik}^j)\right)$$

being

$$\hat{\delta}_i = \frac{\left(Y_i^1 - \hat{\mu}_1(\mathbf{X}_i)\right)\left(Y_i^2 - \hat{\mu}_2(\mathbf{X}_i)\right)}{\hat{\sigma}_1(\mathbf{X}_i)\hat{\sigma}_2(\mathbf{X}_i)}$$

where f_{rk}^j, g_{rk}^j and m_k^j are smooth and unknown functions, α_r, β_r and γ are coefficients, p the number of predictors (covariates), and K the number of basis. Please note that the link functions $H_\sigma(\cdot) = \exp(\cdot)$ and $H_\rho(\cdot) = \tanh(\cdot)$ used in the variance and correlation structures, respectively, ensure that the restrictions on the parameter spaces ($\sigma_r^2(\mathbf{X}) \geq 0$ and $0 \leq \rho(\mathbf{X}) \leq 1$) are maintained. Moreover, in order to guarantee the identification of the model, we assume that all the means of functions f_j, g_j and m_j are zero.

Step 3: Compute the standardized residuals

$$\begin{pmatrix} \hat{\varepsilon}_{i1} \\ \hat{\varepsilon}_{i2} \end{pmatrix} = \hat{\boldsymbol{\Sigma}}^{-1/2}(\mathbf{X}_i) \begin{pmatrix} Y_{i1} - \hat{\mu}_1(\mathbf{X}_i) \\ Y_{i2} - \hat{\mu}_2(\mathbf{X}_i) \end{pmatrix} \quad i = 1, \ldots, n \tag{6}$$

where

$$\hat{\Sigma}(\mathbf{X}_i) = \begin{pmatrix} \hat{\sigma}_1^2(\mathbf{X}_i) & \hat{\sigma}_{12}(\mathbf{X}_i) \\ \hat{\sigma}_{12}(\mathbf{X}_i) & \hat{\sigma}_2^2(\mathbf{X}_i) \end{pmatrix} \tag{7}$$

with $\hat{\sigma}_{12}(\mathbf{X}_i) = \hat{\sigma}_1(\mathbf{X}_i)\hat{\sigma}_2(\mathbf{X}_i)\hat{\rho}(\mathbf{X}_i)$.

Step 4: Obtain the kernel estimation of the bivariate density $\hat{f}(\varepsilon_1, \varepsilon_2)$ given by

$$\hat{f}((\varepsilon_1, \varepsilon_2), \mathbf{H}) = \frac{1}{n}\sum_{i=1}^{n} K_{\mathbf{H}} \begin{pmatrix} \varepsilon_1 - \hat{\varepsilon}_{i1} \\ \varepsilon_2 - \hat{\varepsilon}_{i2} \end{pmatrix} \tag{8}$$

where $K(\cdot)$ is the kernel which is a symmetric probability density function and $\mathbf{H}$ is a 2×2 positive definite matrix. Then, obtain the τth unconditional bivariate uncertainty region on the residual scale as

$$\hat{\varepsilon}_\tau = \{(\varepsilon_1, \varepsilon_2)) \in \mathbb{R}^2 | \hat{f}(\varepsilon_1, \varepsilon_2)) \geq \hat{k}\} \tag{9}$$

$\hat{k}$ being the empirical τ quantile of the values $\hat{f}(\varepsilon_{11}, \varepsilon_{12}), \ldots, \hat{f}(\varepsilon_{n1}, \varepsilon_{n2})$. Finally, for a given $\mathbf{X}$, the conditional bivariate uncertainty region $R_\tau(\mathbf{X})$ is estimated according to (3)

$$\hat{R}_\tau(\mathbf{X}) = \begin{pmatrix} \hat{\mu}_1(\mathbf{X}) \\ \hat{\mu}_2(\mathbf{X}) \end{pmatrix} + \hat{\Sigma}^{1/2}(\mathbf{X})\hat{\varepsilon}_\tau \tag{10}$$

3. Case Study: Joint Forecasting of (SO_2, NO_x) Pollution Episodes

The mathematical model exposed in the previous section was applied to the forecasting of pollution episodes registered at a coal-fired power station located in the northwest of Spain. SO_2 and NO_x are two of the main air pollutants generated by combustion processes, and both have harmful effects on human health. Moreover, it was proven that both pollutants are correlated [46], which is consistent with the model in (1). Fortunately, pollution episodes are not very frequent and the trend is that they will become scarcer as technology advances.

Let t_0 be the present time measured each five minutes, and $\mathrm{SO}(t_0)$ and $\mathrm{NO}(t_0)$ the concentrations obtained respectively by the series of bi-hourly SO_2 and NO_x means at instant t_0. Being t_h the prediction horizon time, the interest is to predict

$$(Y_1, Y_2) = (SO(t_0 + t_h), NO(t_0 + t_h))$$

and provide an uncertainty region for these estimations given a specific value of τ, using the predictive covariates

$$\mathbf{X} = \left(X^1(t), X^2(t), X^3(t), X^4(t)\right) = (\mathrm{SO}(t), \mathrm{NO}(t), \mathrm{SO}'(t), \mathrm{NO}'(t)) \quad \text{with} \quad t \in [t_0 - t_{lag}, t_0]$$

where $(\mathrm{NO}'(t), \mathrm{SO}'(t))$ represents the first derivatives of the functions that approximate the concentrations of both pollutants. These derivatives are obtained from the functional representation of the discrete data, according to Step 1 of the estimation algorithm. Please note that t_{lag} represents the lagged time used in the predictors. In particular,, we are interested in predicting an hour in advance, according to the requirements of current Spanish legislation and, therefore, we will consider $t_h = 12$ (60 min) from now on.

Most of the time, these concentrations times series are low, close to zero, and in order to obtain a reasonably large number of pollution incidents, we took as our sample a historical matrix $\{(\mathbf{X}_i, \mathbf{Y}_i)\}_{i=1}^N$ with pollution data of approximately 12 years, which includes a considerable number of pollution episodes (see [9] for a detailed description of the historical matrix construction). In summary, in the historical matrix not all the data are used, but only part of them, following a quantile-weighted criterion. This means that the larger the concentration, the greater the number of observations of that concentration in the sample. Figure 1 shows a sample of the historical matrix, on top are the curves of both pollutants $(\mathrm{NO}(t), \mathrm{SO}(t))$ measured in $t_{lag} = 20$ discretization points and evaluated in

5 B-spline basis functions of order $p = 4$. On the bottom, the first derivative of the B-spline curves $(NO'(t), SO'(t))$ with order $p - 1$ are represented.

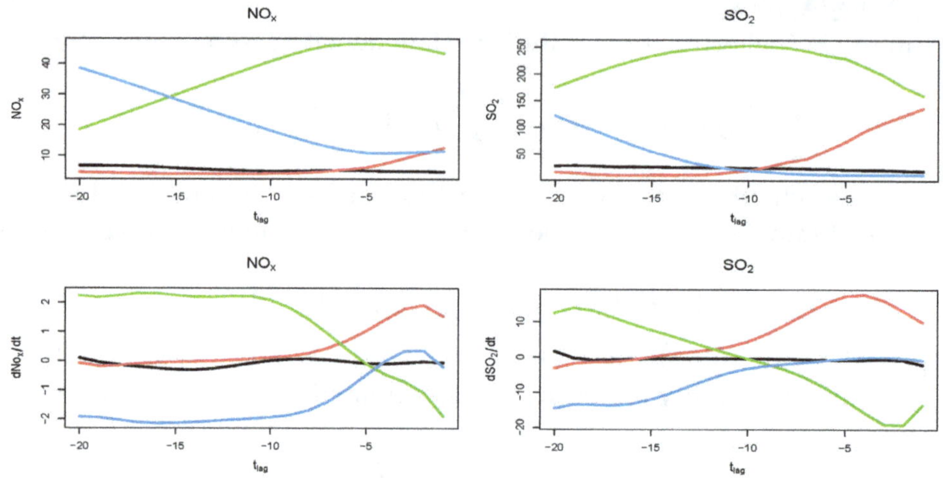

Figure 1. Five curves of the historical matrix for pollutants $(NO(t), SO(t))$ and their corresponding first derivatives $(NO'(t), SO'(t))$, observed in a period of time $t_{lag} = 20$.

In this paper, we tested four models using as predictors different combinations of the covariates $(X^1(t), X^2(t), X^3(t), X^4(t))$ that include the concentrations of both pollutants and their first derivatives. In particular, we will consider models given by:

$$\begin{pmatrix} SO(t_0+12) \\ NO(t_0+12) \end{pmatrix} = \begin{pmatrix} \mu_1(\mathbf{X}^1) \\ \mu_2(\mathbf{X}^2) \end{pmatrix} + \begin{pmatrix} \sigma_1^2(\mathbf{X}^1) & \sigma_{12}(\mathbf{X}^1, \mathbf{X}^2) \\ \sigma_{12}(\mathbf{X}^1, \mathbf{X}^2) & \sigma_2^2(\mathbf{X}^2) \end{pmatrix} \begin{pmatrix} \varepsilon_1 \\ \varepsilon_2 \end{pmatrix} \tag{11}$$

The four considered models, M_1, M_2, M_3 and M_4, are configured in Table 1 where the cross X indicates the covariates included in each model.

Table 1. Selected models from equation in (11) . Cross X indicates the covariates included in each of the four considered models. The derivatives of the functions are indicated with a single quote.

	$\mathbf{X}^1$				$\mathbf{X}^2$			
Model	**SO(t)**	**NO(t)**	**SO′(t)**	**NO′(t)**	**SO(t)**	**NO(t)**	**SO′(t)**	**NO′(t)**
M_1	X					X		
M_2	X	X			X	X		
M_3	X		X			X		X
M_4	X	X	X		X	X	X	X

To validate and compare the four proposed models, we randomly select from the full historical matrix a training set $\mathbf{M}^I = \left\{(\mathbf{X}_i^I, \mathbf{Y}_i^I)\right\}_{i=1}^{n_{train}}$ and a test set $\mathbf{M}^{II} = \left\{(\mathbf{X}_i^{II}, \mathbf{Y}_i^{II})\right\}_{i=n_{train}+1}^{N}$.

The estimates $\hat{\mu}_1$, $\hat{\mu}_2$, $\hat{\mathbf{\Sigma}}$ were obtained from the samples in the first matrix M_t^I. The bivariate uncertainty regions for the values of the covariates on the second matrix M^{II} were obtained using (3).

The estimated coverage $\hat{\tau}$ is given by

$$\hat{\tau} = \frac{1}{n_{test}} \sum_{i=n_{train}+1}^{N} I\{\mathbf{Y}_i^{II} \in \hat{R}_\tau(\mathbf{X}_i^{II})\}; n_{test} = N - n_{train} \tag{12}$$

The performance of the proposed predictors was evaluated in two pollution incidents. A bivariate representation of these episodes is shown in Figure 2. The orientation of the points shows a clear correlation between both pollutants although the range of concentrations is quite different.

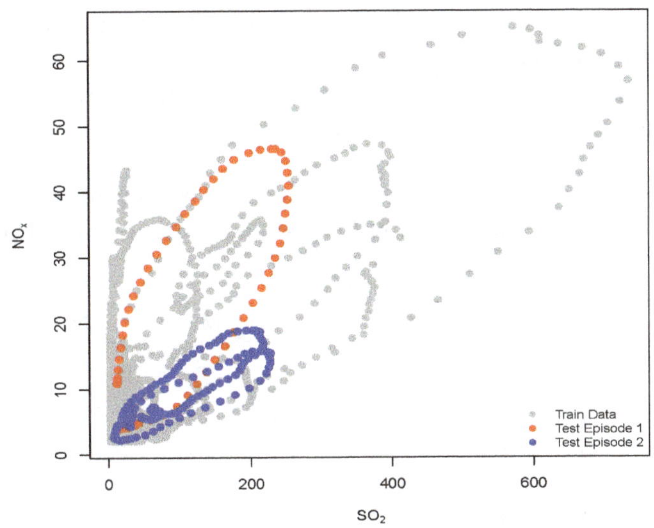

Figure 2. Observed and forecasted concentrations of SO_2 and NO_x for two pollution episodes.

The nominal and the estimated coverages for different time lags and training sample sizes are shown in Table 2. The coverages correspond to the bivariate solution, and were obtained for n_{test} consecutive observations that might or might not correspond to pollution incidents.

Table 2. Nominal τ and estimated $\hat{\tau}$ coverages for each of the four models under study. Two time lags, $t_{lag} = 10$, $t_{lag} = 20$, two sizes of the training sample $n_{train} = 10,000$ and $n_{train} = 4900$, and two numbers of principal components, K = 3 and K = 5, were considered. Results correspond to the test sample.

				$\hat{\tau}$			
τ	t_{lag}	$n^{\bullet}_{train}$	K	M_1	M_2	M_3	M_4
0.50	10	20	3	0.43	0.47	0.45	0.51
			5	0.42	0.48	0.47	0.49
	10	49	3	0.51	0.52	0.52	0.52
			5	0.51	**0.50**	**0.50**	**0.50**
	20	20	3	0.45	0.49	0.44	0.49
			5	0.48	0.46	0.43	0.46
	20	49	3	0.50	0.54	0.51	0.50
			5	0.49	0.49	0.49	0.48
0.75	10	20	3	0.70	0.73	0.72	**0.75**
			5	0.69	0.74	0.73	0.74
	10	49	3	0.76	0.78	0.78	0.78
			5	0.77	0.76	0.76	0.76
	20	20	3	0.70	0.72	0.70	0.72
			5	0.70	0.71	0.69	0.69
	20	49	3	0.77	0.78	**0.75**	0.73
			5	**0.75**	0.74	0.72	0.72
0.90	10	20	3	0.88	0.87	0.87	0.89
			5	0.86	0.88	0.87	0.88
	10	49	3	0.91	**0.90**	**0.90**	**0.90**
			5	**0.90**	**0.90**	**0.90**	**0.90**
	20	20	3	0.87	0.87	0.85	0.86
			5	0.87	0.86	0.86	0.84
	20	49	3	0.90	0.89	0.90	0.86
			5	0.87	0.87	0.88	0.85
0.95	10	20	3	0.93	0.93	0.93	0.93
			5	0.93	0.93	0.93	0.93
	10	49	3	0.96	0.93	0.93	0.93
			5	0.95	0.94	0.94	0.94
	20	20	3	0.93	0.93	0.92	0.92
			5	0.92	0.92	0.93	0.90
	20	49	3	**0.95**	0.94	**0.95**	0.92
			5	0.92	0.92	0.93	0.90

RMSE values for each model are shown in Table 3 , considering an expansion of the functions in three or five principal components, and lags $t_{lag} = 10$ and $t_{lag} = 20$. Please note that this table makes reference to the marginal distributions, and that the range of concentrations for each pollutant is very different, therefore the RMSE values are also different.

Table 3. RMSE values for two pollution episodes and the four models tested, considering curves with two different time lags, size of the training samples and number of principal components.

				Episode 1				Episode 2			
Repsonse	t_{lag}	$n^{\bullet}_{train}$	K	M_1	M_2	M_3	M_4	M_1	M_2	M_3	M_4
NO_x	10	20	3	20.7	25.2	20.0	23.0	1.9	1.7	1.4	1.4
			5	19.6	18.6	20.6	**16.4**	0.9	0.8	0.8	**0.6**
		49	3	20.9	24.7	18.2	23.8	1.8	1.9	1.4	1.4
			5	19.2	17.5	19.3	16.7	0.9	0.9	0.8	0.7
	20	100	3	34.0	24.7	19.9	19.6	3.3	5.5	1.7	3.1
			5	40.9	23.3	46.5	29.2	1.5	2.5	1.0	1.8
		49	3	30.2	18.2	18.7	20.4	3.4	5.3	1.7	2.8
			5	36.2	27.6	39.8	35.6	1.5	2.5	1.0	1.8
SO_2	10	100	3	505.0	841.0	**407.5**	837.3	544.8	531.9	419.8	419.1
			5	914.9	868.6	669.4	516.3	215.6	230.3	184.5	199.1
		49	3	686.0	685.3	515.8	518.3	481.4	484.4	338.0	361.1
			5	991.1	925.5	846.3	682.9	199.9	214.7	**170.6**	179.9
	20	100	3	1463.4	2172.7	825.4	1199.9	1154.5	1133.1	709.5	659.0
			5	1470.6	2531.2	1002.9	1428.5	525.5	482.1	352.0	341.3
		49	3	1.458.7	2485.4	768.4	698.4	1162.6	1125.8	644.8	628.7
			5	1787.6	2811.0	1111.2	951.5	548.1	492.0	352.7	359.3

Note: $n_{train} = 100 \cdot n^{\bullet}_{train}$.

For the two episodes analyzed, Figure 3 shows the observed and the predicted values as well as the quantile for $\hat{\tau} = 0.95$, calculated for the test sample. The results correspond to curves observed in ten points ($t_{lag} = 10$) and represented in a basis expansion in three functional principal components. These univariate confidence intervals were respectively constructed from (11) as $\mu_1(\mathbf{X}^1) + \sigma_1(\mathbf{X}^1)\varepsilon_1^{0.95}$ and $\mu_2(\mathbf{X}^2) + \sigma_2(\mathbf{X}^2)\varepsilon_2^{0.95}$, $\varepsilon_1^{0.95}$ and $\varepsilon_1^{0.95}$ being the 0.95 quantile of the distributions of errors ε_1 and ε_2, respectively.

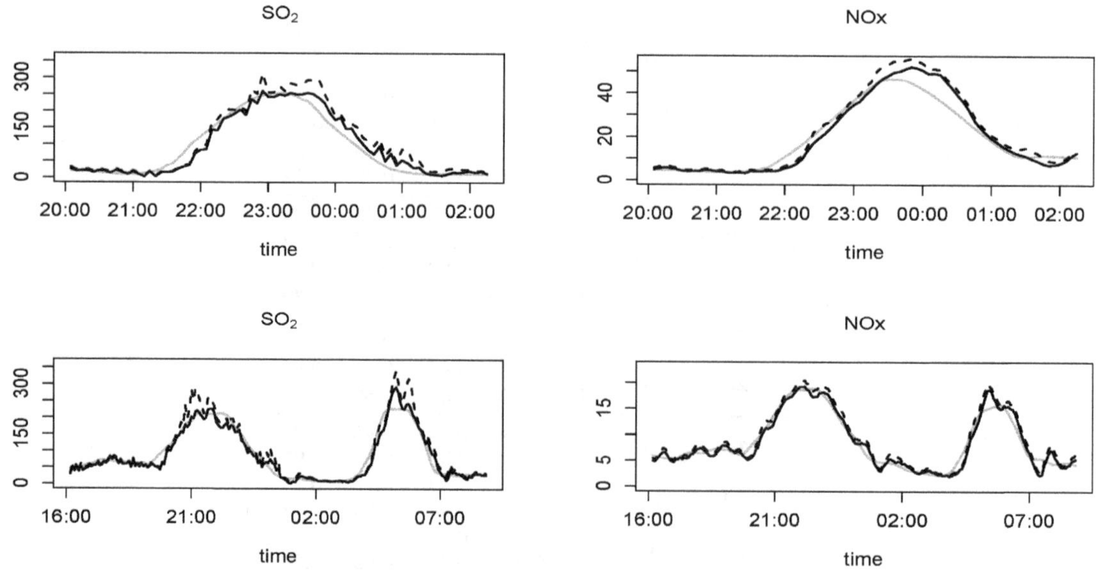

Figure 3. Observations (solid black line), mean (solid gray line) and 0.95*th* quantile estimations (discontinuous line) for both pollutants, SO_2 and NO_x and for two pollution episodes.

Table 4 shows the maximum consumed memory and the runtime (in seconds) for the four models tested and two different dimensions of the submatrices are executed in a Intel Core i7-2600K with 16 GB of RAM.

Table 4. Maximum memory consumption (MB) and computation time (seconds) for the four models tested, M_i, following different strategies concerning the time lag, t_{lag}, the size of the training sample, n_{train}, and the number of basis functions, K.

			Memory (MB)				Runtime (seconds)			
t_{lag}	n_{train}	K	M_1	M_2	M_3	M_4	M_1	M_2	M_3	M_4
10	20	3	652.70	1083.03	1319.57	2266.34	17.98	29.66	35.99	61.62
		5	1090.78	1855.39	2299.14	4075.88	35.55	55.29	63.15	124.7
	49	3	329.94	548.58	669.42	1153.38	10.53	17.42	21.38	37.01
		5	552.74	942.68	1171.67	2089.53	19.79	31.28	46.62	88.20
20	20	3	653.34	1084.15	1320.66	2268.16	18.11	29.63	35.97	61.21
		5	1091.90	1857.26	2300.99	4078.95	32.14	49.51	72.45	124.81
	49	3	330.10	548.84	669.69	1153.82	10.51	17.56	21.49	37.82
		5	553.01	943.13	1172.20	2090.36	17.80	36.44	39.03	76.11

Note: $n_{train} = 100 \cdot n^{\bullet}_{train}$.

4. Discussion

We begin the discussion of the results analyzing Table 2 that show the estimated coverage for the bivariate prediction depending on the time lag, the size of the training sample, the number of principal components and the model used. It can be appreciated that the estimated coverages are generally lower than the theoretical coverages, although very close. Therefore, the mathematical models proposed show a good performance although there is a trend to underestimate the observed values. This effect can also be appreciated in Figure 3, where the mean tends to be under the observed values. Then, in order to be on the safe wide, it would be preferable to use the quantile $\tau = 0.95$, which provides greater guarantee of predicting the highest values of the pollution episodes. Regarding the rest of the parameters, it is not possible to establish a combination of them that provides the best results. However, in general they were obtained for the lowest training size, $n_{train} = 49,000$, and for models that includes one or two derivatives (models M_3 and M_4).

The prediction errors are shown in Table 1, where the best results (minimum RMSE) are marked in bold. As can be seen, they correspond to model M_4 for NO_x and model M_3 for SO_2. In both cases, these models incorporate the derivatives of the original functions. Accordingly, we conclude that the derivatives contribute positively to improve the results, which reinforces the role of the functional approach. However, there is an asymmetry between both pollutants: using the concentrations of SO_2 and their derivatives improves the results for NO_x, but using the concentrations of NO_x and their derivatives is not an advantage in the estimation of SO_2. When SO_2 and NO_x concentrations of both episodes are plotted against time (Figure 4), a slight advance can be seen on the first pollutant compared to the second, which would explain this asymmetry.

With respect to the time lag, the minimum RMSE values were obtained for the shorter period of time $t_{lag} = 10$, so it seems that using 20 observations to predict one our in advance introduced noise into the model instead of adding useful information. This result is in agreement with those obtained for the same data in previous studies of some of the authors, which indicated that only a few observations close to the time of prediction contribute to that prediction. Talking about the size of the training sample, simplifying the original data by removing small values of the concentrations improves the results in most of the cases, so this would be the advisable option.

When the effect of the number of principal components used as basis functions is analyzed, using $K = 5$ is always favorable for episode 2, for both SO_2 and NO_x, but not for episode 1, for which the trend is opposite.

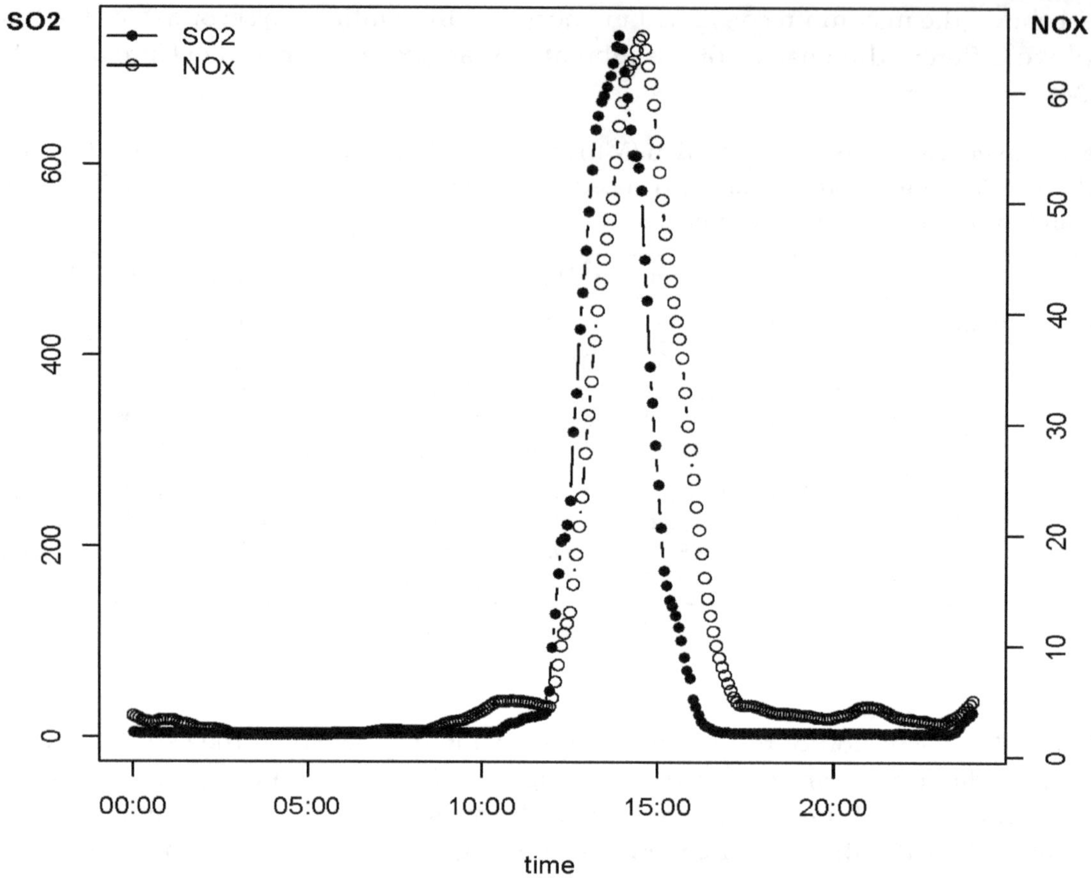

Figure 4. Example of a pollution incident showing SO_2 and NO_x concentrations versus time. Notice that there is an advance in the first pollutant compared to the second.

Although they are not shown in the article, so as not to overstretch it, a comparison of the estimated coverages using 3 or 5 principal components, or 5 B-splines basis functions, tell us that there are not substantial differences among them, so it seems that one or other base functions can be used interchangeably.

Finally, regarding memory consumption and runtime for model training, it is evident, from Table 4 that more complex models consume more resources and requires more computing time. For fixed values of the time lag (t_{lag}) and the size of the training sample (n_{train}), model M_4 is between 3 and 7 times more expensive than model M_1 in terms of memory consumption and runtime. Using time lags $t_{lag} = 10$ or $t_{lag} = 50$ has no effect in terms of computation requirements; and employing 5 principal components instead of 3 principal components implies an approximately double memory consumption and runtime.

To conclude, it is possible to establish that the functional location-scale model proposed were quite a good approach (in terms of coverage and prediction error) to forecast bivariate pollution episodes one hour in advance, as it is required by the Spanish legislation. The best results were obtained when the derivatives of functions adjusted to the observed data are included in the model, when the raw data are filtered and when the shorter period of time is used for the prediction. The size of the training data and the type and number of basis functions are, instead, parameters on which definitive conclusions could not be drawn.

Author Contributions: Conceptualization, J.R.-P. and C.O.; methodology, J.R.-P. and M.O.-d.L.F.; software, J.R.-P. and M.O.-d.L.F.; validation, J.R.-P., C.O. and M.O.-d.L.F.; writing—original draft preparation, J.R.-P. and C.O.; writing—review and editing, J.R.-P., C.O. and M.O.-d.L.F.; supervision, C.O. All authors have read and agreed to the published version of the manuscript.

Acknowledgments: The authors acknowledge financial support from: (1) UO-Proyecto Uni-Ovi (PAPI-18-GR-2014-0014), (2) Project MTM2016-76969-P from Ministerio de Economía y Competitividad—Agencia Estatal de Investigación and European Regional Development Fund (ERDF) and IAP network StUDyS from Belgian Science Policy, (3) Nuevos avances metodológicos y computacionales en estadística no-paramétrica y semiparamétrica—Ministerio de Ciencia e Investigación (MTM2017-89422-P).

References

1. Siew, L.Y.; Ching, L.Y.; Wee, P.M.J. ARIMA and integrated ARFIMA models for forecasting air pollution index in Shah Alam, Selangor. *Malay. J. Anal. Sci.* **2008**, *12*, 257–263.
2. Ibrahim, M.Z.; Roziah, Z.; Marzuki, I.; Muhd S.L. Forecasting and Time Series Analysis of Air Pollutants in Several Area of Malaysia. *Am. J. Enverion. Sci.* **2009**, *5*, 625–632. [CrossRef]
3. Abhilash, M.S.K.; Thakur, A.; Gupta, D.; Sreevidya, B. Time Series Analysis of Air Pollution in Bengaluru Using ARIMA Model. In *Ambient Communications and Computer Systems*; Advances in Intelligent Systems and Computing; Perez, G., Tiwari, S., Trivedi, M., Mishra, K., Eds.; Springer: Singapore, 2018.
4. Liu, P.W.G. Simulation of the daily average PM10 concentrations at Ta-Liao with Box-Jenkins time series models and multivariate analysis. *Atmos. Environ.* **2009**, *43*, 2104–2113. [CrossRef]
5. Nazif, A.; Mohammed, N.I.; Malakahmad, A.; Abualqumboz, M.S. Regression and multivariate models for predicting particulate matter concentration level. *Environ. Sci. Pollut. Res. Int.* **2018**, *25*, 283–289. [CrossRef]
6. Zhao, R.; Gu, X.; Xue, B.; Zhang, J.; Ren, W. Short period PM2.5 prediction based on multivariate linear regression model. *PLoS ONE* **2018**, *13*, e0201011. [CrossRef]
7. Ng, K.Y.; Awang, N. Multiple linear regression and regression with time series error models in forecasting PM10 concentrations in Peninsular Malaysia. *Environ. Monit. Assess.* **2018**, *190*, 63. [CrossRef]
8. Roca-Pardiñas, J.; Gonzàlez Manteiga, W.; Febrero-Bande, M.; Prada-Sànchez, J.M.; Cadarso-Suàrez, C. Predicting binary time series of SO_2 using generalized additive models with unknown link function. *Environmetrics* **2004**, *15*, 729–742. [CrossRef]
9. Martínez-Silva, I.; Roca-Pardiñas, J.; Ordóñez, C. Forecasting SO_2 pollution incidents by means of quantile curves based on additive models. *Environmetrics* **2016**, *27*, 147–157. [CrossRef]
10. Garcia, J.M.; Teodoro F.; Cerdeira, R.; Coelho, L.M.R.; Prashant, K.; Carvalho, M.G. Developing a methodology to predict PM10 concentrations in urban areas using generalized linear models. *Environ. Technol.* **2016**, *37*, 2316–2325. [CrossRef]
11. Roca-Pardiñas, J.; Ordóñez, C. Predicting pollution incidents through semiparametric quantile regression models. *Stoch. Environ. Res. Risk Assess.* **2019**, *33*, 673–685. [CrossRef]
12. Azid, I.A.; Ripin, Z.M.; Aris, M.S.; Ahmad, A.L.; Seetharamu, K.N.; Yusoff, R.M. Predicting combined-cycle natural gas power plant emissions by using artificial neural networks. In Proceedings of the 2000 TENCON Proceedings, Intelligent Systems and Technologies for the New Millennium (Cat. No.00CH37119), Kuala Lumpur, Malaysia, 24–27 September 2000; Volume 3, pp. 512–517.
13. Perez, P.; Trier, A.; Reyes, J. Prediction of PM2.5 Concentrations Several Hours in Advance Using Neural Networks in Santiago, Chile. *Atmos. Environ.* **2000**, *34*, 1189–1196. [CrossRef]
14. Ferretti, G.; Piroddi, L. Estimation of NO_x Emissions in Thermal Power Plants Using Neural Networks. *J. Eng. Gas Turbines Power* **2001**, *132*, 465–471. [CrossRef]
15. Siwek, K.; Osowski, S. Improving the accuracy of prediction of PM10 pollution by the wavelet transformation and an ensemble of neural predictors. *Eng. Appl. Artif. Intell.* **2012**, *25*, 1246–1258. [CrossRef]
16. Muñoz, E.; Martín, M.L.; Turias, I.J.; Jimenez-Come, M.J.; Trujillo, F.J. Prediction of PM10 and SO_2 exceedances to control air pollution in the Bay of Algeciras, Spain. *Stoch. Environ. Res. Risk Assess.* **2014**, *28*, 1409–1420. [CrossRef]
17. He, H.D.; Lu, W.Z.; Xue, Y. Prediction of particulate matters at urban intersection by using multilayer perceptron model based on principal components. *Stoch Environ. Res. Risk. Assess.* **2015**, *29*, 2107–2114. [CrossRef]
18. Antanasijević, D.; Pocajt, V.; Perić-Grujić, A.; Ristić, M. Multiple-input–multiple-output general regression neural networks model for the simultaneous estimation of traffic-related air pollutants. *Atmos. Pollut. Res.* **2018**, *9*, 388–397. [CrossRef]

19. Gilson, M.; Dahmen, D.; Moreno-Bote, R.; Insabato, A.; Helias M. The covariance perceptron: A new paradigm for classification and processing of time series in recurrent neuronal networks. *BioRxiv* **2019**. [CrossRef]
20. Ramsay, J.O.; Silverman, B.W. *Applied Functional Data Analysis: Methods and Case Studies*; Springer: New York, USA, 2002.
21. Ferraty, F.; Vieu, P. *Nonparametric Functional Data Analysis*; Springer: New York, USA, 2006.
22. Febrero-Bande, M.; Galeano, P.; González-Manteiga, W. Outlier detection in functional data by depth measures with application to identify abnormal NOx levels. *Environmetrics* **2008**, *19*, 331–345. [CrossRef]
23. Martinez, J.; Saavedra, Á.; García-Nieto, P.J.; Piñeiro, J.I.; Iglesias, C.; Taboada, J.; Sanchoa, J.; Pastor, J. Air quality parameters outliers detection using functional data analysis in the Langreo urban area (Northern Spain). *Appl. Math. Comput.* **2014**, *241*, 1–10. [CrossRef]
24. Shaadan, N.; Jemain, A.A.; Latif, M.T. Anomaly detection and assessment of PM10, functional data at several locations in the Klang Valley, Malaysia. *Atmos. Pollut. Res.* **2015**, *6*, 365–375. [CrossRef]
25. Ignaccolo, R.; Mateu, J.; Giraldo, R. Kriging with external drift for functional data for air quality monitoring. *Stoch. Environ. Res. Risk Assess.* **2014**, *28*, 1171–1186. [CrossRef]
26. Wang, D.; Zhong, Z.; Kaixu, B.; Lingyun, H. Spatial and Temporal Variabilities of PM2.5 Concentrations in China Using Functional Data Analysis. *Sustainability* **2019**, *11*, 1620. [CrossRef]
27. Aneiros-Pérez, G.; Cardot, H.; Estévez-Pérez, G.; Vieu, P.H. Maximum ozone concentration forecasting by functional non-parametric approaches. *Environmetrics* **2004**, *15*, 675–685. [CrossRef]
28. Fernández de Castro, B.M.; González-Manteiga, W.; Guillas, S. Functional samples and bootstrap for predicting sulfur dioxide levels. *Technometrics* **2005**, *47*, 212–222. [CrossRef]
29. Quintela-del-Río, A.; Francisco-Fernández, M. Nonparametric functional data estimation applied to ozone data: Prediction and extreme value analysis. *Chemosphere* **2001**, *82*, 800–808. [CrossRef]
30. Besse, P.C.; Cardot, H.; Stephenson, D.B. Autoregressive forecasting of some functional climatic variations. *Scand. J. Stat.* **2000**, *27*, 673–687. [CrossRef]
31. Damon, J.; Guillas, S. The inclusion of exogenous variables in functional autoregressive ozone forecasting. *Environmetrics* **2002**, *13*, 759–774. [CrossRef]
32. Ruiz-Medina, M.D.; Espejo R.M. Spatial autoregressive functional plug-in prediction of ocean surface temperature. *Stoch. Environ. Res. Risk. Assess.* **2012**, *26*, 335–344. [CrossRef]
33. Ruiz-Medina, M.D.; Espejo, R.M.; Ugarte, M.D.; Militino A.F. Functional time series analysis of spatio-temporal epidemiological data. *Stoch. Environ. Res. Risk Assess.* **2014**, *28*, 943–954. [CrossRef]
34. Alvarez-Liebana, J.; Ruiz Medina, M.D. Prediction of air pollutants PM10 by ARBX(1) processes. *Stoch. Environ. Res. Risk Assess.* **2019**, *33*, 1721–1736. [CrossRef]
35. Hsu K.J. Time series analysis of the interdependence among air pollutants. *Atm. Environ. Part B Urban Atmos.* **1992**, *26*, 491–503. [CrossRef]
36. Kadiyala, A.; Kumar, A. Vector time series models for prediction of air quality inside a public transportation bus using available software. *Environ. Prog. Sustain.* **2014**, *33*, 337–341. [CrossRef]
37. García-Nieto, P.J.; Sánchez-Lasheras, F.; García-Gonzalo, E.; de Cos Juez F.J. Estimation of PM10 concentration from air quality data in the vicinity of a major steelworks site in the metropolitan area of Avilés (Northern Spain) using machine learning techniques. *Stoch Environ. Res. Risk Assess.* **2018**, *32*, 3287–3298. [CrossRef]
38. Hedeker, D.; Mermelstein, R.J.; Demirtas, H. An Application of a Mixed-Effects Location Scale Model for Analysis of Ecological Momentary Assessment (EMA) Data. *Biometrics* **2008**, *64*, 627–634. [CrossRef]
39. Taylor, J.; Verbyla, A. Joint modelling of location and scale parameters of the t distribution. *Stat. Model.* **2004**, *4*, 91–112. [CrossRef]
40. Pugach, O.; Hedeker, D.; Mermelstein, R. A Bivariate Mixed-Effects Location-Scale Model with application to Ecological Momentary Assessment (EMA) data. *Health Serv. Outcomes Res. Methodol.* **2014**, *14*, 194–212. [CrossRef]
41. He, W.; Lawless, J.F. Bivariate location-scale models for regression analysis, with applications to lifetime data. *J. R. Statist. Soc. B* **2005**, *67 Pt 1*, 63–78. [CrossRef]
42. Jäntschi, L.; Bálint, D.; Bolboaca, S.D. Multiple Linear Regressions by Maximizing the Likelihood under Assumption of Generalized Gauss-Laplace Distribution of the Error. *Comput. Math. Methods Med.* **2016**, *2016*, 8578156. [CrossRef]

43. Rigby, R.A.; Stasinopoulos, D.M. Generalized additive models for location, scale and shape. *J. R. Stat. Soc. Ser. C* **2005**, *54, 507–554.* [CrossRef]
44. Karhunen, K. Zur Spektraltheorie Stochastischer Prozesse. *Annales Academiae Scientiarum Fennicae Series A1 Mathematica-Physica* **1946**, *54*, 1–7. Available online: https://katalog.ub.uni-heidelberg.de/cgi-bin/titel.cgi?katkey=67295489 (accessed on 24 February 2020).
45. Febrero-Bande, M.; Oviedo de la Fuente, M. Statistical Computing in Functional Data Analysis: The R Package fda.usc. *J. Stat. Softw.* **2012**, *51*, 12. [CrossRef]
46. Dogruparmak, S.C.; Özbay, B. Investigating Correlations and Variations of Air Pollutant Concentrations under Conditions of Rapid Industrialization–Kocaeli (1987–2009). *Clean-Soil Air Water* **2011**, *39*, 597–604. [CrossRef]

11

Two-Stage Classification with SIS using a New Filter Ranking Method in High Throughput Data

Sangjin Kim [1,*] and Jong-Min Kim [2]

[1] Department of Mathematical Sciences, University of Texas at El Paso, El Paso, TX 79968, USA
[2] Division of Sciences and Mathematics, University of Minnesota at Morris, Morris, MN 56267, USA; jongmink@morris.umn.edu
* Correspondence: skim10@utep.edu

Abstract: Over the last decade, high dimensional data have been popularly paid attention to in bioinformatics. These data increase the likelihood of detecting the most promising novel information. However, there are limitations of high-performance computing and overfitting issues. To overcome the issues, alternative strategies need to be explored for the detection of true important features. A two-stage approach, filtering and variable selection steps, has been receiving attention. Filtering methods are divided into two categories of individual ranking and feature subset selection methods. Both have issues with the lack of consideration for joint correlation among features and computing time of an NP-hard problem. Therefore, we proposed a new filter ranking method (PF) using the elastic net penalty with sure independence screening (SIS) based on resampling technique to overcome these issues. We demonstrated that SIS-LASSO, SIS-MCP, and SIS-SCAD with the proposed filtering method achieved superior performance of not only accuracy, AUROC, and geometric mean but also true positive detection compared to those with the marginal maximum likelihood ranking method (MMLR) through extensive simulation studies. In addition, we applied it in a real application of colon and lung cancer gene expression data to investigate the classification performance and power of detecting true genes associated with colon and lung cancer.

Keywords: LASSO; SCAD; MCP; SIS; elastic net; accuracy; AUROC; geometric mean

---

1. Introduction

In the last decade, high dimensional data has appeared with the development of high throughput techniques, especially in the research area of machine learning [1,2] and data mining [3,4] in biology. The possibility of finding novel true important variables has potentially become high with a huge amount of data. However, due to limitations of computing capabilities and overfitting issues, two-stage approaches of filtering and variable selection for prediction purpose has been popular. These include methods for microarray [5–8] and RNA-Seq [9,10] data, and genome-wide association studies (GWAS) [11,12]. Filtering methods, which reduce dimensionality and try to retain the most promising features as possible, have long been under development. A number of filtering methods has been proposed to rank features, such as Information gain [13], Markov blanket [14], Bayesian variable selection [15], Boruta [16], Fisher score [17], Relief [18], maximum relevance and minimum redundancy (MRMR) [19], marginal maximum likelihood score (MMLs) [20], among which MMLS is one of the simplest and computationally efficient methods of feature selection with some criteria.

Feature selection methods are divided into two categories of marginal feature ranking and feature subset selection considering relationship among features. Marginal feature ranking methods order individual features by their scores and then drop out irrelevant features with small scores using the desired criteria. [21] utilized the Relief statistical method to rank features. [20] gave a marginal maximum likelihood estimator as a feature ranking method and improved classification accuracy. [22]

also developed a novel method to rank features and then chose the optimal subset of features. Individual ranking methods have been widely used in high throughput data analysis because of their simplicity and computational time efficiency but a predetermined threshold is required before variable selection stage. To overcome this issue, the sure independent screening (SIS) approach [23] was developed to ensure that all true important variables survive after the variable screening with probability tending to one. Feature subset selection methods [24–26] detect an optimal subset of features leading to the best performance of prediction. However, these methods have heavy computational time leading to be NP-hard [27] under a high dimensional setting.

In this paper, we proposed a filter ranking method (PF) utilizing selection probability with an elastic net based on resampling technique with SIS. The selected features are then applied to three popular variable selection algorithms such as least absolute shrinkage and selection operator (LASSO) [28], minimax concave penalty (MCP) [29], and smoothly clipped absolute deviation (SCAD) [30].

The rest of this article is organized as follows. In Section 2, we described three penalized logistic methods of LASSO, MCP, and SCAD, marginal maximum likelihood ranking method, sure independence screening method (SIS), the proposed statistical methods for filter ranking and its algorithm, and metrics of performance including accuracy, area under the receiver operating characteristic (AUROC), and geometric mean of sensitivity and specificity (G-mean). In Section 3, we describe the superior performance of our proposed method compared to an individual ranking method of marginal maximum likelihood logistic regression (MMLR) with SIS through the extensive simulation studies. We next applied the proposed method to the high dimensional colon gene expression data and investigate the biological meaning of selected genes. Finally, in Section 4, we discuss our findings.

2. Materials and Methods

We split this section into several subsections describing the methods used in the study. The section of sparse logistic regression, such as LASSO, adaptive LASSO, SCAD, and MCP, is discussed. A filtering method with SIS used as a reference is briefly described and then our proposed method is explained in detail. The final section considers metrics of the performance including accuracy, AUROC, and G-mean. All simulations and real applications were done with R software and the corresponding codes, results, and data are available at [31].

2.1. Penalized Logistic Regression Method

Binary logistic regression is widely used in the classification of clinical outcomes of cancer using gene expression data to identify the relationship between the outcome and a set of predictors to build prediction models. However, the logistic regression has limited use in high dimensional settings when N << P because the inverted matrix does not exist for the estimation of regression coefficients. Embedded methods such as LASSO, SCAD, and MCP are the most popular methods in gene selection under a high dimensional setting because they are allowed to select a sparse subset of genes by continuously shrinking unimportant covariates' regression coefficients into zero. A number of penalty based embedded methods has been extensively studied and modified in the area of cancer genes selection under high throughput data [32–43].

Let the expression levels of genes in i^{th} individual be denoted as $x_i = (x_{i1}, x_{i2}, \ldots, x_{id})$ for $i = 1, \ldots, n$ and d is a total number of genes. Given a training data set $\{(x_i, y_i)\}_{i=1}^{n}$ where $y_i \in (0,1)$, $y_i = 0$ indicates that i^{th} individual is in normal group and $y_i = 1$ in cancer group. Assuming that $\mathrm{p}(x_i) = p(y_i = 1|x_i)$, the logistic regression is defined as follows:

$$\log\left(\frac{\mathrm{p}(\mathrm{x_i})}{1-\mathrm{p}(\mathrm{x_i})}\right) = \beta_0 + \mathrm{x_i}\beta, \text{ where } \mathrm{i} = 1, \ldots, \mathrm{n} \text{ and } \beta = (\beta_1, \ldots, \beta_d)^{\mathrm{T}}. \quad (1)$$

The following formula is for the maximum log-likelihood estimator of logistic regression (MLR). $\hat{\beta}_{MLR}$ is defined as follows:

$$\hat{\beta}_{MLR} = \underset{\beta}{\operatorname{argmax}}[\log(\mathcal{L}(\beta))] = \underset{\beta}{\operatorname{argmax}}\left[\sum_{i=1}^{n}(y_i \log(\mathrm{p}(x_i)) + (1-y_i)\log(1-\mathrm{p}(x_i)))\right] \tag{2}$$

The estimation of parameters can be calculated by maximizing the above log-likelihood function $\log(\mathcal{L}(\beta))$. The criterion for classification is that if $p(y_i = 1|x_i) \geq 0.5$, then the individual belongs to the cancer group, otherwise, normal group. The penalized logistic regression (PLR) is a combination of logistic regression with penalty function and parameters can be estimated by minimizing the log-likelihood function with penalty function as follows:

$$\hat{\beta}_{PLR} = \underset{\beta}{\operatorname{argmin}}\left[-\sum_{i=1}^{n}(y_i \log(\mathrm{p}(x_i)) + (1-y_i)\log(1-\mathrm{p}(x_i))) + p(\beta)\right], \tag{3}$$

where $p(\beta)$ is a penalty function.

One of the most popular penalty functions is LASSO [12–18]. It forces most of the unimportant genes' regression coefficients into zero. Although it is widely used in high throughput biomedical data, it has the tendency to randomly choose one of the genes with high correlation and then throw out the rest of the genes. The estimation of regression coefficients can be done by minimizing the following likelihood:

$$\hat{\beta}_{lasso} = \underset{\beta}{\operatorname{argmin}}\left[-\sum_{i=1}^{n}(y_i \log(\mathrm{p}(x_i)) + (1-y_i)\log(1-\mathrm{p}(x_i))) + \lambda\sum_{i=1}^{d}|\beta_i|\right]. \tag{4}$$

Another popular sparse logistic regression is SCAD with a concave penalty that complements the limitation of lasso mentioned above. To estimate parameters of regression coefficients, the following log-likelihood can be minimized:

$$\hat{\beta}_{SCAD} = \underset{\beta}{\operatorname{argmin}}\left[-\sum_{i=1}^{n}(y_i \log(\mathrm{p}(x_i)) + (1-y_i)\log(1-\mathrm{p}(x_i))) + \lambda\sum_{i=1}^{d}p_\lambda(\beta_i)\right]. \tag{5}$$

The $p_\lambda(\beta_i)$ is

$$|\beta_j|\mathrm{I}_{(|\beta_j|\leq\lambda)} + \left(\frac{\left\{(a^2-1)\lambda^2 - \left(a\lambda - |\beta_i|\right)_+^2\right\} I\left(\lambda \leq |\beta_i|\right)}{2(a-1)}\right), \text{ for } \lambda \geq 0 \textit{ and } a > 2. \tag{6}$$

The minimax concave penalty (MCP) is also popular as much as SCAD. The estimation of regression coefficients can be achieved by minimizing the following log-likelihood function:

$$\hat{\beta}_{MCP} = \underset{\beta}{\operatorname{argmin}}\left[-\sum_{i=1}^{n}(y_i \log(\mathrm{p}(x_i)) + (1-y_i)\log(1-\mathrm{p}(x_i))) + \lambda\sum_{i=1}^{d}p_\lambda(\beta_i)\right]. \tag{7}$$

$p_\lambda(\beta_i)$ is written as follows.

$$\left(\frac{2a\lambda|\beta_i| - {\beta_i}^2}{2a}\right)\mathrm{I}\left(|\beta_i| \leq \mathrm{a}\lambda\right) + \left(\frac{a\lambda^2}{2}\right)\mathrm{I}\left(|\beta_i| > \mathrm{a}\lambda\right), \textit{ for } \lambda \geq 0 \textit{ and } a > 1. \tag{8}$$

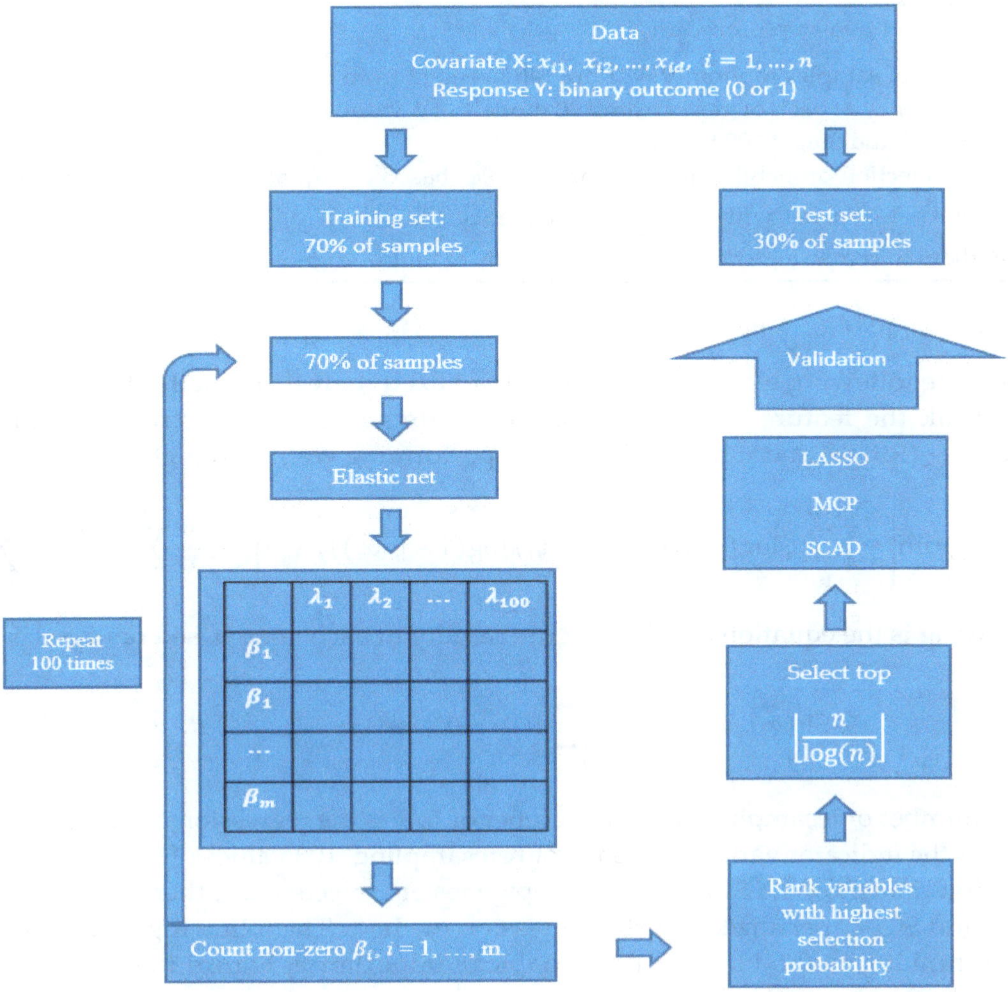

Figure 1. Diagram showing the proposed two-step procedure.

2.2. *Variable Ranking with MMLR*

In practice, gene expression data usually contain irrelevant genes that lead to low classification performance under high dimensional settings. Therefore, the analysis with respect to important variable detection has become a main part of the classification. Filter methods have been paid attention to such a goal. These methods essentially measure the strength of the relationship of each of genes with a binary outcome and then ranks them [44]. They have serval benefits for the analysis of huge amounts of gene expression data. First of all, they reduce high dimension into the appropriate dimension as well as the cost of computation time. Furthermore, they can also help improve the classification performance by increasing the likelihood to choose true important genes. There are a lot of filter methods applied to big data analysis of gene expression. One of the popular ranking methods is a logistic regression as a classifier. The value of maximum marginal likelihood estimator of logistic regression (MMLR) in each gene can be calculated using Equation (2) with a single gene. According to this method, a significant gene should have a large magnitude for its MMLR. Likewise, the list of ranking genes is made by the marginal strength of association with the response. That is, the top-ranked genes considered as most promising features have larger values of MMLR. To make a decision, the threshold of selecting top genes from the list, SIS would be used. It is a simple and effective algorithm which includes the true significant variables with probability tending toward one [43]. The cutoff value to select top-ranked genes is set up with $\left\lfloor \frac{n}{\log(n)} \right\rfloor$. Those filtered genes would be plugged into the sparse logistic regression models such as LASSO, MCP, and SCAD to further evaluate the performance of classification as well as gene selection. The Algorithm 1 describes the procedure of the proposed two-stage approach.

Algorithm 1 Proposed two-step procedure

Step 1: Sample 70% of samples randomly without replacement from the training set.
Step 2: Count frequency of each of genes from 100 models of λ values.
Step 3: Repeat Step 1 and Step 2 100 times.
Step 4: Calculate selection probability for each of variables based on Equation (10) and then rank them.
Step 5: Select top $\left\lfloor \frac{n}{\log(n)} \right\rfloor$ genes with the highest frequency.
Step 6: Apply them to sparse logistic regression methods to build prognostic models.

2.3. The Proposed Variable Ranking Method

We utilize the following elastic net ($\alpha = 0.5$) penalized regression method based on resampling technique to rank the features of importance using frequency. Elastic net is a combination of L_1(LASSO) and L_2(Ridge) and it has the benefit of performing well with highly correlated variables.

$$\hat{\beta}_{elastic\ net} = \underset{\beta}{\text{argmin}}\left[-\sum_{i=1}^{n}(y_i \log(\text{p}(x_i)) + (1-y_i)\log(1-\text{p}(x_i))) + \lambda\left(\frac{1-\alpha}{2}\sum_{i=1}^{d}|\beta_i|^2 + \alpha\sum_{i=1}^{d}|\beta_i|\right)\right]. \quad (9)$$

The following is the equation of selection probability in each gene based on the elastic net.

$$\text{SP}(g_l) = \frac{1}{K}\sum_{i=1}^{K}\frac{1}{L}\sum_{j=1}^{L} I(\beta_i \neq 0), \text{for } l = 1, 2, \ldots, d, \quad (10)$$

where K is the number of resampling, L is the number of λ, β_i is the regression coefficient corresponding gene l, and $I(\,)$ is the indicator variable. In each of K resampling, 100 values of λ are considered to build variable selection models. With SIS approach, top genes are selected and then applied those genes to LASSO, MCP, and SCAD penalized logistic regression method. The following is the algorithm of our proposed filter ranking method to rank the variable of importance. Figure 1 describes the schema of the proposed two-step approach.

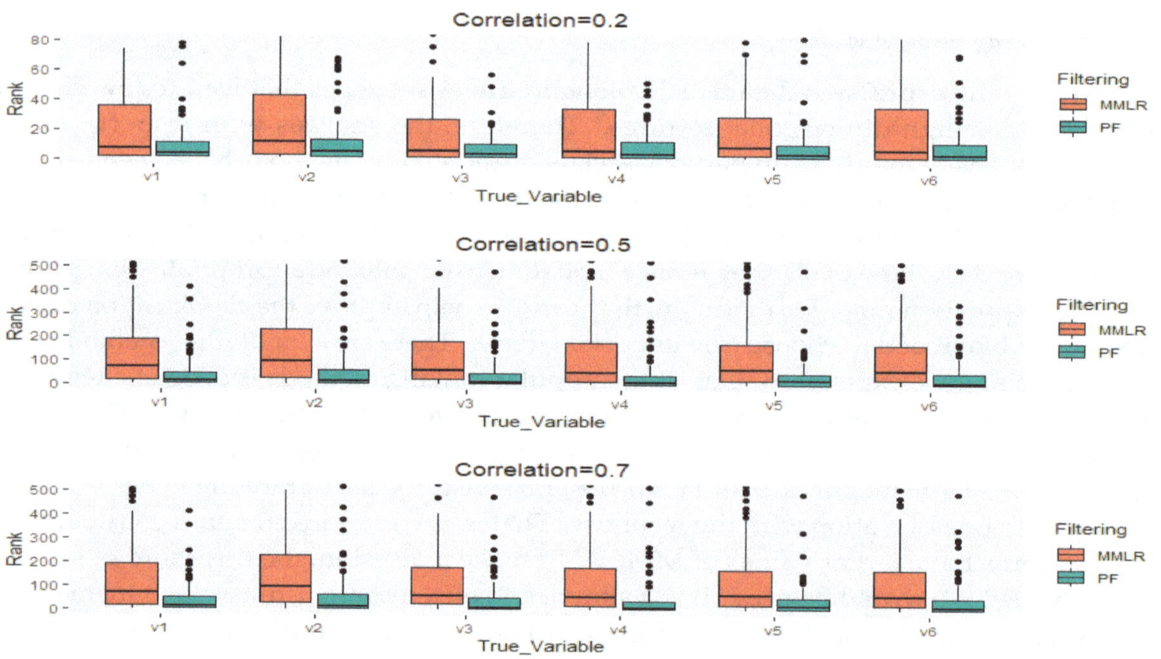

Figure 2. The boxplots of ranking true variables with the proposed filter method (PF) and MMLR method under correlation coefficients 0.2, 0.5, and 0.7 with 100 iterations.

2.4. *Metrics of Performance*

We calculated accuracy, the geometric mean of sensitivity and specificity (G-mean), and area under the receiver operating characteristic curve (AUROC). The accuracy is done with the following equation:

$$\text{Accuracy} = \frac{TP + TN}{TP + FP + TN + FN} \times 100, \tag{11}$$

where TP is the number of true positives, TN the number of true negatives, FP the number of false positives, and FN the number of false negatives.

The geometric mean of sensitivity and specificity was used to check the joint performance. The equation is as follows:

$$\text{Geometric mean} = \sqrt{Sensitivity \times Specificity}. \tag{12}$$

AUROC was also considered to evaluate the overall classification performance of the proposed method. A perfect overall classification produces an AUROC = 1 whereas a random overall classification has an AUROC = 0.5.

3. Results

3.1. *Simulation Results*

The response variable is generated by a sequence of Bernoulli trial with the following probability:

$$\pi_i(y_i = 1|x_i) = \frac{\exp(x_i\beta)}{1 + \exp(x_i\beta)}. \tag{13}$$

Data in each iteration are generated by using a multivariate normal distribution with mean 0 and variance-covariance matrix Σ with compound symmetry correlation structure whose diagonal elements are 1 and off-diagonal elements are $\rho =$ 0.2, 0.5, and 0.7, respectively. The following is the variance-covariance matrix:

$$\Sigma = \begin{pmatrix} 1 & \rho & \cdots & \rho \\ \rho & 1 & \cdots & \rho \\ \vdots & \vdots & \ddots & \vdots \\ \rho & \rho & \cdots & 1 \end{pmatrix}_{d \times d} \tag{14}$$

$x_j \sim N_d(0, \Sigma)$ is the jth row of design matrix X and y_i is a binary outcome generated by a Bernoulli trial with the probability from Equation (13). 100 datasets, where n is 200 and d is 1000, are generated and six true regression coefficients are generated from a uniform distribution with min and max values which are 2 and 4, respectively. The simulation data are applied to PF as well as MMLR as a first stage to show the superiority of performance that true variables are highly ranked. The variable ranking procedure in PF was run 100 times with resampling technique. Then the calculated average selection probabilities of each of the 1000 variables were used to rank them. The result of filtering performance was summarized as boxplots described in Figure 2. As seen in boxplots with three different correlation structures, the ranking of six true important variables is higher than that of MMLR. Under the correlation coefficient of 0.2, the average ranking of the six true variables with the proposed ranking method was at 22nd among 1000 variables, whereas the MMLR method was at 44th. In case of high correlation coefficients of 0.5 and 0.7, the proposed one was 59th and 62nd while MMLR was 132nd and 139th.

In addition, an average number of true variables included in filtered data with SIS is reported in Table 1. As seen in Table 1, the proposed method includes more true variables than MMLR in the various correlation settings. For each correlation setting, we used a paired two-sample *t*-test to check for significance level for the mean difference of the true number of variables between the two methods through 100 iterations, and all three were significant. That is, the proposed method is superior to MMLR for filtering true variables with SIS.

Table 2 shows that the performance of prediction as well as geometric mean with SIS-LASSO, SIS-MCP, and SIS-SCAD based on the proposed filter method are better than that of MMLR. As seen in TP (average number of true positives) of Table 2, all three variable selection methods capture mostly a true number of variable filtered from each of PF and MMLR. However, model size (MS) with the proposed filter ranking method is larger than that of MMLR because the methods with more true variables have a tendency to select unimportant variables highly correlated with the true variables. Figure 3 shows the boxplots of the area under the receiver operating characteristic (AUROC) for each of three methods with both proposed filter ranking and MMLR ranking methods based on SIS under three different correlation coefficients ($\rho = 0.2, 0.5,$ and 0.7). It also demonstrated that the AUROCs of SIS-methods based on the proposed filter ranking method is better performed compared to those of MMLR.

Table 1. An average number of true positives from the proposed PF and MMLR with SIS and a significance level of paired two-sample t-test for the mean difference of the number of true positives between two methods using the number of true positives obtained over 100 iterations.

Filtering Method	Metric	Correlation Coefficient		
		0.2	0.5	0.7
PF	Number of True Positive	5.4 (0.765)	4.21 (1.09)	3.11 (1.09)
MMRL		4.52(0.948)	2.15 (1.26)	0.29 (0.50)
two sample t-test (p value)		1.204×10^{-11}	$< 2.2 \times 10^{-16}$	$< 2.2 \times 10^{-16}$

*(): standard deviation.

The variable selection procedures of SIS-LASSO, SIS-MCP, and SIS-SCAD with both PF and MMRL filtered data were run 100 times using compound symmetry correlation structure with 0.2, 0.5, and 0.7. In each iteration, accuracy, area under the receiver operating characteristic (AUROC), geometric mean (G-mean) for sensitivity and specificity, true positives (TP), and false positives (FP). The results of performance for the variable selection methods with both filter ranking methods are summarized in Table 2.

3.2. *Real Data Analysis*

To test the performance of SIS-LASSO, SIS-MCP, and SIS-SCAD after filtering with the proposed method, we analyzed colon cancer gene expression data. The dataset contains 62 samples, which included 40 colon tumors and 22 normal colon tissue samples and 2000 genes whose gene expression information was extracted from DNA microarray data resulting from preprocessing; all 2,000 genes have unique expressed tags (ESTs) named. We also analyzed lung cancer gene expression data, GSE10072. The dataset includes 107 samples, which are made up of 49 normal lung and 58 lung tumor samples with 22,283 genes. Initially, we calculated the pairwise correlation for the normal and cancer samples combined to check the extent of overall correlation among genes in the colon cancer. The pairwise correlation is summarized in Figure 4 as a histogram with boxplot. The mean correlation between genes is 0.428 with a standard deviation of 0.203. It is clear that there is a high correlation between genes and this falls between the values tested in the simulation studies. In case of the lung cancer, the mean correlation between genes is 0.012 with a standard deviation of 0.246 because we used a full gene expression data unlike the colon gene expression data.

Table 2. Classification performance of proposed filtering (PF) compared to marginal maximum likelihood logistic regression estimator (MMLR) with SIS-LASSO, SIS-MCP, and SIS-SCAD over 100 iterations.

Correlation	Filtering	Methods	Accuracy	G-mean	TP	FP	MS
0.2	PF	SIS-LASSO	0.856(0.047)	0.854(0.049)	5.25(0.757)	0.019(0.002)	24.55(1.971)
		SIS-MCP	0.878(0.054)	0.877(0.056)	5.03(0.937)	0.006(0.003)	11.3(2.805)
		SIS-SCAD	0.878(0.053)	0.876(0.055)	5.18(0.757)	0.012(0.005)	17.24(5.053)
		average	0.871(0.051)	0.869(0.053)	5.153(0.817)	0.012(0.003)	17.697(3.276)
	MMLR	SIS-LASSO	0.847(0.056)	0.844(0.06)	4.3(0.99)	0.015(0.002)	18.73(2.131)
		SIS-MCP	0.86(0.061)	0.858(0.063)	4.21(0.988)	0.006(0.003)	10.32(2.449)
		SIS-SCAD	0.861(0.059)	0.858(0.062)	4.3(0.99)	0.011(0.004)	14.8(3.649)
		average	0.856(0.059)	0.853(0.062)	4.27(0.989)	0.011(0.003)	14.617(2.743)
0.5	PF	SIS-LASSO	0.886(0.041)	0.884(0.042)	3.65(1.266)	0.019(0.003)	22.71(2.267)
		SIS-MCP	0.869(0.055)	0.868(0.057)	2.93(1.409)	0.008(0.003)	10.87(2.058)
		SIS-SCAD	0.884(0.048)	0.883(0.05)	3.57(1.257)	0.017(0.004)	20.06(3.92)
		average	0.88(0.048)	0.878(0.05)	3.383(1.311)	0.015(0.003)	17.88(2.748)
	MMLR	SIS-LASSO	0.865(0.046)	0.863(0.047)	1.84(1.237)	0.015(0.003)	17.02(2.137)
		SIS-MCP	0.858(0.048)	0.857(0.048)	1.66(1.233)	0.008(0.002)	9.89(1.681)
		SIS-SCAD	0.863(0.047)	0.861(0.047)	1.83(1.28)	0.014(0.003)	15.64(2.873)
		average	0.862(0.047)	0.86(0.047)	1.777(1.25)	0.012(0.003)	14.183(2.23)
0.7	PF	SIS-LASSO	0.911(0.037)	0.911(0.038)	2.74(1.16)	0.019(0.003)	21.14(2.274)
		SIS-MCP	0.899(0.042)	0.899(0.043)	1.82(1.158)	0.007(0.002)	8.88(1.981)
		SIS-SCAD	0.907(0.038)	0.907(0.038)	2.68(1.171)	0.016(0.004)	18.88(3.699)
		average	0.906(0.039)	0.906(0.04)	2.413(1.163)	0.014(0.003)	16.3(2.651)
	MMLR	SIS-LASSO	0.887(0.037)	0.886(0.037)	0.26(0.543)	0.014(0.002)	13.72(1.724)
		SIS-MCP	0.881(0.04)	0.88(0.041)	0.21(0.498)	0.008(0.002)	7.75(1.591)
		SIS-SCAD	0.888(0.036)	0.888(0.037)	0.25(0.52)	0.013(0.002)	13.45(2.285)
		average	0.885(0.038)	0.885(0.038)	0.24(0.52)	0.012(0.002)	11.64(1.867)

*(): standard deviation.

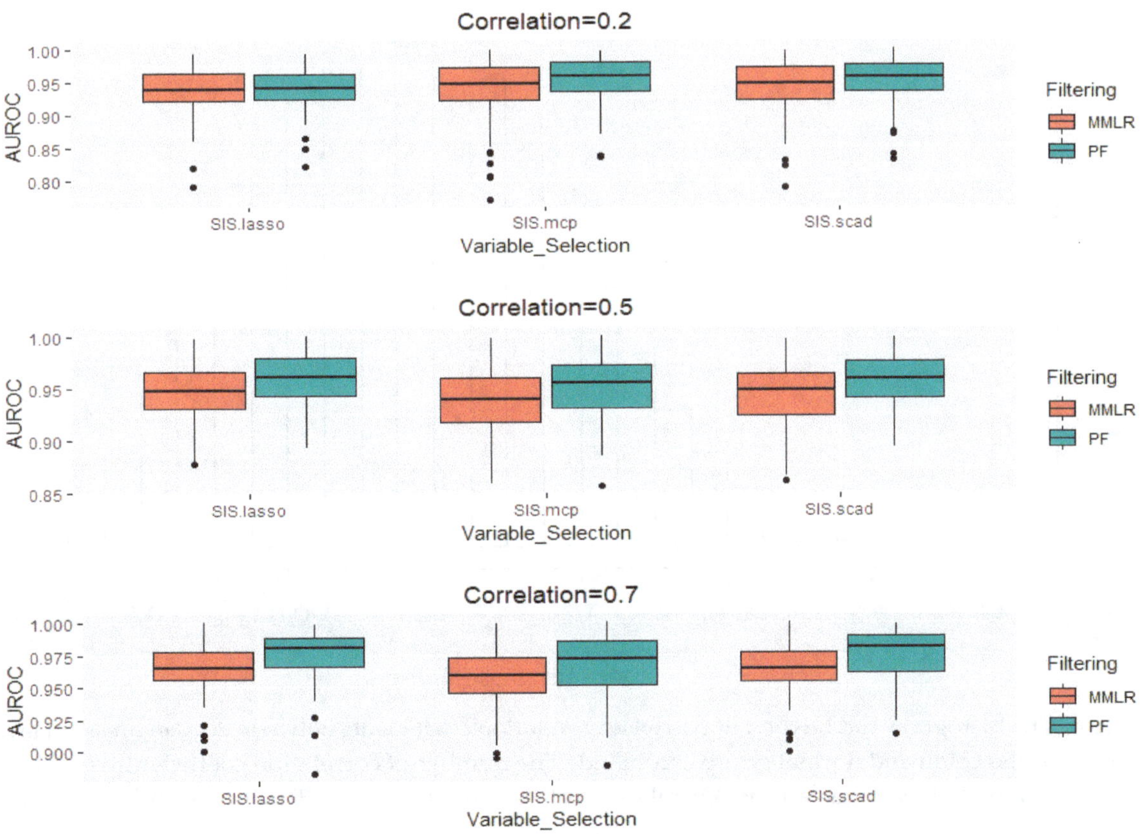

Figure 3. Comparison of area under the receiver operating characteristic (AUROC) with SIS-LASSO, SIS-MCP, and SIS-SCAD after filtering with both proposed filter ranking method and MMLR method under three correlation settings.

To obtain reliable results of the performance of accuracy, AUROC, and G-mean with screened variables, we iterated 100 times of both the colon and lung cancer data with resampling technique. In each iteration, we firstly divided the data into a training set of 70% of samples and a testing set of 30% of samples. Secondly, we select top ranked number of genes with SIS to plug into LASSO, MCP, and SCAD. Finally, we select genes with non-zero coefficients in the model and estimate the performance. We also count genes appeared in the models across three variable selection methods to build lists of ranking genes.

As in the simulation studies, we estimated the average of accuracy, AUROC, G-mean, and model size as the results of using three methods with PF. The results are reported in Table 3. SIS-LASSO with the performance of accuracy and AUROC, each of which is 0.803 and 0.886 with the standard deviations of 0.098 and 0.077 for colon and 0.976 and 0.998 with standard of 0.017 and 0.007, respectively, is relatively better compared to those of other variable selection methods in both datasets. We also presented the top 10 genes selected from each of the three lists of ranking genes across the three variable selection methods based on 100 resampling for the colon cancer and lung cancer data in both Tables 4 and 5. There are eight common genes of G50753, M76378, H08393 H55916, M63391, T62947, R80427, and T71025 among top 10 ranked genes from the results of three methods in the colon data.

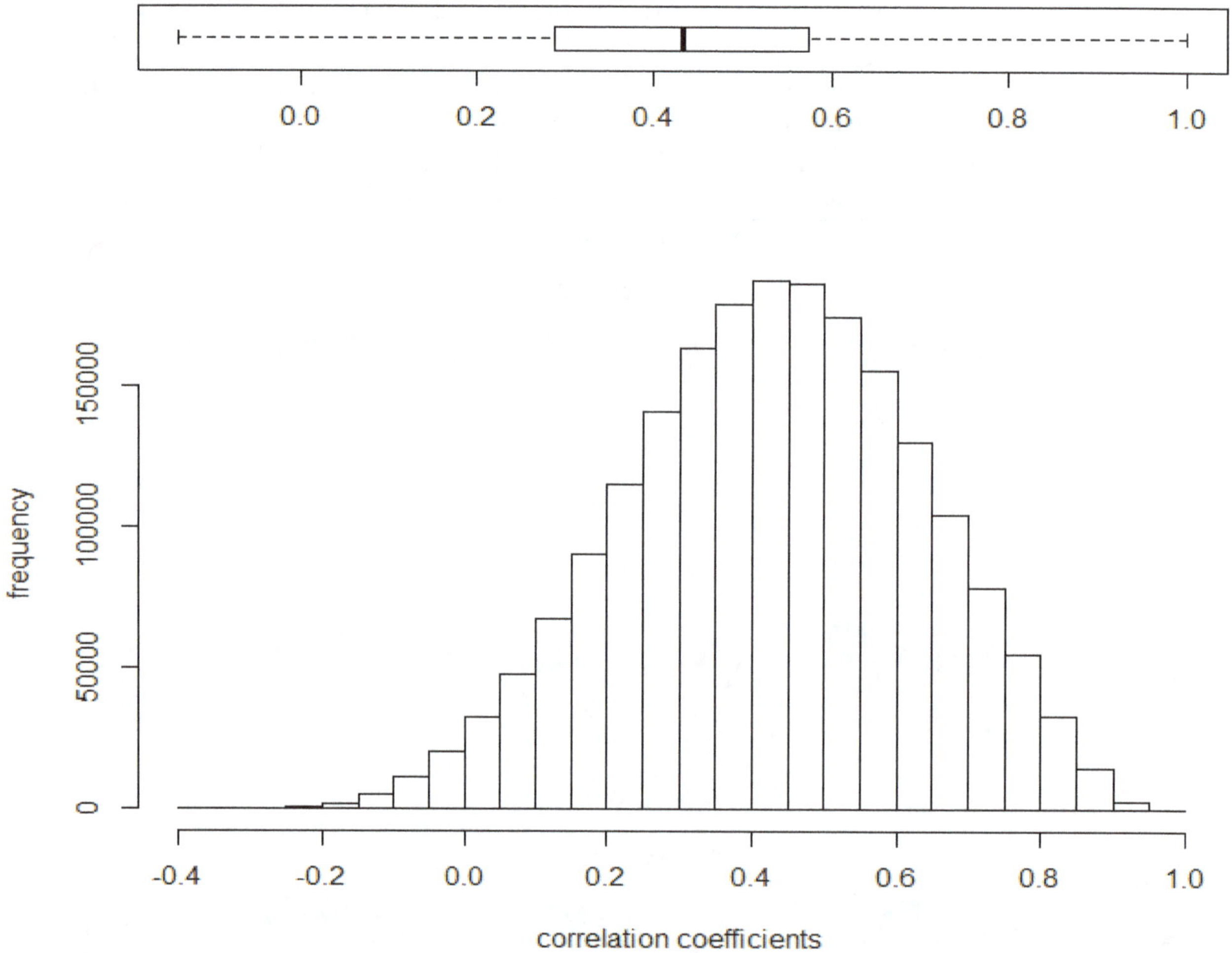

Figure 4. The histogram and boxplot of pairwise correlation coefficients between 2000 expression levels of genes for the colon and normal group combined. The number of correlation coefficients is 1,999,000. Two plots show that average pairwise correlation is 0.428 (median = 0.433) with a standard deviation of 0.203.

The gene of R87126 is common between the results of SIS-LASSO and SIS-MCP, T47377 between SIS-LASSO and SIS-SCAD, and T64012 between SIS-SCAD and SIS-MCP. In particular, G50753, H08393, and H55916 were consistently ranked.

Table 3. Classification performance of the proposed selection method with SIS-LASSO, SIS-MCP, and SIS-SCAD in both colon and lung cancer. It is the average performance resulting from 100 iterations.

Dataset	Method	Accuracy	AUROC	G-Mean	Model Size
Colon	SIS-LASSO	0.803 (0.098)	0.886 (0.077)	0.745 (0.144)	7.8 (1.47)
	SIS-MCP	0.793 (0.097)	0.864 (0.088)	0.748 (0.132)	4.14 (1.054)
	SIS-SCAD	0.798 (0.096)	0.874 (0.082)	0.753 (0.13)	6.73 (1.896)
Lung	SIS-LASSO	0.976 (0.017)	0.998 (0.007)	0.975 (0.019)	9.53 (1.453)
	SIS-MCP	0.952 (0.03)	0.983 (0.017)	0.95 (0.032)	1.09 (0.288)
	SIS-SCAD	0.975 (0.021)	0.997 (0.006)	0.973 (0.023)	8.65 (2.222)

(): standard deviation.

G50753, M63391, and M76378 were reported as significant genes related to colon cancer in [45]. M76378, H08393, H55916, M63391, R87126, and T47377 were also reported as genes associated with colon cancer in [46]. In addition, H08393 (collagen alpha 2(XI) chain) involved in cell adhesion is also known as a gene related to colon carcinoma whose cell has collagen-degrading activity as part of the metastatic process. T62947 has the potential to affect colon cancer by playing a role in controlling cell growth and proliferation through the selective translation of particular classes of mRNA. R80427 is also identified as genes distinguishing colon cancer in [47].

Table 4. Top 10 ranked genes with highest selection frequency from the lists of ranking genes using 100 times resampling approach across three methods of SIS-LASSO, SIS-MCP, and SIS-SCAD on both the colon cancer and the lung cancer gene expression data.

Rank	SIS-LASSO	SIS-MCP	SIS-SCAD
		Gene Accession ID	
1	Hsa.36689 *** (G50753)	Hsa.36689	Hsa.36689
2	Hsa.692.2 *** (M76378)	Hsa.8147	Hsa.692.2
3	Hsa.6814 *** (H08393)	Hsa.6814	Hsa.6814
4	Hsa.1660 *** (H55916)	Hsa.1660	Hsa.1660
5	Hsa.8147 *** (M63391)	Hsa.692.2	Hsa.33268
6	Hsa.5392 *** (T62947)	Hsa.12241 ** (T64012)	Hsa.12241
7	Hsa.37937 ** (R87126)	Hsa.33268	Hsa.5392
8	Hsa.33268 *** (R80427)	Hsa.5392	Hsa.8147
9	Hsa.3016 ** (T47377)	Hsa.8125	Hsa.8125
10	Hsa.8125 *** (T71025)	Hsa.37937	Hsa.3016

***: common genes in all three ranked gene lists, **: common genes in two of the three ranked gene lists.; (): GenBank Accession Number.

Likewise, the top 10 ranked genes in Table 4 from SIS-LASSO, SIS-MCP, and SIS-SCAD with PF were shown to play an important role in colon cancer. Figure 5 shows the boxplots of significantly differentially expressed genes between normal and colon samples on the eight genes found in all three methods. H08393 and H55916 are significantly expressed and downregulated while the other six are upregulated. In case of lung cancer data, there are five common genes of 21957_s_at, 209555_s_at, 209875_s_at, 209074_s_at, and 219213_at among top 10 ranked genes in lung cancer. The genes of 205357_s_at, 203980_at, 208982_at, and 220,170 are common between the results of SIS-LASSO and

SIS-MCP. The gene of 32625_at is common gene between SIS-LASSO and SIS-SCAD. Specially, first top four genes between the results of SIS-LASSO and SIS-SCAD have the same ranking. In addition, there are four unique genes of 209614_at from SIS-SCAD, 206209_s_at, 204271_s_at, 204396_s_at, and 219719_at from SIS-MCP. 219597_s_at (DUOX1) usually is downregulated and associated with lung breast cancer [48,49]. 209555_s_at (CD36) is also related to breast cancer [50] and affects the progression of lung cancer [51]. 209875_s_at (SPP1) is reported as a prognostic biomarker for lung adenocarcinoma [52,53]. 209074_s_at (FAM107A) is also emphasized as a lung cancer biomarker downregulated [54]. Although 219213_at (JAM2) are not directly known as a variant of lung cancer, it is worthwhile to be further investigated as a potential biomarker related to lung adenocarcinoma. We also found that most of five common genes play significant roles in lung cancer. Figure 6 also represents the boxplots of significantly differentially expressed genes between normal and colon samples on the five genes found commonly in the top ten ranked genes in all three methods. Only the gene of 209875_s_at (SPP1) is upregulated while the rest of them are downregulated.

Table 5. Top 10 ranked genes with highest selection frequency from lists of gene ranking using 100 times resampling approach of three methods of SIS-LASSO, SIS-MCP, and SIS-SCAD on the lung cancer gene expression data.

Rank	SIS-LASSO	SIS-MCP	SIS-SCAD
-		Gene Accession ID	
1	219597_s_at ***(DUOX1)	209555_s_at	219597_s_at
2	205357_s_at **	209074_s_at	205357_s_at
3	209555_s_at ***(CD36)	32625_at	209555_s_at
4	209875_s_at ***(SPP1)	206209_s_at *	209875_s_at
5	203980_at **	204271_s_at *	209074_s_at
6	208982_at **	204396_s_at *	219213_at
7	209074_s_at *** (FAM107A)	219213_at	208982_at
8	220170_at **	219597_s_at	220170_at
9	219213_at *** (JAM2)	219719_at *	209614_at *
10	32625_at **	209875_s_at	203980_at

***: common genes in all three ranked gene lists, **: common genes in two of the three ranked gene lists. *: unique genes. (): Gene symbol.

4. Discussion

We explored the feasibility of using the proposed feature ranking method as a filtering stage with Elastic net ($\alpha = 0.5$) based on a resampling approach followed by SIS as screening in conjunction with LASSO, MCP, and SCAD penalized logistic variable selection methods in high dimensional settings to improve the performance of variable selection and classification prediction. One of the currently popular methods achieving such a goal is MMLR. It ranks variables in order from largest to smallest scores of maximum likelihood. It performs poorly with important variables that are marginally weak but jointly and strongly associated with the response since it screens out such variables. The simulation studies demonstrated that the PF method retained more true important variables when compared to MMLR in Table 1. PF method also showed a better performance of retaining a true number of variables as the correlation of the variables was increased than MMRL. It is clear that the elastic net-based PF takes into account correlation among true important variables, while MMLR only considers marginal strength with the outcome variable. However, as seen in the results of using three variable selection methods with SIS based on both filtered data, the proportion of unimportant variables in the models is still high.

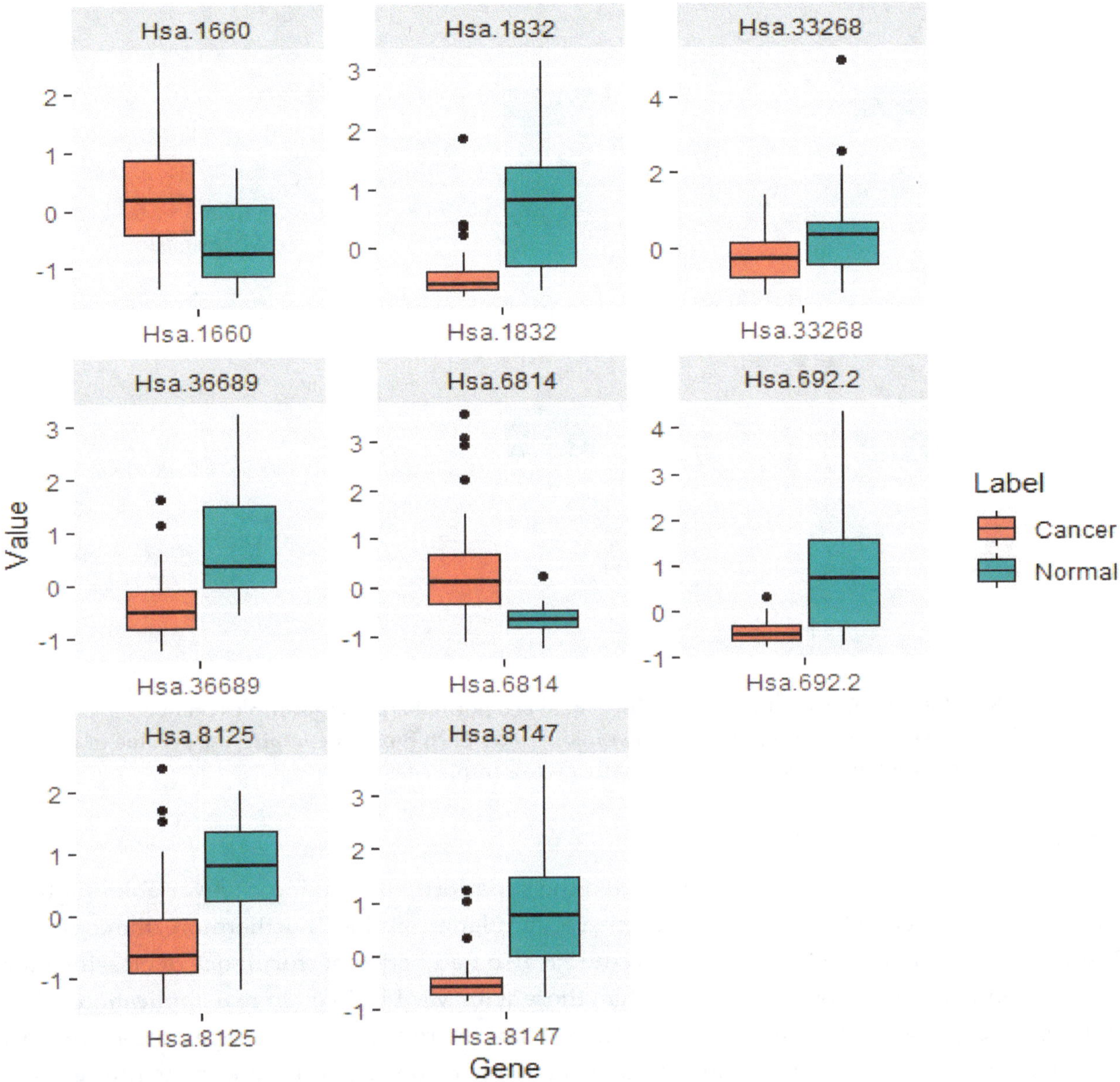

Figure 5. Boxplots of differential expression level between normal and colon samples on eight genes from SIS-LASSO, SIS-MCP, and SIS-SCAD with the ranked data. Each boxplot contains the *p*-value of mean differential expression between two groups with a two-sample *t*-test.

For further confirmation of the PF in selecting the most promising genes for superior classification performance, we applied it to a real example of both colon and lung cancer gene expression data. The SIS-LASSO method produced the best performance scores compared to SIS-MCP and SIS-SCAD. We also selected the top 10 ranked genes with highest selection frequency from the lists of ranking genes generated by the resampling approach in each of three variable selection methods to check gene selection consistency as well as biological significance connecting to colon and lung cancer. There were eight and five overlapped genes among top 10 ranked genes from the results of three methods in Tables 4 and 5, respectively. Most of the genes are reported as significant genes related to colon and lung carcinoma. In addition, some of the genes was consistently highly ranked across the three methods.

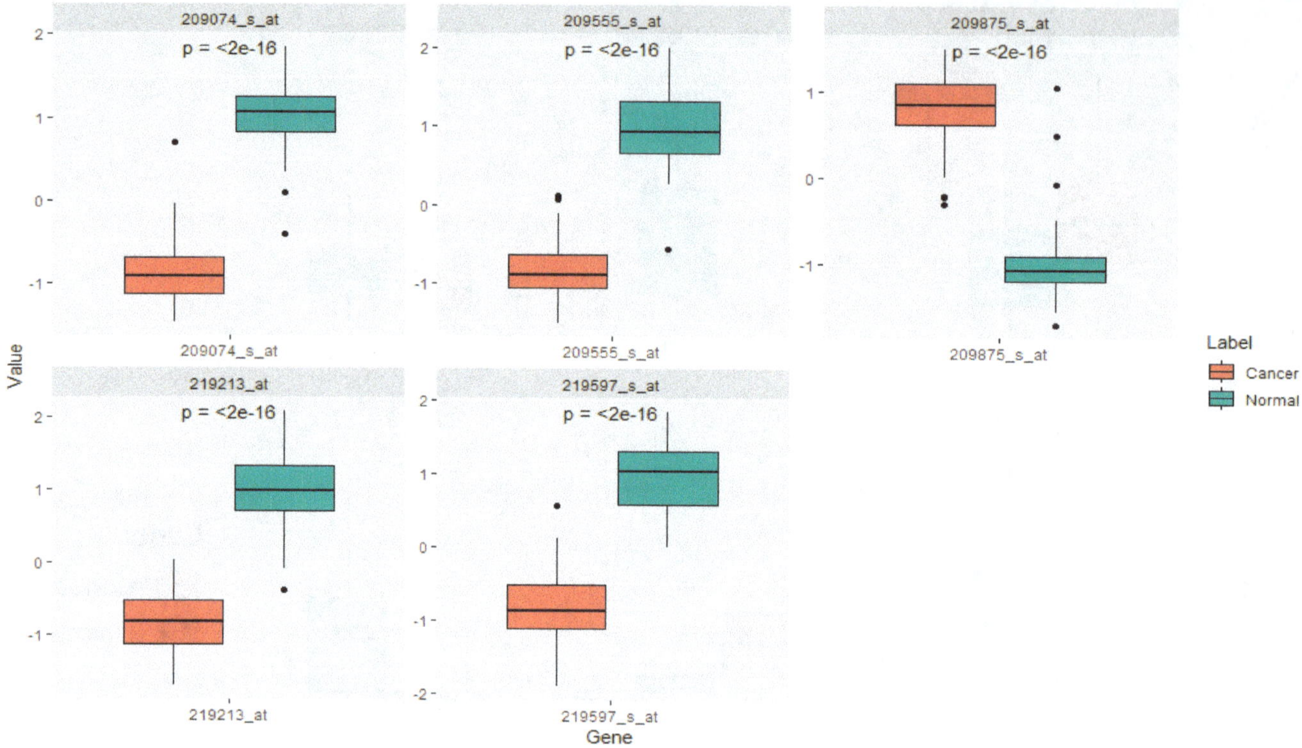

Figure 6. Boxplots of differential expression level between normal and lung samples on five genes from SIS-LASSO, SIS-MCP, and SIS-SCAD with the ranked data. Each boxplot contains the p-value of mean differential expression between two groups with a two-sample t-test.

5. Conclusions

In this study, the proposed PF demonstrated the superiority of ranking true variables highly as a filtering stage compared to MMLR through extensive simulation studies. Furthermore, the combination of SIS-LASSO, SIS-MCP, and SIS-SCAD with the PF also had better performance of classification as well as detection of true important variables than those with MMLR. Even in real applications of colon and lung gene expression data, it was demonstrated that the proposed two-stage procedure with PF consistently captures the most promising features related to colon and lung cancer. As future research, we plan to develop the methodology of variable selection with PF to increase the power of detecting true important variables as well as prediction of classification.

Author Contributions: All authors drafted the manuscript, and read and approved the final manuscript.

Acknowledgments: We thank the referees for reading the manuscript very carefully and making a number of valuable and kind comments that improve the presentation of the manuscript.

References

1. Sangjin, K.; Susan, H. High Dimensional Variable Selection with Error Control. *Biomed. Res. Int. Vol.* **2016**, *2016*. [CrossRef]
2. Shuangge, M.; Jian, H. Penalized feature selection and classification in bioinformatics. *Brief. Bioinform.* **2008**, *9*, 392–403.
3. Abhishek, B.; Shailendra, S. Gene Selection Using High Dimensional Gene Expression Data: An Appraisal. *Curr. Bioinform.* **2018**, *13*, 225–233. [CrossRef]

4. Hassan, T.; Elf, E.; lan, W. An efficient approach for feature construction of high-dimensional microarray data by random projections. *PLoS ONE* **2018**, *13*, e0196385. [CrossRef]
5. Bourgon, R. Independent filtering increases detection power for high-throughput experiments. *Proc. Natlacad. Sci.* **2010**, *107*, 9546–9951. [CrossRef] [PubMed]
6. Bourgon, R.; Gentleman, R.; Huber, W. Reply to Talloen et al.: Independent filtering is a generic approach that needs domain-specific adaptation. *Proc. Natl Acad. Sci. USA* **2010**, *107*, E175. [CrossRef]
7. Lu, J.; Peddada, S.D.; Bushel, P.R. Principal component analysis-based filtering improves detection for Affymetrix gene expression arrays. *Nucleic Acids Res.* **2011**, *e86*, 39. [CrossRef]
8. Jiang, H.; Doerge, R.W. A two-step multiple comparison procedure for a large number of tests and multiple treatments. *Stat. Appl. Genet. Mol. Biol.* **2006**, *5*. [CrossRef]
9. Ramskold, E.; Kerns, R.T. An abundance of ubiquitously expressed genes revealed by tissue transcriptome sequence data. *PLoS Comput. Biol.* **2009**, *5*, e1000598. [CrossRef]
10. Sultan, M.; Schulz, M.H.; Richard, H.; Magen, A.; Klingenhoff, A.; Scherf, M.; Seifert, M.; Borodina, T.; Soldatov, A.; Parkhomchuk, D.; et al. A global view of gene activity and alternative splicing by deep sequencing of the human transcriptome. *Science* **2008**, *321*, 956–960. [CrossRef]
11. Calle, M.L.; Urrea, V.; Malats, V.N.; Steen, K.V. Improving strategies for detecting genetic patterns of disease susceptibility in association studies. *Stat. Med.* **2008**, *27*, 6532–6546. [CrossRef] [PubMed]
12. Li, L.; Kabesch, M.; Bouzigon, E.; Demenais, F.; Farrall, M.; Moffatt, M.F.; Lin, X.; Liang, L. Using eQTL weights to improve power for genome-wide association studies: A genetic study of childhood asthma. *Fron. Genet.* **2013**, *4*, 103. [CrossRef] [PubMed]
13. Taqwa, A.A.; Siraj, M.M.; Zainal, A.; Elshoush, H.T.; Elhaj, F. Feature Selection Using Information Gain for Improved Structural-Based Alert Correlation. *PLoS ONE* **2016**, *11*, e0166017. [CrossRef]
14. Tan, Y.; Liu, Z. Feature selection and prediction with a Markov blanket structure learning algorithm. *BMC Bioinform.* **2013**, *14*, A3. [CrossRef]
15. Kakourou, A.; Mertens, B. Bayesian variable selection logistic regression with paired proteomic measurements. *Biom. J.* **2018**. [CrossRef] [PubMed]
16. Kursa, M.B.; Rudnicki, W.R. Feature Selection with the Boruta Package. *J. Stat. Softw.* **2010**, *36*, 1–13. [CrossRef]
17. Okeh, U.M.; Oyeka, I.C.A. Estimating the fisher's scoring matrix formula from the logistic model. *Am. J. Theor. Appl. Stat.* **2013**, *2*, 221–227.
18. Urbanowicz, R.J.; Meekerb, M.; La Cavaa, W.; Olsona, R.S.; Moorea, J.H. Relief-based feature selection: Introduction and review. *J. Biomed. Inform.* **2018**, *85*, 189–203. [CrossRef]
19. Milos, R.; Mohamed, G.; Nenad, F.; Zoran, O. Minimum redundancy maximum relevance feature selection approach for temporal gene expression data. *BMC Bioinform. BMC Ser.* **2017**, *18*, 9. [CrossRef]
20. Algamal, Z.Y.; Lee, M.H. A two-stage sparse logistic regression for optimal gene selection in high-dimensional microarray data classification. *Adv. Data Anal. Classif.* **2018**, 1–19. [CrossRef]
21. Le, T.T.; Urbanowicz, R.J.; Moore, J.H.; McKinney, B.A. Statistical Inference Relief (STIR) feature selection. *Bioinformatics* **2018**, *788*. [CrossRef] [PubMed]
22. Abdel-Aal, R.E. GMDH-based feature ranking and selection for improved classification of medical data. *J. Biomed. Inf.* **2005**, *38*, 456–468. [CrossRef] [PubMed]
23. Fan, J. Sure Independence screening for ultrahigh dimensional feature space. *J. R. Stat. Soc. B* **2008**, *70*, 849–911. [CrossRef]
24. Dizler, G.; Morrison, J.C.; Lan, Y.; Rosen, G.L. Fizzy: Feature subset selection for metagenomics. *BMC Bioinform.* **2015**, *1*, 358. [CrossRef]
25. Peng, H.; Long, F.; Ding, C. Feature selection based on mutual information: Criteria of max-dependency, max-relevance, and min-redundancy. *IEEE Trans. Pattern Anal. Mach. Intell.* **2005**, *27*, 1226–1238. [CrossRef]
26. Wei, M.; Chow, T.W.S.; Chan, R.H.M. Heterogeneous feature subset selection using mutual information based feature transformation. *Neurocomputing* **2015**, *168*, 706–718. [CrossRef]
27. Su, C.-T.; Yang, C.-H. Feature selection for the SVM: An application to hypertension diagnosis. *Expert Syst. Appl.* **2008**, *34*, 754–763. [CrossRef]

28. Tibshirani, R. Regression Shrinkage and Selection via the Lasso. *J. R. Stat. Soc. Ser. B* **1996**, *58*, 267–288. [CrossRef]
29. Zhang, C.-H. Nearly unbiased variable selection under minimax concave penalty. *Ann. Stat.* **2010**, *38*, 894–942. [CrossRef]
30. Fan, J.; Li, R. Variable Selection via Nonconcave Penalized Likelihood and its Oracle Properties. *J. Am. Stat. Assoc.* **2001**, *96*, 1348–1360. [CrossRef]
31. Two-Stage-Resources-2019. Available online: https://sites.google.com/site/sangjinkim0716/data-repository/two-stage-resources-2019 (accessed on 29 May 2019).
32. Pappua, V.; Panagopoulosb, O.P.; Xanthopoulosb, P.; Pardalosa, P.M. Sparse proximal support vector machines for features selection in high dimensional datasets. *Expert Syst. Appl.* **2015**, *42*, 9183–9191. [CrossRef]
33. Liao, J.G.; Chin, K.-V. Logistic regression for disease classification using micro data: Model selection in a large p and small n case. *Bioinformatics* **2007**, *23*, 1945–1951. [CrossRef] [PubMed]
34. Park, M.Y.; Hastie, T. Penalized logistic regression for detecting gene interactions. *Biostatistics* **2008**, *9*, 30–50. [CrossRef] [PubMed]
35. Bielza, C.; Robles, V.; Larrañaga, P. Regularized logistic regression without a penalty term: An application to cancer classification with microarray data. *Expert Syst. Appl.* **2011**, *38*, 5110–5118. [CrossRef]
36. Bootkrajang, J.; Kabán, A. Classification of mislabelled microarrays using robust sparse logistic regression. *Bioinformatics* **2013**, *29*, 870–877. [CrossRef] [PubMed]
37. Cawley, G.C.; Talbot, N.L.C. Gene selection in cancer classification using sparse logistic regression with Bayesian regularization. *Bioinformatics* **2006**, *22*, 2348–2355. [CrossRef] [PubMed]
38. Li, J.; Jia, Y.; Zhao, Z. Partly adaptive elastic net and its application to microarray classification. *Neural Comput. Appl.* **2012**, *22*, 1193–1200. [CrossRef]
39. Sun, H.; Wang, S. Penalized logistic regression for high-dimensional DNA methylation data with case-control studies. *Bioinformatics* **2012**, *28*, 1368–1375. [CrossRef] [PubMed]
40. Zhu, J.; Hastie, T. Classification of gene microarrays by penalized logistic regression. *Biostatistics* **2004**, *5*, 427–443. [CrossRef]
41. Liang, Y.; Liu, C.; Luan, X.-Z.; Leung, K.-S.; Chan, T.-M.; Xu, Z.-B.; Zhang, H. Sparse logistic regression with an L1/2 penalty for gene selection in cancer classification. *BMC Bioinform.* **2013**, *14*, 198–211. [CrossRef]
42. Huang, H.H.; Liu, X.Y.; Liang, Y. Feature selection and cancer classification via sparse logistic regression with the hybrid L1/2 + 2 regularization. *PLoS ONE* **2016**, *11*, e0149675. [CrossRef] [PubMed]
43. Algamal, Z.Y.; Lee, M.H. Penalized logistic regression with the adaptive LASSO for gene selection in high-dimensional cancer classification. *Expert Syst. Appl.* **2015**, *42*, 9326–9332. [CrossRef]
44. Ben Brahim, A.; Limam, M. A hybrid feature selection method based on instance learning and cooperative subset search. *Pattern Recogn. Lett.* **2016**, *69*, 28–34. [CrossRef]
45. Wang, Y.; Yang, X.-G.; Lu, Y. Informative Gene Selection for Microarray Classification via Adaptive Elastic Net with Conditional Mutual Information. *Appl. Math. Model.* **2019**, *71*, 286–297. [CrossRef]
46. Patrick, M.; John, S.; Rebecca, W. Methods for Bayesian Variable Selection with Binary Response Data using the EM algorithm. *arXiv* **2016**, arXiv:1605.05429.
47. Castellanos-Garzon, J.A.; Ramos-Gonzalez, J. A Gene Selection Approach based on Clustering for Classification Tasks in Colon Cancer. *Adv. Distrib. Comput. Artif. Intell. J.* **2015**, *4*. [CrossRef]
48. Fortunato, R.S.; Gomes, L.R.; Munford, V.; Pessoa, C.F.; Quinet, A.; Hecht, F.; Kajitani, G.S.; Milito, C.B.; Carvalho, D.P.; Martins Menck, C.F. DUOX1 Silencing in Mammary Cell Alters the Response to Genotoxic Stress. *Oxid. Med. Cell. Longev.* **2018**, *2018*. [CrossRef] [PubMed]
49. Little, A.C.; Sham, D.; Hristova, M.; Danyal, K.; Heppner, D.E.; Bauer, R.A.; Sipsey, L.M.; Habibovic, A.; van der Vliet, A. DUOX1 silencing in lung cancer promotes EMT, cancer stem cell characteristics and invasive properties. *Oncogenesis* **2016**, *5*. [CrossRef] [PubMed]
50. Liang, Y.; Han, H.; Liu, L.; Duan, Y.; Yang, X.; Ma, C.; Zhu, Y.; Han, J.; Li, X.; Chen, Y. CD36 plays a critical role in proliferation, migration and tamoxifen-inhibited growth of ER-positive breast cancer cells. *Oncogenesis* **2018**, *7*, 98. [CrossRef] [PubMed]

51. Sun, Q.; Zhang, W.; Guo, F. Hypermethylated CD36 gene affected the progression of lung cancer. *Genetics* **2018**, *678*, 395–406. [CrossRef] [PubMed]
52. Zhang, W.; Fan, J.; Chen, Q.; Lei, C.; Qiao, B.; Liu, Q. SPP1 and AGER as potential prognostic biomarkers for lung adenocarcinoma. *Oncol. Lett.* **2018**, *15*, 7028–7036. [CrossRef] [PubMed]
53. Ioanna, G.; Vasilieios, P.; Ioannis, L.; Nikolaos, K.; Theodora, A.; Georgios, S. Tumor cell-derived osteopontin promotes lung metastasis via both cell-autonomous and paracrine pathways. *Eur. Respir. J.* **2016**, *48*. [CrossRef]
54. Pastuszak-Lewandoska, D.; Czarnecka, K.H.; Nawrot, E.; Domanska, D.; Kiszalkiewicz, J. Decreased FAM107A Expression in Patients with Non-small Cell Lung Cancer. *Adv. Exp. Med. Biol.* **2015**, *852*, 39–48. [PubMed]

12

Constructing a Control Chart using Functional Data

Miguel Flores [1], Salvador Naya [2], Rubén Fernández-Casal [3], Sonia Zaragoza [4], Paula Raña [3] and Javier Tarrío-Saavedra [5,*]

1 MODES Group, Department of Mathematics, Escuela Politécnica Nacional, 170517 Quito, Ecuador; miguel.flores@epn.edu.ec
2 MODES Group, CITIC, ITMATI, Department of Mathematics, Escola Politécnica Superior, Campus Industrial, Universidade da Coruña, Mendizábal s/n, 15403 Ferrol, Spain; salva@udc.es
3 MODES Group, CITIC, Department of Mathematics, Faculty of Computer Science, Campus de Elviña, Universidade da Coruña, 15008 A Coruña, Spain; ruben.fcasal@udc.es (R.F.-C.); paula.rana@udc.es (P.R.)
4 PROTERM Group, Department of Naval and Industrial Engineering, Campus Industrial, Universidade da Coruña, Mendizábal s/n, 15403 Ferrol, Spain; szaragoza@udc.es
5 MODES Group, CITIC, Department of Mathematics, Escola Politécnica Superior, Campus Industrial, Universidade da Coruña, Mendizábal s/n, 15403 Ferrol, Spain
* Correspondence: javier.tarrio@udc.es

Abstract: This study proposes a control chart based on functional data to detect anomalies and estimate the normal output of industrial processes and services such as those related to the energy efficiency domain. Companies providing statistical consultancy services in the fields of energy efficiency; heating, ventilation and air conditioning (HVAC); installation and control; and big data for buildings, have been striving to solve the problem of automatic anomaly detection in buildings controlled by sensors. Given the functional nature of the critical to quality (CTQ) variables, this study proposed a new functional data analysis (FDA) control chart method based on the concept of data depth. Specifically, it developed a control methodology, including the Phase I and II control charts. It is based on the calculation of the depth of functional data, the identification of outliers by smooth bootstrap resampling and the customization of nonparametric rank control charts. A comprehensive simulation study, comprising scenarios defined with different degrees of dependence between curves, was conducted to evaluate the control procedure. The proposed statistical process control procedure was also applied to detect energy efficiency anomalies in the stores of a textile company in the Panama City. In this case, energy consumption has been defined as the CTQ variable of the HVAC system. Briefly, the proposed methodology, which combines FDA and multivariate techniques, adapts the concept of the control chart based on a specific case of functional data and thereby presents a novel alternative for controlling facilities in which the data are obtained by continuous monitoring, as is the case with a great deal of process in the framework of Industry 4.0.

Keywords: functional data analysis; statistical process control; control chart; data depth; nonparametric control chart; energy efficiency

---

1. Introduction

Generally, univariate and multivariate control charts are applied to identify anomalies in the industry and control the quality of products and services. However, the specific characteristics of data obtained by continuous monitoring (which has become the main trend due to advancements in sensoring and communications in the framework of Industry 4.0), require the use of increasingly sophisticated tools that take into account the presence of autocorrelation in critical to quality (CTQ) variables for the process under study.

Recently, diverse solutions have been provided that propose the modification and application of exponentially weighted moving average chart (EWMA) control charts [?]; control charts based on the fitting of autoregressive integrated moving average (ARIMA) models [?]; and the use of control charts for profiles, which is understood as the control of the parameters that define the relationship between two different CTQ variables [? ? ? ?]. Additionally, there are solutions that propose the use of techniques based on the control chart concept for anomaly detection such as machine learning techniques (neural networks and support vector machines, among others) and time series [? ? ? ? ? ? ? ?]. The increasingly common use of these tools is attributed to the fact that they adapt very well to the new paradigm of data (in the framework of Industry 4.0) defined by the continuous monitoring of numerous variables. Traditional control charts cannot be applied to many of the new cases due to their non-compliance with their starting hypotheses. In the domain of statistical quality control, the use of the concept of latent variables along with multivariate control charts is one of the most popular and useful alternatives to solve the problem of process control in the Industry 4.0. domain [? ? ?].

Many of these new data, usually curves, can be studied as functional data. This is the case of data in the fields of energy consumption, indoor and machine temperatures, relative humidity, amount of CO_2, among other variables, measured in all types of buildings. These new data, which currently characterize industrial processes, require innovative solutions, developed by researchers in the field of statistical quality control, based on the application of functional data analysis (FDA) techniques. Until now, few works have been published (in relation to the importance of the topic and the frequent monitoring of this type of data) on controlling the quality of processes when the data to be monitored are functional CTQ variables. Among the most outstanding research in this context, Reference [?] compared the performances of different control charts used for monitoring functional data; these charts are often identified as profiles in the statistical quality control literature. According to References [? ?], in the case of the monitoring of profiles, a set of statistical techniques (usually multivariate) is applied for controlling processes when they are defined by the functional relationship between two variables. However, a control chart for monitoring functional data, which is based on bootstrap resampling, is also proposed in Reference [?], whereas, in References [? ?], two complete monographs are proposed in which the concepts of control charts for profiles are adapted to the context of the functional data. Moreover, recently, new alternatives for outlier detection based on interlaboratory studies and those with application in industrial anomaly detection have been performed in the FDA framework [? ? ?].

This study intends to provide an alternative solution to the detection of out-of-control anomalies or states in the area of energy efficiency in buildings, specifically in commercial areas such as different stores of textile companies. Therefore, an important aim of this work is to solve the problem of such companies, specifically the ones providing lighting facilities; office automation; and heating, ventilation and air conditioning (HVAC). One such company is Fridama SL. The group comprises Fridama's facilities, Σqus (web platforms for big data management) and Nerxus (statistical consultancy for data analysis in the energy sector); the solution to the problem raised by this company is defined by continuously monitored data over time that can be treated as profiles or functional data. Based on this need, the present study proposes methodologies to build control charts that allow us to control the aforementioned processes. in order to test the applied statistical methodologies for anomaly detection properly, these companies have provided a database on a real HVAC installation, whose anomalies and assignable causes are identified by its maintenance personnel.

These energy efficiency data are non-Gaussian and autocorrelated, as the main component of continuously monitored data in the Industry 4.0 framework. Thus, the present proposal provides alternative control charts for Phases I and II of the statistical process control that can deal with non-Gaussian and autocorrelated data. As is well-known, using control charts, the process can be controlled in the following two phases: Phase I involves the stabilization or the calibration of the process and Phase II focuses on process monitoring. For Phase I, a control chart is proposed based on the depth measurement of a functional datum and the idea of atypical detection; in Phase II, a nonparametric range control chart is proposed, based on the calculation of the functional data depth, to monitor the process of interest. The procedure of building control charts corresponding to Phases I and II are explained in the next section; it also shows the results of their performance through a simulation study including diverse scenarios. The proposed methodology has been developed and programmed in R through different functions integrated in the "Quality Control Review" package, qcr, which can be freely and easily accessed by the practitioners.

Additionally, it provides a way to visualize the control charts for functional data, including the original data and the curves corresponding to the estimated control limits. This visual tool allows users and maintenance managers to relate each anomaly to an intuitively assignable cause. The results of the simulation study and its application to the real data show the usefulness of this control chart methodology in detecting anomalies when the process is defined by functional data, specifically the daily curves of energy consumption in commercial areas.

1.1. Alternatives of the Statistical Quality Control When the Basic Assumptions of the Control Charts Are Not Met

This control chart approximation for functional data is based on nonparametric multivariate control charts, which are useful when the assumptions of Gaussianity are not met. Therefore, a brief introduction of nonparametric control charts and statistical tools for dealing with autocorrelated data are presented in the following section.

Recently, analyses of the robustness of different control charts against the non-compliance of Gaussianity hypothesis have been developed. Particularly, several nonparametric control charts have been proposed. In this domain, we should highlight the works of Regina Liu [? ? ?], who has developed the r, Q and S control charts based on the data depth and rank concepts and the rank or ranges. In this regard, it is important to highlight the strong influence of Liu's works; it must be noted that several nonparametric alternatives for control charts are based on the concepts of data depth and rank. Moreover, it should be emphasized that one of the most important lines of research of the SQC, the profiles' control charts, are based, in many cases, on the application of nonparametric or semiparametric regression models [? ?]. Additionally, it is also important to highlight the use of resampling techniques for calculating natural control limits for different types of control charts [? ?]. The work in Reference [?] constitutes a complete monograph on the current trends for the development of control charts.

Despite their important advantages (no assumptions on the probability distribution of CTQ variables), nonparametric control charts are not predominantly used in the industrial and business framework. As noted in Reference [?], this can be attributed to several factors such as the lack of specific software, both commercial and free; the lack of general training in nonparametric statistics that generates insecurity and distrust in users and the lack of contrasted reference texts for the application of nonparametric methods in SQC. However, the research activity in this field is growing, as illustrated by the work in Reference [?].

Another starting hypothesis of control charts is the independence of the observations. The data continuously monitored over time by different sensors usually show a variable level of autocorrelation (the greater the correlation the closer the observations are in time). The application of standard techniques in the case of the violation of the independence hypothesis often results in the detection of an unacceptable number of false alarms [?]. Therefore, the development and analysis of techniques that remove the sample autocorrelation is fully justified. Within these techniques, the most widespread is the application of time series models (e.g., autoregressive moving average (ARMA) and ARIMA) to remove the correlation between observations and the subsequent monitoring of the error variable (difference between actual values and those estimated by the model) using control charts [? ? ?]. Moreover, References [? ?] propose the combination of control charts with adjustment algorithms. Finally, the monographs of [? ?] describe the most relevant lines of research for controlling and monitoring autocorrelated data.

It is important to note that this type of data is related to the control charts for functional data. The use of FDA techniques allows us to consider the autocorrelation of the data as well as, by means of resampling techniques, to circumvent parametric assumptions about the trend and dependence.

1.2. Control Charts for Phases I and II

The studied process is stabilized in the framework of the Phase I control charts, that is, the process is left under control [?]. This implies that there are no other assignable causes in the process, except for those present due to randomness itself (non assignable causes). This is equivalent to stating that the process remains stable—the parameters of the probability distribution of the CTQ characteristic remain unchanged. This allows us to estimate the natural control limits for the variable that describes the quality of the process. The natural control limits are estimated using a preliminary or calibration sample. Thus, in this stage, the main assignable causes of variation are identified and rectified through corrective actions. After the variations are removed, we estimate the natural control limits of the CTQ variable corresponding to the process under control. While most of the literature in the field has referred to Phase II control charts, within the new data paradigm for Industry 4.0, it would be essential to develop methodologies for Phase I [?]. In this regard, Reference [?] presents an overview of the recent contributions to the development of Phase I control charts. Among the latest proposals for control charts, the work of Grasso et al. [?] is especially interesting taking into account that is a Phase I control chart proposal for profiles based on the use of functional data depth concept (as well as the case of the present proposal). In fact, they proposed a Phase I control chart methodology for profiles belonging to the "multi-modelling framework", that includes the following stages: (1) a classification stage of new profiles into different operating modes or profile patterns using functional classification techniques from functional data depth measures (maximum depth approach based on Mode depth); (2) even the identification of a novel operation mode is included; (3) Once the operation mode corresponding to a new profile is identified, this is assigned to the corresponding control chart and consequently a suitable control charting method is applied to determine if the process remained in control over the period of time where those data were collected.

In this study, the atypical detection procedure presented in Reference [?] is adapted to develop a control chart for Phase I; this approach allows us to obtain a calibration sample from an under-control process that can be monitored through new measurements using a Phase II rank control chart. In fact, in Phase II, it is assumed that the process is under control [?]; additionally, in each new sample (monitored sample), the statistical rank is obtained and it is represented in the control chart with the estimated lower control limit.

2. Methodology

In this section, methodologies of control charts for functional random variables, **X**, which take values in a functional space $\mathcal{E} = L^2(T)$, with $T \subset \mathbb{R}$, are developed.

Based on observations of the functional variable **X**, we obtain a sample each of calibration and monitoring; they are functional datasets of sizes n and m, respectively, which allow us to build control charts for Phase I (in the case of the calibration sample) and Phase II (from the monitoring sample).

In the case of designing the control charts, any unstable or out-of-control process is refereed to the assignable causes of variation emerging from unusual and avoidable events that interrupt the process, that is, when they cause a change in the parameters of the underlying model of the profile or functional data [?]; these variations can be eliminated from the data by identifying and acting on the cause; this approach will avoid such variations in the future [?].

Concerning a method for building quality control charts, the probability of process instability (level of significance) presents a measure of its performance. This probability, provided that it is within the H_0, allows the derivation of at least one measure (observed value of the statistic) outside the control limits [?]. Phase I involves the development of a method and the estimation of the level of significance;

however, in the case of Phase II, is the process is assumed to be under control, that is, the level of significance is fixed.

2.1. Procedure for Building a Control Chart for Phase I (Stabilization)

As mentioned above, in Phase I, the retrospective data corresponding to a calibration sample of size n is analyzed for evaluating the stability of the process whose quality is ascertained over time and for estimating the parameters of the control chart [?].

In Phase I, a control chart is used to test the hypothesis that there is no change in the distribution of observations of the variable ordered with respect to time $\{\mathcal{X}_1(t), \mathcal{X}_2(t), \ldots, \mathcal{X}_n(t)\}$.

These changes can be punctual (freaks or bunches) or they can be related to a change in process that is evaluated (observable through patterns of sudden or gradual change in the mean of the process). Concerning isolated changes, it refers to the occurrence of at least one observation of the observed variable that deviates from the distribution of the other observations [?]. The hypothesis tested in Phase I is:

$$\begin{aligned} H_0 &: \mathcal{X}_i(t) \stackrel{d}{=} \mathcal{X}_j(t), \forall i, j \in \{1, \ldots, n\} \\ H_a &: \mathcal{X}_i(t) \stackrel{d}{\neq} \mathcal{X}_j(t), \text{for some } i, j \in \{1, \ldots, n\}. \end{aligned} \tag{1}$$

The stabilization phase of a process consists of applying an iterative method that allows the detection and elimination of those observations (in this context curves) that have a deviation with respect to the shape or magnitude of most of the observed curves. In other words, a curve is an atypical value if it has been generated by a different stochastic process or there is a change in the trend or variability of the stochastic process with respect to that corresponding to the remaining data [?]. The quantity of outliers is assumed to be unknown, although small.

In this study, the proposed control charts for Phase I are based on the adaptation of outlier detection methodologies from functional data, based on the data depth calculations [?].

The method of the detection of outliers for functional data [?] considers an atypical curve if its depth is less than a specific quantile of the distribution of depths estimated by bootstrap. In other words, an atypical curve will have a significantly low depth.

This procedure can be used with different types of functional depths. In the library `fda.usc` [?], the following alternatives are offered: Fraiman and Muniz (FM) [? ?], mode depth [?] and random projections depth (RP) [? ?]. In this work, the outlier detection procedure proposed by Febrero et al. `fda.usc` has been adapted to estimate a specific quantile of the depth distribution that plays the role of the lower control limit (LCL) for a Phase I control chart.

The control chart proposed for Phase I is estimated and plotted from a depth measurement (FM, RP or mode) and only the lower control limit (LCL) is considered to detect if the process is out-of-control (the depth of a curve is less than the LCL). In addition to this representation, an additional chart showing the original curves is proposed to provide an intuitive idea about the cause behind the identified anomaly (by checking its shape or magnitude) and thus to identify assignable causes. For instance, in the case of the HVAC energy consumption in buildings, anomalies may include stopping air handler, failure of the counter or sensor, a change in the regulation of the machines and adverse weather conditions.

In Phase I, we consider the functional random variable $\mathcal{X}$, from which a random sample is drawn—$\{\mathcal{X}_1(t), \mathcal{X}_2(t), \ldots, \mathcal{X}_n(t)\}$. These data are used to formulate the following steps to build the control chart for Phase I:

1. Calculate the depth corresponding to each observation of the dataset, $D(\mathcal{X}_i)_{i=1}^n$ and make a control chart based on the depth of each datum with respect to time.
2. Choose the parameter LCL according to the significance level of the control chart, that is, the percentage of false alarms (observations under control but erroneously detected as

out-of-control) is small (for example, $\alpha = 1\%$). The following procedures are used to estimate the LCL.

- Bootstrap procedure based on Trimmed:
  - Reorder the curves according to their depths in a decreasing way. $\mathcal{X}_{(1)}, \ldots, \mathcal{X}_{(n)}$.
  - It is assumed that, at most, $\alpha\%$ of the sample can be considered outliers.
  - B samples are obtained by a smoothed bootstrap procedure from the dataset resulting from discarding the $\alpha\%$ of the less depth curves. Let $\mathcal{X}_i^{*b}, i = 1, \ldots, n, b = 1, \ldots, B$ these bootstrap samples. To obtain each bootstrap resample:
    * A uniform sampling is done, i^* of $1, \ldots, [n(1-\alpha)]$.
    * Z_{i*} is generated as a Gaussian process with zero mean and variance and covariance matrix.

      $\gamma\Sigma_{\mathcal{X}}$ with $\gamma \in [0,1]$, where $\Sigma_{\mathcal{X}}$ is the variance and covariance matrix of the observations $\mathcal{X}_{(1)}, \ldots, \mathcal{X}_{([n(1-\alpha)])}$.
    * Finally, you get $\mathcal{X}_i^{*b} = \mathcal{X}_{(i*)} + Z_{i*}$.
  - For each $b = 1, \ldots, B$, we obtain C^b, which is the $\alpha\%$ empirical quantile corresponding to the distribution of the depths, $D(\mathcal{X}_i^{*b})$. The output value, C = LCL, is the median of the values $C^b, b = 1, \ldots, B$.
- Bootstrap procedure based on weighing:
  - The depths of the $\mathcal{X}_1, \ldots, \mathcal{X}_n$ curves are obtained.
  - B samples are obtained through a smoothed bootstrap from the original dataset weighted by their depths. Let $\mathcal{X}_i^{*b}, i = 1, \ldots, n, b = 1, \ldots, B$ these bootstrap samples. These replicas would be obtained:
    * Weighted sampling is performed, with i^* of $1, \ldots, n$ and with probability proportional to $D(\mathcal{X}_1), \ldots, D(\mathcal{X}_n)$.
    * Z_{i*} is generated as a Gaussian process with zero mean and variance-covariance matrix $\gamma\Sigma_{\mathcal{X}}$, with γ $[0,1]$, where $\Sigma_{\mathcal{X}}$ is the variance and covariance matrix of the observations $\mathcal{X}_1, \ldots, \mathcal{X}_n$.
    * Finally, we get $\mathcal{X}_i^{*b} = \mathcal{X}_{i*} + Z_{i*}$
  - For each $b = 1, \ldots, B$, we get C^b, which is the empirical quantile corresponding to the $\alpha\%$ of the distribution of the depths, $D(\mathcal{X}_i^{*b})$. The final value C = LCL is the quantile β of the values C^b, with $b = 1, \ldots, B$.

3. If there is any curve such that $D(\mathcal{X}_i) \leq$ LCL for a given LCL, then it would be considered atypical and the process would be out-of-control.
4. Additionally, a control chart including the original curves and the functional envelope obtained from 99% of the deeper bootstrap replicas is also developed.

Moreover, once the atypical curves are detected, they are removed and the procedure is repeated until the process becomes stable (under control), namely, defined by a total absence of atypical data.

2.2. Procedure for Building a Control Chart for Process Monitoring (Phase II)

Phase II deals with process monitoring; it involves quick detection of changes from the calibrated sample stabilized in Phase I [?]. For the scalar and multivariate cases, the process is monitored by taking the estimated control limits in Phase I [?] as a reference. In this phase, the average run length (ARL) is used to evaluate the performance of the control charts [?].

In this context, we test if there are deviations between the data obtained in Phase II, also called monitoring sample, $\{\mathcal{X}_{n+1}(t), \mathcal{X}_{n+2}(t), \ldots, \mathcal{X}_m(t)\}$ and the reference data $\{\mathcal{X}_1(t), \mathcal{X}_2(t), \ldots, \mathcal{X}_n(t)\}$ or calibration sample, taking into account its distribution.

In Phase II, in the univariate case, an F distribution for the under-control process is estimated from a calibration sample or reference data. It is assumed that F is the distribution of the CTQ variable of an under-control process (Phase I). This distribution is used to establish control limits that will be used to monitor the process in Phase II. The limits comprise an interval that will cover new observations of the process with a high probability, assuming that the process is under control. In Phase II, a sample of the G distribution is monitored. Therefore, in this stage, the methods for constructing control charts are based on contrasting the hypothesis:

$$\begin{aligned} H_0: &\quad F = G \\ H_1: &\quad F \neq G. \end{aligned} \tag{2}$$

In the FDA context, we do not have a density function for a functional random variable $\mathcal{X}$ that allows us to perform different tests corresponding to Phases I and II. Alternatively, we can estimate the distribution of the depth corresponding to the curves that belong to a sample of functional data. Thus, for Phase II, the use of the rank control charts [?] is proposed in an FDA context. The calculation of depths for functional data is proposed, which facilitate the calculation of ranks. These form the basis for developing r control charts, called rank charts.

The adaptation of r control charts involves the calculation of the rank statistic from functional depth measurements. The r chart plots the rank statistic as a function of time. The central control line CL = 0.5 serves as a reference point for observing possible patterns or trends. The lower limit is LCL = α, where α is the false alarm rate.

For the practical application of this proposal, the `qcr` [?] and `fda.usc` [?] R packages are used. The `fda.usc` package provides tools for the calculation of functional data depth, while the rank control chart, among other nonparametric charts proposed in Reference [?], is applied using the `qcr` package, which was developed by the authors.

As mentioned above, in Phase II, the curves corresponding to the calibration sample of Phase I, $\{\mathcal{X}_1(t), \mathcal{X}_2(t), \ldots, \mathcal{X}_n(t)\}$, are used for detecting changes or deviations with respect to the behavior of the process described in Phase I. The curves of the monitoring sample, $\{\mathcal{X}_{n+1}(t), \mathcal{X}_{n+2}(t), \ldots, \mathcal{X}_{n+m}(t)\}$, are collected; additionally, we test the hypothesis of each new curve belonging to the same distribution that corresponds to the calibration sample.

The procedure for estimating control charts for Phase II follows the same scheme presented in Reference [?]. Particularly, we assume that the rank statistic follows a uniform asymptotic distribution. This result is applicable to the functional case because of the way the rank corresponding to each observation is calculated (percentage of less deep curves than the observed ones). This fact provides a computational advantage in the monitoring of continuous processes since it eliminates the need to estimate the LCL. However, it is set as the quantile of a uniform distribution at a significance level α.

The procedure to develop the rank control chart for the functional univariate case (process defined by one functional variable) is detailed below, which can be easily generalized to the functional multivariate case (process defined by more than one functional variable).

1. From the reference sample $\{\mathcal{X}_1(t), \mathcal{X}_2(t), \ldots, \mathcal{X}_n(t)\}$, get the depths of the dataset, $D(\mathcal{X}_i)_{i=1}^{n}$ and the depths of the curves that make up the monitoring sample, $D(\mathcal{X}_j)_{j=n+1}^{n+m}$. The depth of each curve corresponding to the monitoring sample is calculated from the n curves of the calibration sample, that is, with respect to $n+1$ cases.
2. The rank statistic is calculated for each curve of the monitoring sample, $r_G(\mathcal{X}_{n+1}), \ldots, r_G(\mathcal{X}_{m+n})$, considering the calibration sample $\{\mathcal{X}_1(t), \mathcal{X}_2(t), \ldots, \mathcal{X}_n(t)\}$ as sample of reference.

$$r_G(\mathcal{X}) = \frac{\#\{\mathcal{X}_i | D(\mathcal{X}_i) \leq D(\mathcal{X}), i = 1, \ldots, n\}}{n}.$$

3. The values of the rank statistic, the lower control limit LCL = α and central line CL = 0.5 (the expected value of the rank statistic) are plotted, thereby generating the control chart.
4. Proceed to monitor the process. If at least the rank of a curve, $\mathcal{X}$, is such that $r_G(\mathcal{X}) \leq$ LCL, then the process is considered out-of-control.
5. A functional control chart is developed. This is a graphical tool that allows us to identify the possible assignable causes of the out-of-control states. The original curves are included, those correspond to the reference and monitoring samples, in addition to the functional envelope obtained from $(1-\alpha)\%$ of the deepest curves of the calibration sample.

3. Data Collection: Case Study of HVAC Installations in Commercial Areas

Here, the case study of HVAC installations' control are considered for a clothing store of a commercial area in the Panama City [?]. The data stream has been obtained by using the Σqus energy web platform. Sixteen CTQ variables are measured taking into account their ability to provide information about the energy efficiency, air quality and the thermal comfort of the store environment—indoor temperatures, overall energy consumption, HVAC energy consumption, CO_2 content in the air (ppm), relative humidity (%), temperatures of impulsion and return temperatures of the chillers in different areas of the store (see Figure ??).

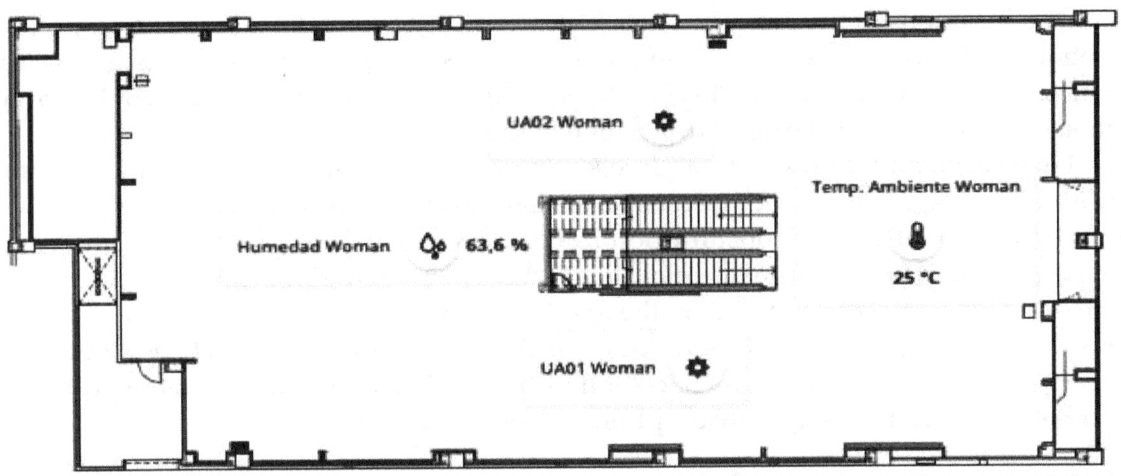

Figure 1. Plan of the case study store located in the Panama City.

Hourly measurements are obtained from 1 August 2017 to 31 October 2018. The operations of the HVAC facilities of the store start at 9:00 a.m. or 10:00 a.m. At start-up, the energy consumption peaks due to the characteristics of the HVAC installation. From 12:00 p.m., the consumption remains relatively constant the store closure at 20:00 p.m., 21:00 p.m. or 22:00 p.m. The shutdown takes about 1 or 2 h, with consumption falling at a constant rate of change. The resulting data can be considered functional data and thus FDA techniques can be applied. It is also important to note that this case study is a controlled study in which the anomalies and their assignable causes have been previously detected for the maintenance staff.

The data were obtained in the framework of a controlled environment where anomalies were identified by the maintenance staff. They are briefly described as follows:

- On 11 September, there was a decline in the air conditioning consumption at about midday.
- On 21, 22 and 30 September, the shopping center was closed and thus there was no energy consumption and the temperatures remained high.
- On 27 September, several maintenance tests were applied to the store's HVAC installations.
- On 29 September, the store's HVAC installations were stopped one hour earlier.

- Additionally, on 19 September, the air conditioning was stopped half an hour before the usual time. Particularly, there was a regulation change in the HVAC system.
- In mid-October, there was a leak in the air conditioning circuit. From that moment, energy consumption began to rise.
- On 1 November, repairing activities were performed. Consequently, the consumption decreased and the start-up consumption peak was removed.
- Between 17 and 20 November, the energy consumption in HVAC was again increased.

It is important to note that only working days have been studied in this work to evaluate the performance of the proposed control charts.

4. Application to Real Data

This section shows the usefulness and performance of the new graphical methodology for quality control using functional data, which is evaluated in the case study on the detection of energy efficiency anomalies of an HVAC installation. Specifically, the case study considers a commercial area of a well-known Galician apparel brand located in the Panama City. In this controlled case study, the anomalies and their assignable causes were detected by the maintenance personnel.

The following section shows the need to develop and apply FDA methodologies for control charts, considering the observations of the data of the present case study, particularly those corresponding to August. As mentioned above, in August, no event destabilized the process due to assignable causes. However, by using a methodology for scalar data (ignoring the autocorrelation between the variables), an unacceptable number of false alarms could be detected.

In the scalar case, boxplot [?] is commonly used to detect anomalous or atypical data. Figure **??** shows a traditional scalar approach for detecting outliers using boxplot. The left panel shows the boxplot for each variable of energy consumption in HVAC systems per hour, while the right panel shows the curves of daily energy consumption in HVAC systems, highlighting curves detected as outliers by the descriptive procedure based on the application of boxplots to each hourly consumption. In the usual procedure, atypical curves are those in which at least one point has been detected as an outlier in some boxplot; however, the drawback of this approach is that it increases the probability of type I error. It detects 12 daily energy consumption curves as outliers.

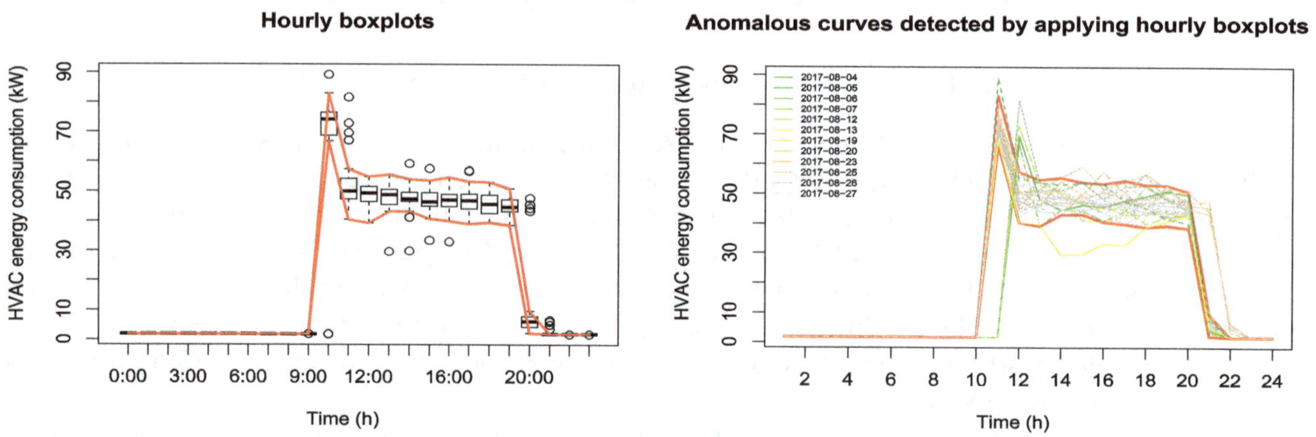

Figure 2. Detecting outliers in the heating, ventilation and air conditioning (HVAC) energy consumption by developing a boxplot for energy consumption per h.

Based on the information described in the previous section, first, we apply the data depth control chart for Phase I and, subsequently, the rank control chart to monitor the process during Phase II. The application of these two statistical techniques, together with the contribution of an intuitive graphic tool (to facilitate the detection of assignable causes for the anomalies), constitutes the new proposed procedure of control charts for functional data. Generally, the procedure can be summarized as follows:

- The energy consumption curves in HVAC during August and September accounts for the calibration sample. A broad temporal range has been taken to estimate the natural control limits of energy consumption in HVAC accurately.
- For Phase I, the reference or calibration sample is stabilized by using the control chart developed from the depth measures of the curves. Particularly, curves detected as atypical by FDA outlier data detection methods are removed from the calculations to estimate the natural control limits.
- Data corresponding to October and November are monitored to confirm that there is no deviation in the HVAC processes. If there is a change, that is, there are days with out-of-control HVAC energy consumption, then it was recommended that the possible assignable causes should be sought, which, when detected, should be removed in order to remove the corresponding process variations.

In Figure **??**, the black curves correspond to August (23 curves), whereas the gray curves account for the HVAC energy consumption in September (21 curves). The days from Monday to Friday are used in this study, taking into account that the work schedule is different on Saturday and Sunday. In September, the actual anomalies were detected (see Figure **??**, curves in red) using the Phase I control chart based on the functional data depth outlier detection methods. Particularly, the anomalies corresponding to 11, 21, 22, 27 and 29 September were identified (refer to Section **??** for more information on the assignable causes of these anomalies).

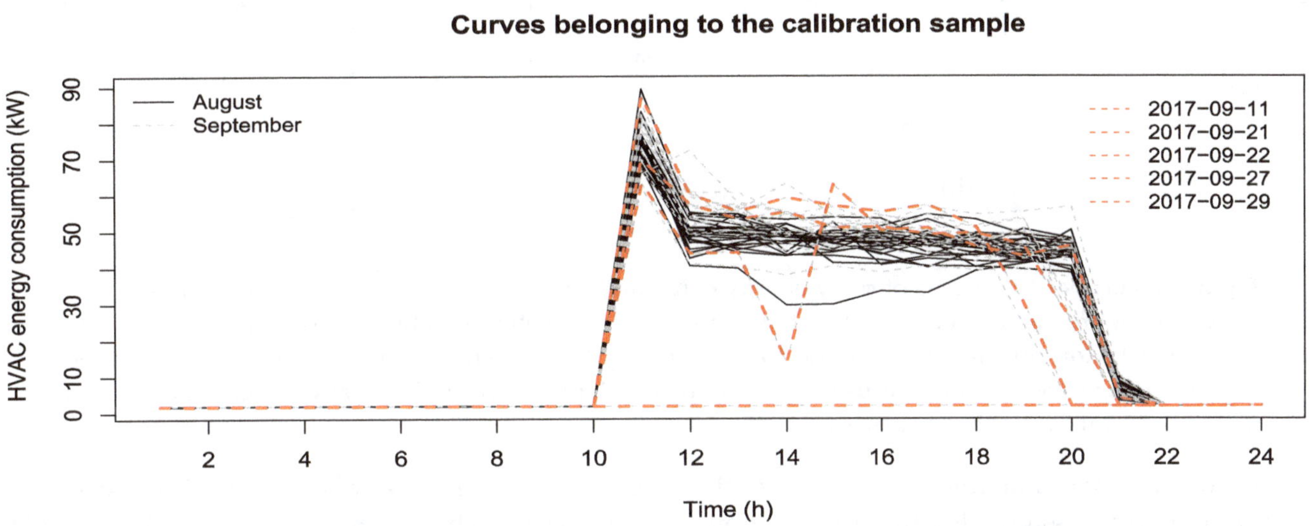

Figure 3. Daily curves of energy consumption in HVAC facilities of the Panama City's store. The natural control limits are estimated from the curves belonging to the calibration sample, which have been shown which (Phase I).

From this reference sample and based on the simulation study corresponding to the control chart based on the functional depths (see Section **??**), the mode functional depth and the weighed method for outlier detection are used. Additionally, a significance level of $\alpha = 0.025$ is used for the estimation of the LCL from $B = 500$ bootstrap resamples, a smoothing coefficient $\gamma = 0.8$, and a percentage of Trimming $trim = 0.05$, which allows obtaining an envelope of 95% of the deepest curves. The advantage of this procedure of detecting atypical curves in the Phase I control chart based on functional data is its flexibility to adapt to a wide variety of real problems, by regulating its parameters (γ and $trim$).

The left panel of the Figure **??** shows the original gray curves, the estimate of the median (blue curve), the functional trimmed mean (red curve) and the envelope corresponding to the 95% of the deepest curves, which is plotted in red. The curves detected as anomalous are shown in gray and dotted in black. After the mentioned atypical energy consumption curves are identified, they are removed from any calculation related to the estimation of the natural control limits, given that their assignable causes have been identified (causes apart from the randomness of the data). Subsequently, the process is repeated according to an iterative scheme until atypical curves associated with assignable causes are not identified.

Figure **??** shows the functional approximation of the rank control chart, where each datum or point accounts for a daily curve of energy consumption in HVAC. The first 23 points represent the HVAC energy consumption curves for August, whereas the next 21 points represent the energy consumption curves of September.

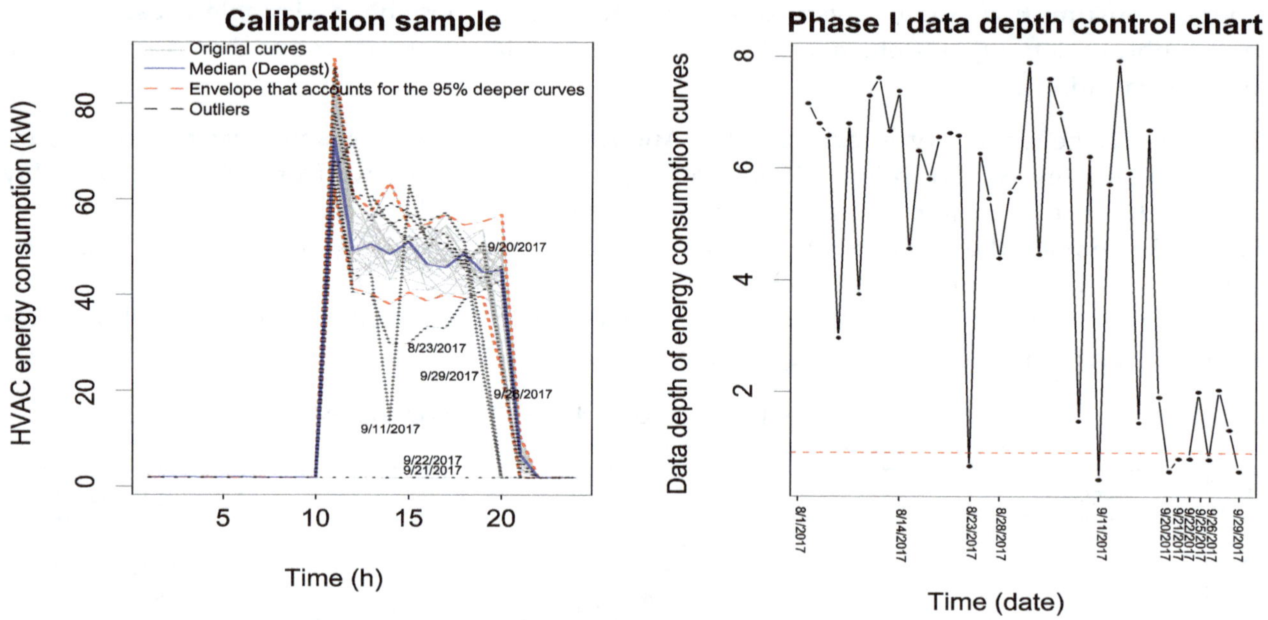

Figure 4. Left panel: Curves of daily energy consumption in HVAC facilities of the Panama City's store. The figure identifies curves detected as anomalies as well as functional position measurements. **Right** panel: Control chart corresponds to Phase I; they facilitate the detection of anomalous curves, thus eliminating the assignable causes of variation. The depths of each of the daily consumption curves and the natural control limit are shown.

In this first iteration, the previously identified curves are detected as anomalous. Their structure corresponds to an assignable cause of variation (see Section **??**). They correspond to 11, 21, 22 and 29 September. However, in this iteration, the anomalous curve corresponding to 27 September has been not detected. Additionally, the curves corresponding to 23 August, 20 and 26 September are detected. Moreover, the data depth corresponding to the curves from 19 September are very small—they are outside the lower control limit and thus they are identified as out-of-control states. The assignable cause corresponding to this behavior is that the air conditioning was being shut off half an hour earlier than usual.

In the next iteration, the following daily consumption curves are detected as out-of-control or anomalous: 4 August, 15, 19, 25 and 27 September. In this iteration, the HVAC energy consumption curve of 27 September and those corresponding to the last days of September are identified; the timing of HVAC shutdown was changed for the latter part of September. Finally, energy consumption curves are not detected as anomalous after the second iteration (see Figure **??**).

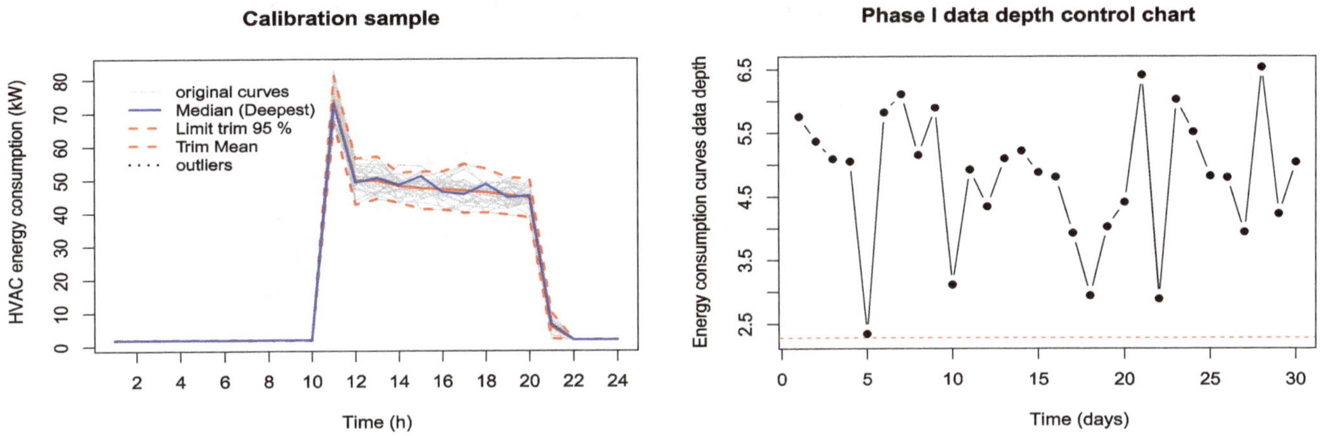

Figure 5. Under-control process (HVAC energy consumption is stabilized). **Left** panel: daily HVAC energy consumption curves in the Panama City store, showing the envelope corresponding to 95% of deeper curves. **Right** panel: control chart for Phase I based on functional data depth.

The process began with 44 curves, of which 9 (in the first iteration) and 5 (in the second iteration) curves were identified as anomalous. Of these 14 energy consumption curves, 2 and 12 curves correspond to August and September, respectively. Hence, the calibration sample comprises days in August and September, until 18 September. The current performance of energy efficiency of HVAC installation has been characterized through this reference or calibration sample. The next step is to detect changes (from the reference HVAC performance) in the HVAC system. This task belongs to Phase II control charts.

After Phase I, the Phase II control charts based on functional data, also called monitoring phase, is performed. Therefore, the sample to be monitored comprises days corresponding to October in which the HVAC facilities of the apparel store were operated at regular times. This is in consideration of the fact that a leak in the air conditioning circuit was recorded in mid-October, especially after the HVAC consumption began to rise. This behavior can be observed in Figure **??**; it plots the monitored sample as well as the curves of the calibration sample and their 95% envelope, which is estimated in Phase I.

Figure **??** shows that the HVAC energy consumption, the CTQ variable for the energy efficiency of HVAC installations, is out of control. The assignable cause is the leak in the air conditioning system. In order to rectify the anomalous operation of the HVAC installations, on 1 November, a provisional repair was performed. These actions produced a decrease in the HVAC energy consumption and attenuated peak energy consumption corresponding to the start-up (see the energy consumption curves before 11:00). Consequently, the consumption of the HVAC facilities in November is slightly different from that of August and September, which account for the calibration sample. However, this energy consumption resumes its rise between 17 and 20 October.

These changes can be observed in Figure **??** wherein it is observed that the HVAC energy consumption corresponding to the monitored sample is characterized by a greater variability than that corresponding to the calibration or reference sample. The results of the application of the rank control chart show that the process is out-of-control both for the monitoring samples of October and November. In October, the assignable cause is the fault in HVAC facilities, whereas the assignable cause in November is the change in the process due to repair activities that deals on a decreasing in HVAC energy consumption.

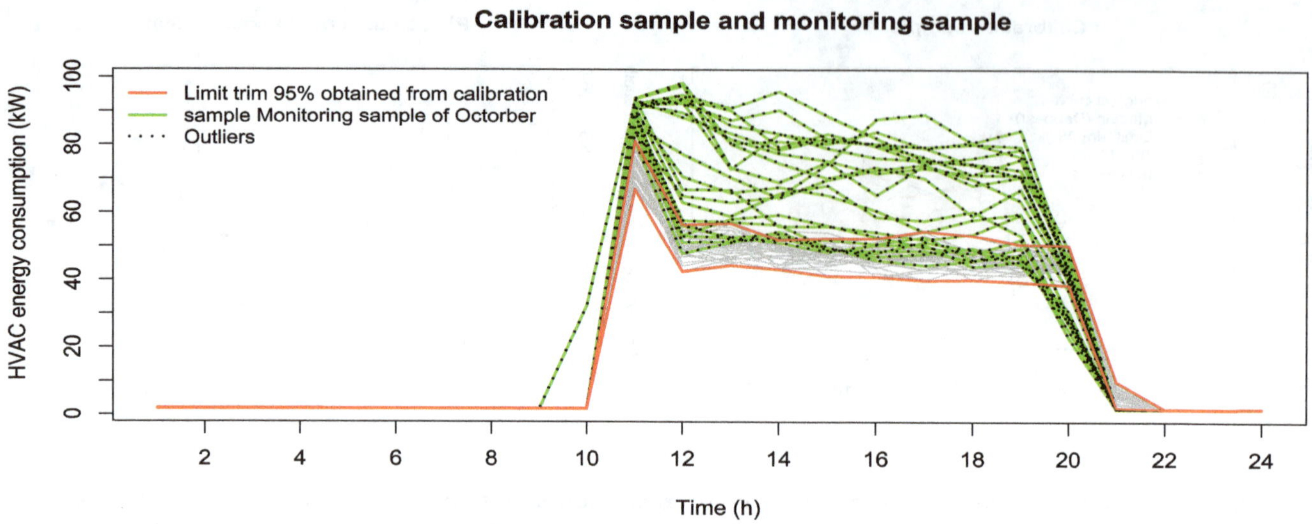

Figure 6. Reference of calibration and monitored samples corresponding to October.

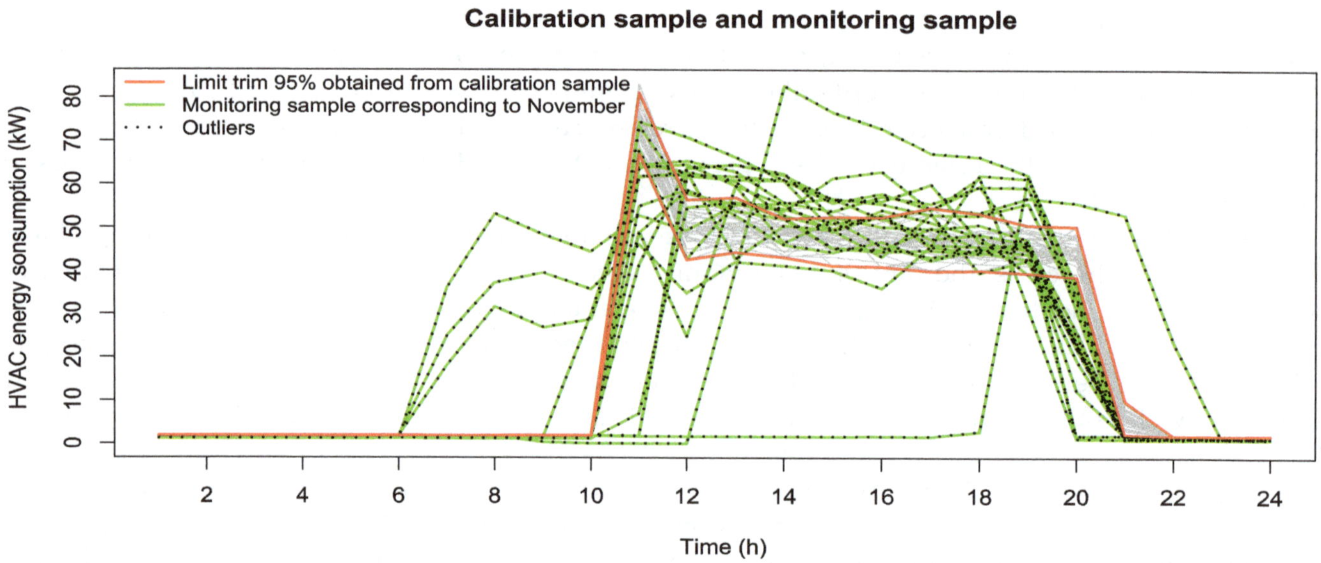

Figure 7. Reference (August and September) and monitoring samples corresponding to November.

Given that the process has changed, a new reference sample should be obtained and studied. The stabilization and monitoring process should begin for the next year, starting December. Figure **??** shows the rank control chart for the calibration (August and September) and monitoring samples of November. In practice, monitoring can be efficiently performed using the proposed control charts because you do not need to estimate the LCL through resampling procedures. Additionally, the rank control chart can be adapted for monitoring multivariate functional data, that is, when there are different types of curves that define the quality of a specific system or process simultaneously. For example, a direct case of application considers their first derivative along with the original curves.

The application of rank control charts also allows us to monitor the process by studying more than one functional variable. Particularly, it not only allows us to study the energy consumption curves but also the curves of daily temperature, daily relative humidity, daily CO_2 concentration, among others; they completely characterize the energy efficiency of the facilities and their thermal comfort and ambient air quality. Precisely, this procedure would allow us to perform a functional multivariate monitoring.

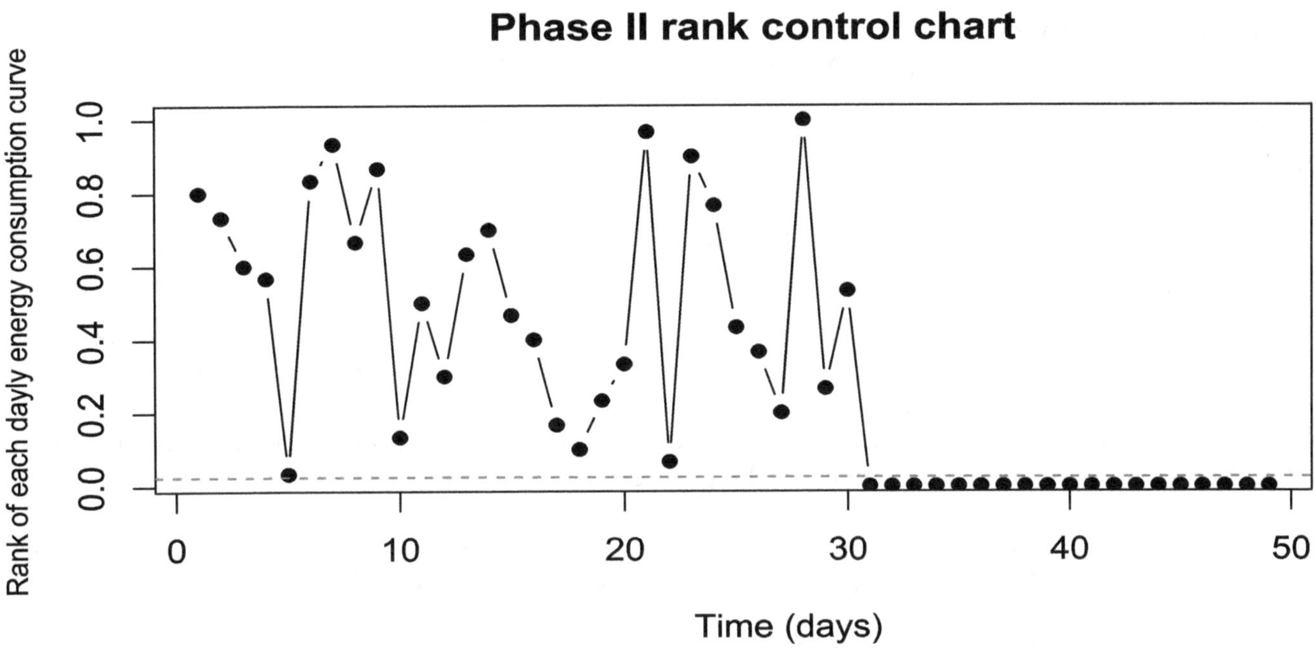

Figure 8. Rank control chart for Phase II of quality control. Both reference (corresponding to August and September) and monitoring (corresponding to November) samples are shown.

5. Simulation Study

The control chart performance is many times measured by observing its power $(1-\beta)$. It is defined as the probability of identifying an out-of-control state when the process state is actually out-of-control [?]. Moreover, assuming that the process is actually under control, the type I error (α) or false alarm rate will be defined as the probability of detecting an out-of-control signal [?].

In the case of the process is under control, the probability of identifying an out-of-control observation should be small enough to prevent an unacceptable number of false alarms. Otherwise, if the process is effectively out of control, the power should be high enough to detect the process change as quickly as possible [?].

Another common index for measuring the performance of a control chart is Average Run Length (ARL), which is defined as the average number of observations plotted before a signal is out-of-control. The ARL is equal to $\frac{1}{p}$ (if we can assume that the signals are independent, that the run length distribution is geometric), where p is the probability of having an out-of-control signal [?].

To evaluate the performance of a control chart in Phase II, the ARL_0 and ARL_1 are often used, which are the average number of observations until the first out of control is detected, in cases where the process is really under control ($ARL_0 = \frac{1}{\alpha}$) or actually is not ($ARL_1 = \frac{1}{1-\beta}$) [?]. The ARL_1 should be at its lowest to increase the probability of quickly identifying events $(1-\beta)$, power of a test that lead to the process being out-of-control [?].

Given that the F distribution is not known, a Monte Carlo simulation is designed to calculate the control charts power. The simulated scenarios allow us to estimate and compare the power of the control charts for different data depth measurements and for the case of independent and dependent functional data.

In this section, the performance of the control graphics proposed for Phases I and II will be evaluated. First, the simulation scheme designed in Febrero et al. [?] will be used to evaluate the performance of the control chart proposal for Phase I. Realizations of a Gaussian stochastic process have been proposed following the expression below [? ?]:

$$\mathcal{X}(t) = \mu(t) + \sigma(t) \cdot \epsilon(t), \tag{3}$$

whereby $\sigma^2(t) = 0.5$ and

$$\mu(t) = \mathbf{E}\left(\mathcal{X}(t)\right) = 30t(1-t)^{3/2}, \tag{4}$$

whereas $\epsilon(t)$ is a Gaussian process $\epsilon(t) \sim GP(\mathbf{0}, \Sigma)$ with $\mathbf{0}$ mean and variance-covariance matrix equal to

$$\mathbf{E}\left[\epsilon(t_i) \times \epsilon(t_j)\right] = e^{-\frac{|t_i - t_j|}{0.3}}.$$

Additionally, Reference [?] developed an alternative model to generate atypical curves, $\mu(t) = 30t^{3/2}(1-t)$. In Figure ??a, the two functional means are presented. The black curve accounts for the process mean without atypical curves, while the red curve is the mean of the process that generates the atypical curves.

The control charts proposed in this work have been designed to monitor the functional mean to detect two events—change in the mean of the process in terms of the magnitude and shape—which reveal that the process is not under control. For designing control charts for Phase II, it is assumed that the process is under control, that is, outliers are not detected. To generate simulation scenarios for each of these events, the following functional means have been considered:

- Mean of the model with a change in the magnitude:

$$\mu(t) = 30t(1-t)^{3/2} + \delta, \tag{5}$$

  by which δ denotes the change that goes from 0.4 to 2 in steps of 0.4.
- Mean of the model with a change in form:

$$\mu(t) = (1-\eta) \cdot 30t(1-t)^{3/2} + \eta \cdot 30t^{3/2}(1-t), \tag{6}$$

  where η is the change from 0.2 to 1 in steps of 0.2.

In Figure ??b, the green curve accounts for the functional mean of a process when there is a change in the magnitude ($\delta = 0.7$), while the curve of blue denotes the mean of a process when there is a change in the shape ($\eta = 0.3$).

In Febrero et al. [?], the functional data $\mathcal{X}_1, \ldots, \mathcal{X}_n$ denote realizations of a stochastic process $X(\cdot)$, assuming continuous trajectories in the $[a, b] = [0, 1]$ period and independence between the curves. However, simulation scenarios in which the simulated curves are defined by a variable degree of dependence have also been considered. This is because several practical applications of this type of chart are related to continuously monitored data with respect to time, forming functional time series, such as the curves of daily energy consumption in commercial areas. In this way, dependent curves are generated from the model $\tilde{Y}_i(t) = \mu(t) + \sigma(t) \cdot \tilde{\epsilon}(t)$, with $\tilde{\epsilon}(t) = \rho \cdot \tilde{\epsilon}_{i-1}(t) + (1-\rho) \cdot \epsilon_i(t)$, where ρ is the correlation measure between curves and $\sigma(t) = 0.5$ and both $\epsilon(t)$ and $\tilde{\epsilon}(t)$ are Gaussian processes [?].

In order to compare the results of the simulations in the scenarios defined by independence and dependence between the curves, the variance of ϵ is rescaled (we define the variance of the error $\tilde{\epsilon}$ to be one). Specifically, considering $\sigma^2_{\epsilon} = \frac{(1-\rho^2)}{(1-\rho)^2} = \frac{(1+\rho)}{(1-\rho)}$, you have $\sigma^2_{\tilde{\epsilon}} = 1$.

In Figure ??, different scenarios are presented considering the changes in the functional mean of the process, in the shape and magnitude, in the cases of independence and dependence between curves. The gray curves show the realizations of the process when it is under control (whose mean is the Equation (??)). However, the red curve in each graph accounts for the scenarios in which the presence of events that destabilize the process is considered, that is, the process is not under control. In Figure ??a,b, the cases of independence between curves and the presence of events defined by changes in the functional mean in terms of the magnitude and in shape, respectively, are shown.

However, Figure ??c,d show two cases defined by the presence of dependence between curves, including changes in the magnitude of the mean (panel c), with respect to its shape (panel d).

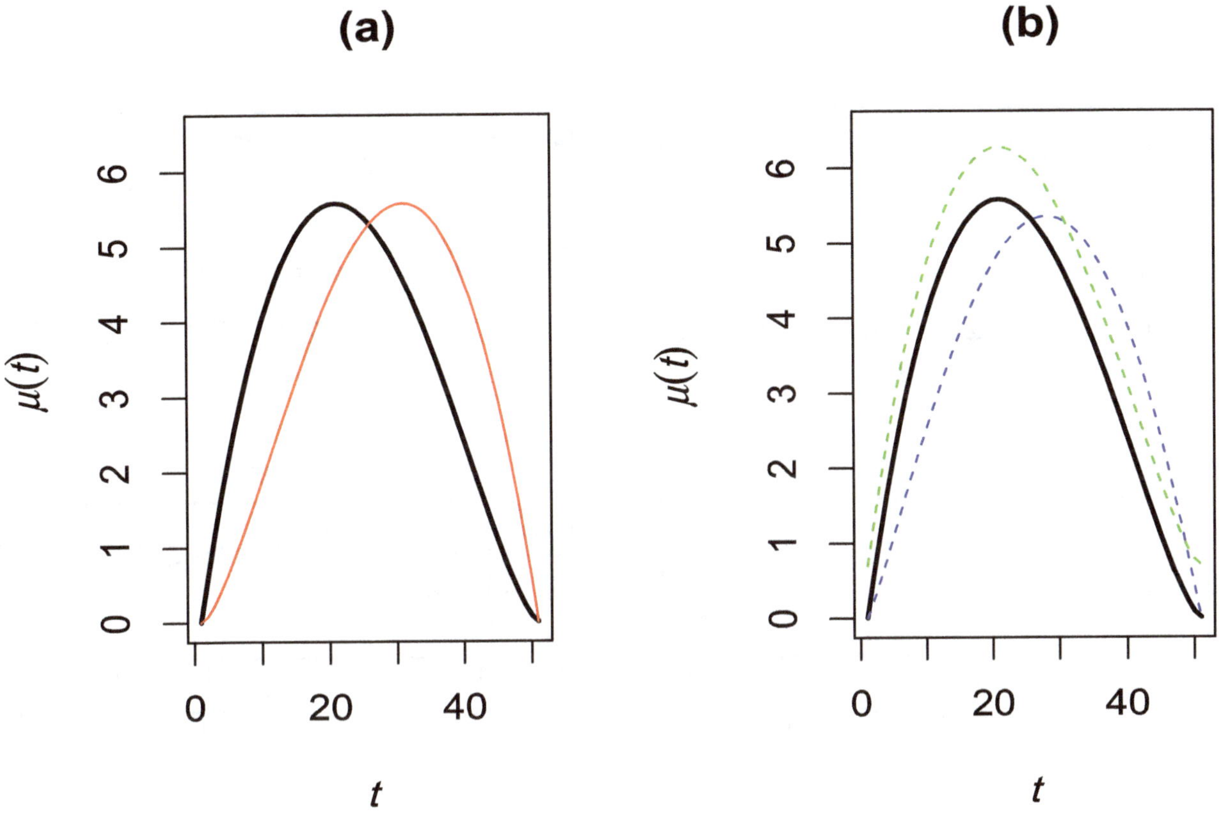

Figure 9. (**a**) Functional means and (**b**) changes in the shape and the magnitude in the mean of the process.

In the building energy efficiency domain, the out of control signals of energy consumption, temperature, CO_2 proportion and humidity, among others, can be defined by a change of shape and/or magnitude with respect to the under control signals analogous to those analyzed in the simulation study. The change in magnitude is related with a change in the scale of the studied process, for example, the increasing of energy consumption, temperature, humidity or concentration of CO_2 in all the hours corresponding of a specific day with respect to the otherwise normal pattern. This type of change is accounted by the addition of the δ term to the Equation (**??**) in order to obtain the Equation (**??**). Thus, the amount of that change is controlled by δ parameter. The green curve of Figure **??** accounts for an example of magnitude change. If we compare with the black one, we could realize that the two curves have the same shape and the only difference is the scale. On the other hand, changes in the shape of the curves are introduced by modifying the Equation (**??**) by the η parameter resulting in the Equation (**??**). In the energy efficiency domain, this type of change can be related to a change in the HVAC facilities programing (changes in temperature regulation, changes in the time schedule), in a failure of the HVAC in just one interval of the day, an extremely high or low level of occupation in a building and extreme changes in the weather, among other causes. The proposed simulation study has performed taking into account the specific domain where the FDA control chart approach is applied, namely energy efficiency in buildings. Thus, the shape of these types of profiles is similar to that corresponding to CO_2, temperature, energy consumption and humidity curves in buildings. In order to measure the performance of the proposed control charts for very different types of profiles, new studies may be necessary.

In the following section, a simulation study is performed to determine conditions under which it can be verified that the smooth bootstrap procedure works when there is independence and dependence between curves.

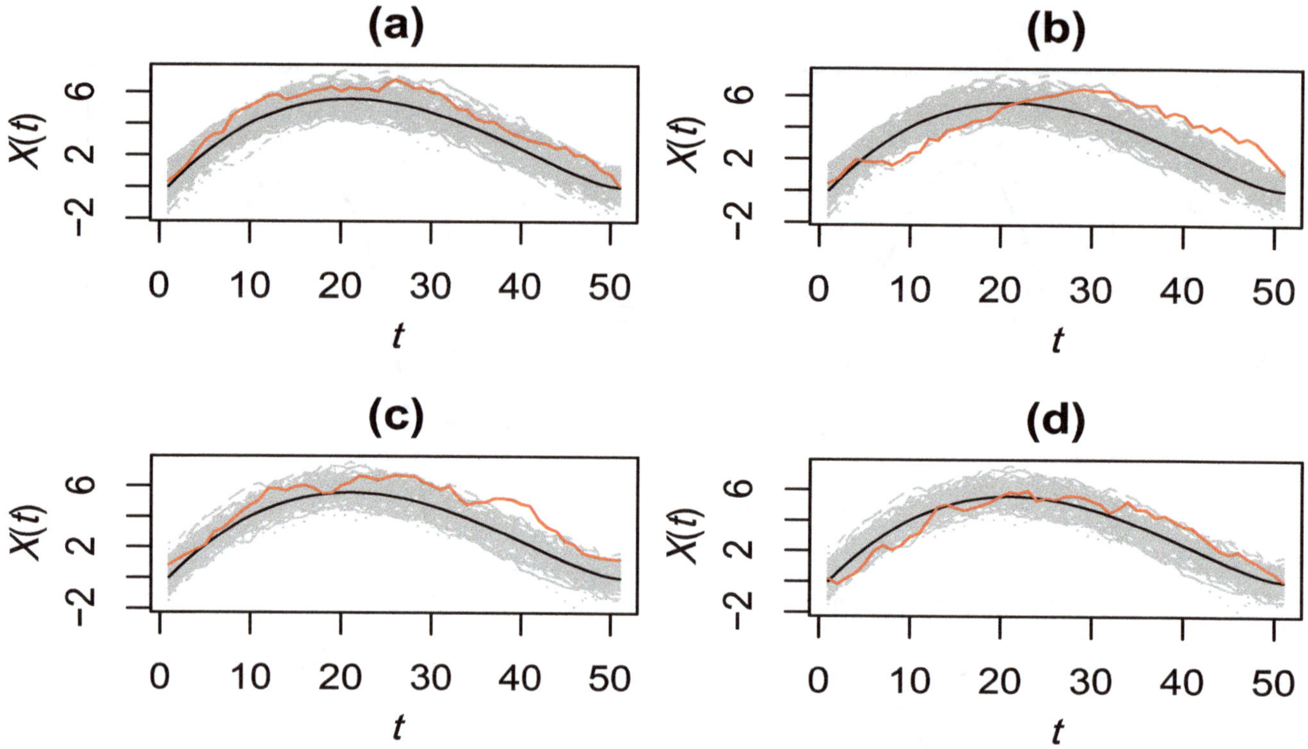

Figure 10. Scenarios in which independence between curves is studied. Changes in the functional mean with respect to its magnitude (**a**) and shape (**b**) are shown. In the case of dependence, panels (**c**) and (**d**) show the simulation scenarios in which changes in the magnitude and shape, respectively, are observed in the functional mean.

5.1. Measurement and Comparison of the Performance of the Control Chart Proposed for Phase I

The performance of the control chart is estimated and compared from the generation of calibrated samples of size 50 and 100 (curves). For each sample, different functional depth measurements described in Section **??** are calculated and the outlier detection robust procedures (weighted and trimmed) are applied for the estimation of type I error when the process is under control and the power of the test when the process is out-of-control. For the estimation of type I error, each scenario is replicated 1000 times ($n = 50$, 100, assuming independence and dependence between curves). When the power of the test is estimated, in each scenario (assuming independence and dependence), a curve within the alternative hypothesis is generated; this procedure is also repeated 1000 times.

Following the scheme described in Reference [?], curves observed at equidistant points are considered; the number of points that define each curve is 51 in the interval $[0, 1]$. From 1000 resamples ($B = 1000$) and with a 2.5% trimming procedure (removing less deep curves), a smoothing bootstrap procedure defined by a smoothing factor $\gamma = 0.05$ is applied to estimate the $C = 0.01$ quantile representing the LCL.

First, a simulation study is performed to estimate and compare the type I error ($\alpha = 0.01$ is fixed) of the proposed control chart, assuming scenarios with independence and dependence between curves. Subsequently, a similar study is carried out to estimate and evaluate the power of the control chart to detect out-of-control signals in different situations (independence, dependence, different sample size and a change in the shape or magnitude).

In Table **??**, the results of the estimation of the false alarm rate (type I error) in the independence scenario are shown. The average of the percentage of false out-of-control signals (type I error) detected by the procedure shown above are very close to the nominal 1% for the two considered sample sizes. Furthermore, it can be observed that, when n increases, the type I error percentages are closer to the nominal level. In general, for a sample of size $n = 100$, the results of applying the weighted method are closer to α, especially when using the mode depth measurement. The results obtained in the simulations are similar to those presented in Reference [**?**].

Table 1. Estimation of the false alarm rate (%) for the case of independence between curves using a nominal type I error of 1%.

n	Method	FM	RP	Mode
50	Weighted	1.94	1.89	1.49
	Trimmed	1.34	1.95	1.36
100	Weighted	1.55	1.75	1.25
	Trimmed	1.67	2.33	1.76

In any process, type I error increases the production cost. Hence, it is essential not to overestimate this error rate when managing the quality.

Table **??** shows the results of the simulation to evaluate the ability of the control chart to detect a change in the shape or magnitude of the functional mean of the process through the estimation of its power $(1-\beta)$. The percentage of out-of-control signals (outliers) correctly detect when the population defined by the Equation (**??**) is contaminated with curves belonging to the M_1 model (Equation (**??**)) and M_2 model (Equation (**??**)); it is denoted by p_c; however, the percentage of false alarms (false states out of control) is p_f. These parameters have been estimated, in all the scenarios assuming the independence of curves, using the average of the corresponding empirical values, $\hat{p}_c$ and $\hat{p}_f$.

Table **??** shows that a better performance is achieved when the curves of model M_1 (where changes in the magnitude of the proposed control chart are simulated) are studied. Precisely, $\hat{p}_f$ and $\hat{p}_c$ are closer to the nominal α and $(1-\beta)$. When identifying changes in the shape of the process average, M_2, the mode depth provides the highest percentages of correctly detected out-of-control signals. However, in the case of the M_1 model, the use of RP depth provides percentages of the detection of the true out-of-control states lower than those corresponding to the use of FM and mode depths. With respect to a robust method for outlier detection, the performance is similar in all the scenarios. However, an exception is the case wherein the RP depth is used; it reveals the low performance of the control chart in detecting observations corresponding to actual out-of-control states.

Briefly, the detection rate of false out-of-control signals for the independence scenario is close to 1%. However, when using the trimmed method, the detection rate of false out-of-control signals is overestimated but this percentage decreases when the sample size increases.

The results of the false alarm rate (type I error) for scenarios defined by dependence between curves are shown in Table **??**. It is important to note that, for different values of ρ, very similar results with respect to the independence scenarios have been obtained. Precisely, the average of the percentages of false out-of-control signals are close to the nominal 1% in the two studied sample sizes. Additionally, when n increases, the type I error percentages are closer to the nominal level. However, some differences are observed when the RP data depth measure is used to develop the control chart. In this case, there is an overestimation of the percentage of false out-of-control signals.

Table 2. Percentages of $\hat{p}_c$ and $\hat{p}_f$ for the cases of curves simulated with M_1 (Equation (??)) and M_2 (Equation (??)) models, assuming independence between curves.

			δ	0.4		0.8		1.2		1.6		2	
Stage	**Displays**	**Depth**	**Method**	$\hat{p}_f$	$\hat{p}_c$	$\hat{p}_f$	$\hat{p}_c$	$\hat{p}_f$	$\hat{p}_c$	$\hat{p}_f$	$\hat{p}_c$	$\hat{p}_f$	$\hat{p}_c$
M_1	50	FM	Weighted	1.80	8.30	1.63	27.80	1.47	56.50	1.48	85.30	1.60	95.70
			Trimmed	1.25	7.70	1.19	26.80	1.19	56.00	1.22	83.00	1.20	95.50
		RP	Weighted	1.88	6.80	1.68	20.10	1.49	45.10	1.42	71.90	1.42	87.70
			Trimmed	1.88	7.00	1.89	23.20	2.08	49.30	2.14	75.70	2.20	91.90
		Mode	Weighted	1.41	5.70	1.27	19.90	1.09	45.10	1.08	76.70	1.17	94.30
			Trimmed	1.33	6.80	1.32	22.20	1.33	48.60	1.38	77.50	1.39	95.20
	100	FM	Weighted	1.51	5.70	1.40	23.30	1.34	53.50	1.34	82.00	1.38	96.20
			Trimmed	1.63	7.40	1.51	25.90	1.32	56.10	1.13	83.60	1.00	96.70
		RP	Weighted	1.68	5.00	1.57	18.60	1.50	45.40	1.45	71.40	1.52	90.10
			Trimmed	2.28	7.40	2.22	23.70	2.23	54.20	2.13	78.70	2.08	93.50
		Mode	Weighted	1.22	4.20	1.15	17.90	1.08	46.80	1.06	78.90	1.12	94.70
			Trimmed	1.70	5.80	1.62	22.90	1.46	54.20	1.28	83.50	1.13	95.90
			η	0.2		0.4		0.6		0.8		1	
M_2	50	FM	Weighed	1.91	1.50	1.85	3.40	1.70	11.00	1.46	26.30	1.25	55.00
			Trimmed	1.28	1.90	1.23	4.10	1.19	11.60	1.13	26.70	1.18	50.60
		RP	Weighted	1.94	2.40	1.90	3.30	1.75	8.00	1.64	16.90	1.42	32.20
			Trimmed	1.89	2.30	1.83	4.60	1.73	10.10	1.84	22.50	1.92	41.40
		Mode	Weighted	1.43	2.80	1.35	10.70	1.19	31.00	1.06	65.30	1.12	91.70
			Trimmed	1.33	4.00	1.31	12.30	1.31	33.70	1.42	68.00	1.48	92.10
	100	FM	Weighted	1.53	2.00	1.51	2.70	1.45	12.20	1.35	27.90	1.25	53.00
			Trimmed	1.66	2.50	1.63	4.30	1.57	14.40	1.51	33.40	1.43	58.00
		RP	Weighted	1.69	2.70	1.66	4.10	1.63	8.20	1.55	17.30	1.48	34.70
			Trimmed	2.25	2.90	2.19	5.60	2.15	13.60	2.15	25.40	2.21	45.50
		Mode	Weighted	1.22	3.30	1.18	11.90	1.12	34.10	1.07	66.80	1.09	89.60
			Trimmed	1.72	4.60	1.65	15.70	1.57	41.70	1.41	72.70	1.22	92.90

Table 3. Results for the scenarios with dependence between curves. The estimates of false alarm rate (type I error) for values of ρ between 0.3 and 0.7 are shown.

ρ	n	Method	FM	RP	Mode
0.3	50	Weighted	1.85	1.88	1.41
		Trimmed	1.31	1.90	1.33
	100	Weighted	1.54	1.71	1.20
		Trimmed	1.65	2.23	1.71
0.5	50	Weighted	1.80	1.81	1.28
		Trimmed	1.26	1.82	1.29
	100	Weighted	1.46	1.68	1.15
		Trimmed	1.61	2.21	1.65
0.7	50	Weighted	1.66	1.73	0.93
		Trimmed	1.25	1.74	1.06
	100	Weighted	1.42	1.61	0.99
		Trimmed	1.56	2.15	1.57

Tables ??–?? show the results of the empirical estimation of p_c and p_f, assuming different values of ρ (from 0.3 to 0.7). The power (estimated by $\hat{p}_c$) of the control chart proposed for the model M_1 (Equation (??)) performs better when the weighted method is applied and if the sample size is increased. It is also observed that the performance of the control chart tends to be the same, independent of the type of data depth measurement used. Certainly, the performance of control charts in detecting real changes in the process, related to differences in the shape and mean, is better when the mode depth is used.

Table 4. Empirical values of $\hat{p}_f$ and $\hat{p}_c$, with $\rho = 0.3$ (assuming dependence between curves).

			δ	0.4		0.8		1.2		1.6		2	
Model	**Sample Size**	**Depth**	**Method**	$\hat{p}_f$	$\hat{p}_c$	$\hat{p}_f$	$\hat{p}_c$	$\hat{p}_f$	$\hat{p}_c$	$\hat{p}_f$	$\hat{p}_c$	$\hat{p}_f$	$\hat{p}_c$
M_1	50	FM	Weighted	1.58	8.90	1.22	33.60	0.98	69.80	1.07	91.70	1.19	99.30
			Trimming	1.12	9.20	0.75	26.60	0.37	47.40	0.09	54.30	0.02	49.20
		RP	Weighted	1.75	7.10	1.38	26.30	0.97	53.80	0.74	77.60	0.67	90.70
			Trimmed	1.72	7.60	1.42	25.90	0.89	45.90	0.49	60.90	0.20	68.70
		Mode	Weighted	1.21	5.80	0.90	24.90	0.67	59.80	0.70	88.30	0.86	98.50
			Trimmed	1.21	6.40	0.89	24.00	0.50	49.40	0.19	73.50	0.04	89.80
	100	FM	Weighted	1.47	7.30	1.23	29.60	1.07	67.80	1.13	92.30	1.24	99.20
			Trimmed	1.53	7.60	1.29	31.20	1.04	68.70	0.96	91.30	0.92	99.10
		RP	Weighted	1.69	5.50	14.67	24.40	1.21	58.00	1.06	83.90	1.02	95.90
			Trimmed	2.19	8.80	1.99	27.40	1.88	64.80	1.92	89.30	1.85	97.30
		Mode	Weighted	1.17	4.60	0.98	23.00	0.86	62.40	0.89	91.00	1.01	98.50
			Trimming	1.62	6.90	1.38	27.90	1.15	67.10	1.07	91.70	1.06	98.90
			η	0.2		0.4		0.6		0.8		1	
M_2	50	FM	Weighted	1.75	1.30	1.61	1.95	1.31	7.70	0.94	16.85	0.64	26.05
			Trimmed	1.25	1.40	1.15	2.35	0.97	7.70	0.73	15.95	0.51	20.50
		RP	Weighted	1.86	1.20	1.75	3.00	1.43	5.50	1.24	10.85	0.96	16.55
			Trimmed	1.79	1.65	1.68	3.40	1.58	7.15	1.46	13.10	1.16	19.40
		Mode	Weighted	1.31	3.70	1.08	16.00	0.78	44.90	0.62	79.30	0.81	96.90
			Trimmed	1.22	4.70	1.11	17.50	0.74	42.20	0.29	65.00	0.07	82.50
	100	FM	Weighted	1.53	0.90	1.46	2.05	1.30	7.45	1.07	19.50	0.92	32.35
			Trimmed	1.65	1.20	1.57	2.70	1.44	9.40	1.22	22.75	1.12	35.35
		RP	Weighted	1.76	0.60	1.73	2.20	1.57	5.10	1.41	12.10	1.27	21.80
			Trimmed	2.23	1.10	2.18	3.30	2.09	8.25	1.91	17.45	1.97	29.35
		Mode	Weighted	1.17	3.55	1.07	14.70	0.91	47.20	0.86	83.10	0.98	97.50
			Trimmed	1.64	4.70	1.48	20.50	1.25	52.60	1.13	85.10	1.12	98.20

Table 5. Empirical values of $\hat{p}_f$ and $\hat{p}_c$, with $\rho = 0.5$ (assuming dependence between curves).

			δ	0.4		0.8		1.2		1.6		2	
Model	**Sample Size**	**Depth**	**Method**	$\hat{p}_f$	$\hat{p}_c$	$\hat{p}_f$	$\hat{p}_c$	$\hat{p}_f$	$\hat{p}_c$	$\hat{p}_f$	$\hat{p}_c$	$\hat{p}_f$	$\hat{p}_c$
M_1	50	FM	Weighted	1.46	11.50	1.04	43.90	0.84	80.80	1.06	97.40	1.14	99.90
			Trimmed	1.02	10.80	0.63	33.40	0.23	50.70	0.28	53.80	0.00	45.60
		RP	Weighted	1.68	9.10	1.21	31.10	0.80	63.00	0.65	85.60	0.60	94.60
			Trimmed	1.73	9.60	1.26	30.70	0.66	52.60	0.28	66.40	0.11	68.80
		Mode	Weighted	1.10	7.50	0.76	34.40	0.55	73.00	0.69	95.20	0.82	99.90
			Trimmed	1.05	8.10	0.72	31.40	0.32	62.60	0.06	86.50	0.01	96.40
	100	FM	Weighted	1.20	10.00	1.08	39.60	1.01	79.40	1.10	97.30	1.18	99.80
			Trimmed	1.30	10.00	1.18	41.60	0.96	80.20	0.89	97.00	0.84	99.80
		RP	Weighted	1.40	10.00	1.39	32.30	1.09	69.20	0.97	90.60	0.97	97.90
			Trimmed	1.80	10.00	1.93	38.20	1.84	77.10	1.80	94.70	1.73	98.80
		Mode	Weighted	0.90	10.00	0.86	32.10	0.79	75.80	0.91	96.00	0.98	99.90
			Trimmed	1.30	10.00	1.30	38.30	1.10	79.40	1.06	97.20	1.02	99.90
			η	0.2		0.4		0.6		0.8		1	
M_2	50	FM	Weighted	1.63	1.40	1.43	3.55	1.02	10.80	0.68	21.60	0.50	33.10
			Trimmed	1.16	1.50	1.02	3.60	0.80	11.20	0.54	18.00	0.32	21.70
		RP	Weighted	1.89	1.75	1.65	3.30	1.32	7.30	0.98	14.30	0.75	22.55
			Trimmed	1.79	2.30	1.62	3.80	1.54	9.05	1.26	16.40	0.95	22.90
		Mode	Weighted	1.13	5.40	0.86	22.60	0.56	60.50	0.59	91.50	0.85	99.80
			Trimmed	1.10	6.30	0.83	23.40	0.46	54.00	0.11	77.20	0.02	93.80
	100	FM	Weighted	0.01	0.01	0.01	0.03	1.13	11.90	0.90	28.25	0.77	38.90
			Trimmed	0.02	0.01	0.01	0.04	1.28	15.25	1.11	30.90	1.04	39.90
		RP	Weighted	0.02	0.01	0.02	0.03	1.42	7.65	1.26	16.65	1.08	27.55
			Trimmed	0.03	0.02	0.02	0.05	2.02	12.60	1.91	23.25	1.96	35.00
		Mode	Weighted	0.01	0.05	0.01	0.23	0.82	63.00	0.87	92.50	0.99	99.80
			Trimmed	0.02	0.07	0.01	0.28	1.21	68.40	1.12	93.40	1.12	99.60

Table 6. Empirical values of $\hat{p}_f$ and $\hat{p}_c$, with $\rho = 0.7$ (assuming dependence between curves).

			δ	0.4		0.8		1.2		1.6		2	
Model	**Sample Size**	**Depth**	**Method**	$\hat{p}_f$	$\hat{p}_c$	$\hat{p}_f$	$\hat{p}_c$	$\hat{p}_f$	$\hat{p}_c$	$\hat{p}_f$	$\hat{p}_c$	$\hat{p}_f$	$\hat{p}_c$
M_1	50	FM	Weighted	1.32	16.00	0.87	57.70	0.85	91.40	1.01	99.30	1.01	100.00
			Trimmed	0.87	13.40	0.42	40.30	0.07	53.60	0.00	47.50	0.00	35.80
		RP	Weighted	1.41	13.00	0.97	43.40	0.61	76.70	0.56	92.00	0.53	98.00
			Trimmed	1.51	14.00	0.97	40.70	0.39	61.40	0.14	68.20	0.04	71.30
		Mode	Weighted	0.71	14.00	0.44	52.20	0.41	88.90	0.51	99.30	0.53	100.00
			Trimmed	0.82	14.30	0.43	47.40	0.09	80.20	0.01	96.00	0.00	99.90
	100	FM	Weighted	1.29	12.00	0.99	54.90	1.05	90.40	1.14	99.50	1.13	100.00
			Trimmed	1.36	13.80	1.08	56.00	0.89	90.20	0.84	99.20	0.76	100.00
		RP	Weighted	1.50	9.90	1.20	48.30	0.97	80.80	0.93	96.40	0.96	99.50
			Trimmed	2.00	13.90	1.80	53.10	1.80	87.30	1.72	98.30	1.60	99.90
		Mode	Weighted	0.90	9.80	0.68	51.00	0.69	90.00	0.81	99.40	0.83	99.90
			Trimmed	1.43	13.40	1.14	58.10	1.01	92.40	0.97	99.60	0.92	99.90
			η	0.2		0.4		0.6		0.8		1	
M_1	50	FM	Weighted	1.52	1.80	1.21	6.05	0.76	17.90	0.49	30.70	0.41	39.95
			Trimming	1.07	2.10	0.85	6.15	0.56	15.30	0.31	22.30	0.18	24.05
		RP	Weighted	1.62	2.10	1.36	5.25	0.98	11.40	0.68	19.95	0.56	28.30
			Trimmed	1.59	2.40	1.41	5.80	1.19	13.85	0.92	21.05	0.70	28.15
		Mode	Weighted	0.75	8.60	0.47	39.90	0.37	82.80	0.50	99.10	0.61	100.00
			Trimmed	0.87	10.30	0.51	38.30	0.14	73.10	0.01	93.80	0.00	99.30
	100	FM	Weighted	2.02	0.00	1.24	6.35	0.99	21.20	0.81	36.45	0.75	45.10
			Trimmed	3.03	0.00	1.36	7.80	1.16	24.30	1.05	37.95	1.03	44.55
		RP	Weighted	3.69	0.00	1.50	5.25	1.25	13.00	1.08	24.50	0.99	34.85
			Trimmed	6.28	0.00	1.98	7.20	1.84	18.50	1.89	31.80	1.89	41.60
		Mode	Weighted	1.63	0.00	0.72	37.90	0.69	82.70	0.80	98.60	0.90	100.00
			Trimmed	3.48	0.00	1.24	46.70	1.10	87.10	1.05	99.00	1.04	100.00

With respect to the false out-of-control rate $\hat{p}_f$, in the scenarios corresponding to the use of M_1 model, when the trimmed method is also used, a lower rate is obtained. In the case of the M_2 model, there are similar results on the scenarios defined by independence between curves, that is, the $\hat{p}_f$ is lower when the trimmed method for outlier detection is used.

In Reference [?], new methods for the detection of outliers were proposed for the case in which there is dependence between curves. From the simulation studies carried out in this study, at different degrees of dependence, we can say that the outlier detection method proposed in Reference [?] was relatively robust against the presence of dependence between curves. The simulation study performed in this section supports the results obtained in the work in Reference [?] and, in conclusion, justifies the use of this method within the new control charts proposed for Phase I, even in scenarios with dependence between curves.

Although the application of the weighed outlier detection method to Mode data depth has generally provided Phase I control charts with best performance, if the false alarm rate of Tables **??** and **??** are observed, the use of trimmed outlier detection method tends to provide values of $\hat{p}_f$ slightly closer to $\alpha = 1\%$ (with respect to the weighted method) when the process is actually under control, the curves are independent and the sample is relatively small ($n = 50$). In addition, if the process is out of control, the curves are independent and the outliers are generated by the Equation (**??**) (changes in magnitude), the trimmed method applied to FM data depth provide a $\hat{p}_f$ close to $\alpha = 1\%$ and the highest $\hat{p}_c$, as shown in Table **??**. In all the remaining scenarios, the use of weighed method applied to Mode data depth tends to provide the closest to $\alpha\%$ $\hat{p}_f$ and the highest $\hat{p}_c$ (see Tables **??**–**??**).

5.2. Measurement and Comparison of the Performance of the Control Chart Proposed for Phase II

For Phase II, the monitoring stage, the use of the rank control chart has been proposed. The application of the rank control chart allows simultaneous monitoring of changes in the mean and variability of a process. In the functional case, in order to calculate the rank statistic, the functional FM, RP and mode depths are used.

An $ARL_0 = \frac{1}{\alpha=0.025}$ (the monitoring sample is assumed under control) is assumed to evaluate the performance of the control chart. The power of the control chart is estimated and compared for an under-control process, based on the generation of a calibration sample of size $n = 50$ by a Monte Carlo procedure.

Following the simulation scheme of Phase I, curves observed at equidistant points are assumed; they are composed of 51 points at $[0, 1]$ interval. A smoothed bootstrap with a smoothing factor $\gamma = 0.05$, 1000 resamples ($B = 1000$), using a 2.5% trimmed procedure (removing the shallowest curves), is applied to estimate and compare the power of the control chart to detect out-of-control signals when a significance level of $\alpha = 0.025$ is assumed. Additionally, in the same way as in Phase I, the simulation of scenarios with independence and dependence between curves are assumed.

Table **??** shows the estimates of the power (%) of the control charts for the scenario of independence between curves, whereas Table **??** shows power of the control chart for the scenario with dependence. In both cases, the ability of the control chart to detect a change in the magnitude of the functional mean of the process is evaluated by the estimation of its power $(1 - \beta)$.

Table 7. Power of the control chart, $1 - \beta$, for the case M_1 (Equation (**??**)) model in the scenario of independence between curves.

δ	FM	RP	Mode
0.5	14.5	14.8	14.8
1	47.6	47.2	48.0
1.5	83.3	80.8	83.0
2	97.8	97.5	97.7

From the results of the Table **??**, any depth measure can be used to detect a shift in the process mean, since the same performance, in terms of power, is obtained.

Table 8. Power of the control chart, $1 - \beta$, for the scenarios defined by the M_1 (**??**) model, assuming dependence between curves.

ρ	δ	FM	RP	Mode
0.3	0.5	20.28	23.73	21.84
	1	62.95	64.75	64.06
	1.5	92.98	92.33	93.64
	2	99.56	99.16	99.75
0.5	0.5	25.59	29.41	28.41
	1	73.84	75.12	76.24
	1.5	97.15	96.56	97.94
	2	99.91	99.77	99.98
0.7	0.5	35.47	40.23	43.55
	1	86.14	86.87	90.38
	1.5	99.48	99.11	99.78
	2	100.00	99.99	100.00

The results of the detection of a shift in the process mean are shown in Table **??**. A similar performance is observed when using any depth measure for different values of ρ. Apparently, the control chart for Phase II is robust against the existence of dependence between curves.

As observed in the simulation study and in the analysis of the case study with real data, the present proposal of control charts for functional data, including Phase I and II control charts, can be useful to detect anomalies in diverse scenarios. In the case of its application to real data, the set of proposed techniques is being examined for implementation in the web platform Σqus and for its use by the company Nerxus for detecting false alarms in facilities in commercial areas. The present control chart methodology can be used for control tasks, monitoring, anomaly detection and continuous improvement in diverse industrial processes, monitoring of environmental variables, chemical industry and, in general, any process involving continuous monitoring of functional data over time.

Regarding the use of our methodology in more complex case studies defined by different operation modes, the application of the multi-modelling framework methodology in combination with our proposal could be useful. Indeed, in the building energy efficiency domain, there are many different operation modes of installations, each one defined by a specific operation pattern. Namely, the HVAC installations can be operated in heating or ventilation modes (there are even different modes within ventilation or heating). The automatic classification of each profile in the corresponding profile pattern could be very useful in the building energy efficiency field and previously to the application of our control chart proposal for Phase I and Phase II. With respect to the work of Grasso et al. [**?**], it is also interesting to mention that the proposed profile monitoring control charting scheme is that based on functional PCA and described in Colosimo and Pacella [**?**].

6. Conclusions

A new alternative of control charts has been proposed when CTQ variables of the process are functional. The proposal includes alternatives to develop the the Phase I and II control charts for stabilizing and monitoring the processes, respectively. In order to develop Phase I control charts based on functional data, outlier detection methods are used based on a method of smooth bootstrap resampling and the depth calculation of functional data is proposed. However, in order to implement Phase II, the use of rank-type nonparametric control charts based on the concept of functional data depth is proposed. This Phase II control chart is directly estimated assuming that the asymptotic distribution of the rank statistic is a uniform distribution. The application of the control charts to the two-process control phases and the development of a new graphic tool for visualizing functional data (including an envelope with 95% of the deepest curves that facilitate the identification of the assignable cause of each anomaly) give rise to the proposed methodology. It has been successfully applied in real case studies belonging to the framework of anomaly detection in building energy efficiency. Additionally, a simulation study is conducted to measure the performance (as the percentage of rejection when the null hypothesis is not met) of the control charts, depending on the functional data depth used, the sample size, the presence of dependence between curves and the use of different FDA procedures for outlier detection.

In the simulation study, the use of different types of functional depths has been compared to develop Phase II of the proposed control chart. In case of the univariate functional data (single type of curves), for the three scenarios, a better performance is obtained with the mode depth measurement combined with the weighted outlier detection method and moderately large samples. Additionally, one of the final observations of the simulation study is that the control chart methodology is robust against the presence of dependence between curves. Thus, this alternative tool can be applied to the framework of continuously monitored data streams.

Generally, the authors recommend using the weighed method and the Mode functional data depth for the case of Phase I taking into account the values of $\hat{p}_f$ and $\hat{p}_c$. Thus, when the Phase I control chart is evaluated, both weighted method and Mode data depth are generally the best options in those scenarios defined by under control assumption and even in those where out of control curves are simulated. In the latter, when the change in magnitude or shape is very small, the corresponding $\hat{p}_c$ tends to be not higher to those obtained by the use of other combinations of data depth measure and outlier detection method. Nevertheless, when the change in magnitude (δ) or shape (η) increases, the power of the combination of weighted method and Mode depth tends to be higher than those corresponding to the other combinations. Moreover, regarding the Phase II control chart and taking into account the higher values of power estimates included in Tables **??** and **??**, we also recommend the use of Mode data depth for Phase II control charts.

The present proposal has been verified by its application in a real case study dealing with the detection of energy efficiency anomalies in buildings. Specifically, all the previously identified real anomalies (by the maintenance personnel) have also been successfully identified by the application of this functional approximation of control charts for Phases I and II of process control. Additionally, the proposed graphical tool helps to intuitively identify the assignable causes corresponding to each anomaly.

This procedure can be used in different industrial and scientific domains in which the control procedures are defined by functional CTQ variables.

Author Contributions: Conceptualization, S.N., M.F. and J.T.-S.; methodology, M.F., R.F.-C. and S.N.; software, M.F. and R.F.-C.; validation, M.F., R.F.-C. and J.T.-S.; formal analysis, M.F., R.F.-C., S.N. and J.T.-S.; investigation, M.F., R.F.-C., S.N., J.T.-S., S.Z. and P.R.; resources, S.N., S.Z. and P.R.; data curation, S.Z. and P.R.; writing—original draft preparation, M.F., J.T.-S., S.N. and R.F.-C.; writing—review and editing, J.T.-S., R.F.-C., M.F., S.N. and S.Z.; visualization, Miguel Flores, R.F.-C. and S.N.; supervision, S.N., J.T.-S. and R.F.-C.; project administration, S.N.; funding acquisition, S.Z. and S.N. All authors have read and agreed to the published version of the manuscript.

Acknowledgments: The authors strongly thank, on the one hand, Fridama, Σqus, and Nerxus companies, and on the other hand CITIC, Campus Industrial and MODES group their valuable help and support.

References

1. Lu, C.W.; Reynolds, M.R, Jr. EWMA control charts for monitoring the mean of autocorrelated processes. *J. Qual. Technol.* **1999**, *31*, 166. [CrossRef]
2. Alwan, L.C.; Roberts, H.V. Time-series modeling for statistical process control. *J. Bus. Econ. Stat.* **1988**, *6*, 87–95.
3. Qiu, P.; Zou, C.; Wang, Z. Nonparametric profile monitoring by mixed effects modeling. *Technometrics* **2010**, *52*, 265–277. [CrossRef]
4. Shiau, J.J.; Huang, H.L.; Lin, S.H.; Tsai, M.Y. Monitoring nonlinear profiles with random effects by nonparametric regression. *Commun. Stat. Methods* **2009**, *38*, 1664–1679. [CrossRef]
5. Noorossana, R.; Saghaei, A.; Amiri, A. *Statistical Analysis of Profile Monitoring*; John Wiley & Sons: Hoboken, NJ, USA, 2011; Volume 865.
6. Chipman, H.; MacKay, R.; Steiner, S. Comment on Nonparametric profile monitoring by mixed effects modeling. *Technometrics* **2010**, *52*, 280–283. [CrossRef]
7. Lavin, A.; Ahmad, S. Evaluating Real-Time Anomaly Detection Algorithms–The Numenta Anomaly Benchmark. In Proceedings of the 2015 IEEE 14th International Conference on Machine Learning and Applications (ICMLA), Miami, FL, USA, 9–11 December 2015; pp. 38–44.
8. Kroll, B.; Schaffranek, D.; Schriegel, S.; Niggemann, O. System modeling based on machine learning for anomaly detection and predictive maintenance in industrial plants. In Proceedings of the 2014 IEEE Emerging Technology and Factory Automation (ETFA), Barcelona, Spain, 16–19 September 2014; pp. 1–7.
9. Ahmad, S.; Lavin, A.; Purdy, S.; Agha, Z. Unsupervised real-time anomaly detection for streaming data. *Neurocomputing* **2017**, *262*, 134–147. [CrossRef]
10. Meshram, A.; Haas, C. Anomaly detection in industrial networks using machine learning: A roadmap. In *Machine Learning for Cyber Physical Systems*; Springer: Berlin/Heidelberg, Germany, 2017; pp. 65–72.
11. Basu, S.; Meckesheimer, M. Automatic outlier detection for time series: An application to sensor data. *Knowl. Inf. Syst.* **2007**, *11*, 137–154. [CrossRef]
12. Mosallam, A.; Medjaher, K.; Zerhouni, N. Nonparametric time series modelling for industrial prognostics and health management. *Int. J. Adv. Manuf. Technol.* **2013**, *69*, 1685–1699. [CrossRef]
13. Talagala, P.D.; Hyndman, R.J.; Smith-Miles, K.; Kandanaarachchi, S.; Muñoz, M.A. Anomaly detection in streaming nonstationary temporal data. *J. Comput. Graph. Stat.* **2019**, pp. 1–28. [CrossRef]
14. Hyndman, R.J.; Liu, X.A.; Pinson, P. Visualizing big energy data: Solutions for this crucial component of data analysis. *IEEE Power Energy Mag.* **2018**, *16*, 18–25. [CrossRef]
15. Golshan, M.; MacGregor, J.F.; Bruwer, M.J.; Mhaskar, P. Latent Variable Model Predictive Control (LV-MPC) for trajectory tracking in batch processes. *J. Process Control* **2010**, *20*, 538–550. [CrossRef]
16. Kourti, T.; MacGregor, J.F. Multivariate SPC methods for process and product monitoring. *J. Qual. Technol.* **1996**, *28*, 409–428. [CrossRef]
17. Ferrer-Riquelme, A. 1.04-Statistical Control of Measures and Processes. In *Comprehensive Chemometrics*; Elsevier: Amsterdam, The Netherlands, 2009; pp. 97–126.

18. Colosimo, B.M.; Pacella, M. A comparison study of control charts for statistical monitoring of functional data. *Int. J. Prod. Res.* **2010**, *48*, 1575–1601. [CrossRef]
19. Megahed, F.; Jones-Farmer, L. Statistical Perspectives on "Big Data". In *Frontiers in Statistical Quality Control 11*; Springer: Berlin, Germany, 2015; pp. 29–47.
20. Woodall, W.; Montgomery, D. Some current directions in the theory and application of statistical process monitoring. *J. Qual. Technol.* **2014**, *46*, 78. [CrossRef]
21. Sheu, S.H.; Ouyoung, C.W.; Hsu, T.S. Phase II statistical process control for functional data. *J. Stat. Comput. Simul.* **2013**, *83*, 2144–2159. [CrossRef]
22. García, D.R. Cartas de Control Para Datos Funcionales. Master's Thesis, Centro de Investigación en Matemáticas (CIMAT), Guanajuato, Mexico, 2011.
23. Rodrigo, O.P. Monitoreo de Datos Funcionales. Master's Thesis, Centro de Investigación en Matemáticas (CIMAT), Guanajuato, Mexico, 2013.
24. Flores, M.; Tarrio-Saavedra, J.; Fernandez-Casal, R.; Naya, S. Functional extensions of Mandel's h and k statistics for outlier detection in interlaboratory studies. *Chemom. Intell. Lab. Syst.* **2018**, *176*, 134–148. [CrossRef]
25. Flores, M.; Fernández-Casal, R.; Naya, S.; Tarrío-Saavedra, J.; Bossano, R. ILS: An R package for statistical analysis in Interlaboratory Studies. *Chemom. Intell. Lab. Syst.* **2018**, *181*, 11–20. [CrossRef]
26. Flores, M.; Naya, S.; Tarrío-Saavedra, J.; Fernández-Casal, R. Functional data analysis approach of Mandel's h and k statistics in Interlaboratory Studies. In *Functional Statistics and Related Fields*; Springer: Berlin/Heidelberg, Germany, 2017; pp. 123–130.
27. Liu, R.Y. Control Charts for Multivariate Processes. *J. Am. Stat. Assoc.* **1995**, *90*, 1380–1387. [CrossRef]
28. Liu, R.Y.; Tang, J. Control charts for dependent and independent measurements based on bootstrap methods. *J. Am. Stat. Assoc.* **1996**, *91*, 1694–1700. [CrossRef]
29. Liu, R. On a notion of data depth based on random simplices. *Ann. Stat.* **1990**, *18*, 405–414. [CrossRef]
30. Chatterjee, S.; Qiu, P. Distribution-free cumulative sum control charts using bootstrap-based control limits. *Ann. Appl. Stat.* **2009**, *3*, 349–369. [CrossRef]
31. Jones, L.; Woodall, W. The performance of bootstrap control charts. *J. Qual. Technol.* **1998**, *30*, 362. [CrossRef]
32. Chakraborti, S. Nonparametric (Distribution-Free) Quality Control Charts. In *Encyclopedia of Statistical Sciences*; John Wiley & Sons, Inc.: Hoboken, NJ, USA, 2011.
33. Qiu, P.; Li, Z. On nonparametric statistical process control of univariate processes. *Technometrics* **2011**, *53*, 390–405. [CrossRef]
34. Barbeito, I.H.; Zaragoza, S.; Tarrío-Saavedra, J.; Naya, S. Assessing thermal comfort and energy efficiency in buildings by statistical quality control for autocorrelated data. *Appl. Energy* **2017**, *190*, 1–17. [CrossRef]
35. Montgomery, D. *Introduction to Statistical Quality Control.*; John Wiley & Sons: New York, NY, USA, 2009.
36. Ledolter, J.; Bisgaard, S. Challenges in constructing time series models from process data. *Qual. Reliab. Eng. Int.* **2011**, *27*, 165–178. [CrossRef]
37. Box, G.E.; Luceño, A.; Paniagua-Quinones, M.D.C. *Statistical Control by Monitoring and Adjustment*; John Wiley & Sons: Hoboken, NJ, USA, 2011; Volume 700.
38. Del Castillo, E. Statistical process adjustment: A brief retrospective, current status, and some opportunities for further work. *Stat. Neerl.* **2006**, *60*, 309–326. [CrossRef]
39. Knoth, S.; Schmid, W. Control charts for time series: A review. *Front. Stat. Qual. Control* **2004**, *7*, 210–236.
40. Prajapati, D.; Singh, S. Control charts for monitoring the autocorrelated process parameters: a literature review. *Int. J. Prod. Qual. Manag.* **2012**, *10*, 207–249. [CrossRef]
41. Tsai, T.R.; Hsieh, Y.W. Simulated Shewhart control chart for monitoring variance components. *Int. J. Reliab. Qual. Saf. Eng.* **2009**, *16*, 1–22. [CrossRef]
42. Jones-Farmer, L.; Woodall, W.; Steiner, S.; Champ, C. An overview of phase I analysis for process improvement and monitoring. *J. Qual. Technol.* **2014**, *46*, 265. [CrossRef]
43. Grasso, M.; Colosimo, B.M.; Tsung, F. A phase I multi-modelling approach for profile monitoring of signal data. *Int. J. Prod. Res.* **2017**, *55*, 4354–4377. [CrossRef]
44. Febrero-Bande, M.; Galeano, P.; González-Manteiga, W. Outlier detection in functional data by depth measures, with application to identify abnormal NOx levels. *Environmetrics* **2007**, *19*, 331–345. [CrossRef]
45. Kazemzadeh, R.B.; Noorossana, R.; Ayoubi, M. Change point estimation of multivariate linear profiles under linear drift. *Commun. Stat.-Simul. Comput.* **2015**, *44*, 1570–1599. [CrossRef]

46. Woodall, W.H.; Spitzner, D.J.; Montgomery, D.C.; Gupta, S. Using control charts to monitor process and product quality profiles. *J. Qual. Technol.* **2004**, *36*, 309–320. [CrossRef]
47. Lanhede, D. Statistical Process Control: Evaluation and Implementation of Methods for Statistical Process Control at GE Healthcare. Master's Thesis, Umeå University, Umeå, Sweden, 2015.
48. Flores Sánchez, M. Nuevas Aportaciones del Análisis de Datos Funcionales en el Control Estadístico de Procesos. Ph.D. Thesis, Universidade da Coruña, A Coruña, Spain, 2019.
49. Febrero-Bande, M.; Oviedo de la Fuente, M. Statistical computing in functional data analysis: The R package fda.usc. *J. Stat. Softw.* **2012**, *51*, 1–28. [CrossRef]
50. Fraiman, R.; Muniz, G. Trimmed means for functional data. *Test* **2001**, *10*, 419–440. [CrossRef]
51. López-Pintado, S.; Romo, J. On the concept of depth for functional data. *J. Am. Stat. Assoc.* **2009**, *104*, 718–734. [CrossRef]
52. Cuevas, A.; Febrero, M.; Fraiman, R. Robust estimation and classification for functional data via projection-based depth notions. *Comput. Stat.* **2007**, *22*, 481–496. [CrossRef]
53. Cuesta-Albertos, J.A.; Nieto-Reyes, A. The random Tukey depth. *Comput. Stat. Data Anal.* **2008**, *52*, 4979–4988. [CrossRef]
54. Flores, M.; Fernández-Casal, R.; Naya, S.; Tarrío-Saavedra, J. qcr: Quality Control Review. R Package Version 1.0. 2016. Available online: https://cran.r-project.org/web/packages/qcr/index.html (accessed on 26 December 2019).
55. Hyndman, R.J.; Shang, H.L. Rainbow plots, bagplots, and boxplots for functional data. *J. Comput. Graph. Stat.* **2010**, *19*, 29–45. [CrossRef]
56. Raña, P. Pointwise Forecast, Confidence and Prediction Intervals in Electricity Demand and Price. Ph.D. Thesis, Universidade da Coruña, A Coruña, Spain, 2016.
57. Colosimo, B.M.; Pacella, M. On the use of principal component analysis to identify systematic patterns in roundness profiles. *Qual. Reliab. Eng. Int.* **2007**, *23*, 707–725. [CrossRef]

13

Robust Linear Trend Test for Low-Coverage Next-Generation Sequence Data Controlling for Covariates

Jung Yeon Lee [1], Myeong-Kyu Kim [2] and Wonkuk Kim [3,*]

1 Department of Psychiatry, New York University School of Medicine, New York, NY 10016, USA; JungYeon.Lee@nyulangone.org
2 Department of Neurology, Chonnam National University Medical School, Gwangju 61469, Korea; mkkim@jnu.ac.kr
3 Department of Applied Statistics, Chung-Ang University, Seoul 06974, Korea
* Correspondence: wkim@cau.ac.kr.

Abstract: Low-coverage next-generation sequencing experiments assisted by statistical methods are popular in a genetic association study. Next-generation sequencing experiments produce genotype data that include allele read counts and read depths. For low sequencing depths, the genotypes tend to be highly uncertain; therefore, the uncertain genotypes are usually removed or imputed before performing a statistical analysis. It may result in the inflated type I error rate and in a loss of statistical power. In this paper, we propose a mixture-based penalized score association test adjusting for non-genetic covariates. The proposed score test statistic is based on a sandwich variance estimator so that it is robust under the model misspecification between the covariates and the latent genotypes. The proposed method takes advantage of not requiring either external imputation or elimination of uncertain genotypes. The results of our simulation study show that the type I error rates are well controlled and the proposed association test have reasonable statistical power. As an illustration, we apply our statistic to pharmacogenomics data for drug responsiveness among 400 epilepsy patients.

Keywords: allele read counts; low-coverage; mixture model; next-generation sequencing; sandwich variance estimator

---

1. Introduction

Genome-wide association study (GWAS) is a powerful tool for screening a high-dimensional genome data set and selecting candidate genetic variants such as single nucleotide polymorphisms (SNPs) in genetic association studies. Next-generation sequencing (NGS) technology is widely used to produce a large amount of genetic information in a fast way. In the past decade, there have been numerous studies using NGS data such as rare variants association study [1,2], pharmacogenomics [3,4], machine learning and deep learning applications [5,6], and big data analysis [7,8]. Many NGS experiments are based on low-coverage sequencing with a large sized sample since there is a trade-off between sample size and sequencing depth in the NGS experiments [9,10]. For the low-coverage NGS data, a high uncertainty of the inferred genotypes is common; however, it causes biased and unreliable results on genetic association analyses. In genetic research based on NGS data, therefore, it is important to obtain accurate genotypes to perform an association analysis.

A number of researchers have worked on the effects of genotype misclassification in genetic association studies. There are two types of genotype misclassifications: differential and non-differential misclassifications, determined by whether the misclassification mechanism differs in the case and control groups or not. In summary, non-differential misclassifications result in a loss of statistical power and differential misclassifications distort type I error rates in a genetic case-control association study [11–14].

While there have been many research on improving the accuracy of genotypes such as the joint genotype calling algorithms across all samples were suggested to increase the accuracy of genotype calls [15–17], several researchers have tried to develop new association statistics accounting for the genotype errors. Their approaches are based on the raw measurements rather than inferred genotypes. In statistical genetics literature, Kim et al. [18] extended a chi-squared test of independence and developed a mixture likelihood based association test using the continuous measurements for copy number polymorphisms. Barnes et al. [19] proposed a mixture model linear trend test for the continuous copy number measurements. In NGS experiments, a likelihood ratio test based on allele read counts of pooled samples was proposed to test independence of genetic variants with a binary phenotype [20]. Gordon et al. [21] proposed a likelihood ratio test of the binomial mixture model of allele read counts with known error parameters. Kim et al. [13,22] proposed an extended version of Cochran–Armitage (CA) trend test and a multi-variant linear trend test for next-generation sequences data by using binomial mixture models. For a case-parent trio design, the binomial mixture model was applied to develop extended transmission disequilibrium tests (TDTs) based on read counts and read depths and to provide power analysis and sample size formulas [23]. All these approaches do not require genotype calls that can be highly uncertain when the read depth or coverage is low. However, none of these previous research has addressed how to include covariates in their mixture-based association studies.

When the covariates are independent of the latent genotypes, the extension of the mixture model based association tests is straightforward. However, if the latent genotype variable is associated with other covariates, then a likelihood based approach requires a model specification between the genotype variable and the other covariates as opposed to the previous research [16–23]. To our knowledge, this is the first study that investigates a genetic case-control association test controlling for covariates in low-coverage NGS experiments. Since we do not know the true model, we apply a sandwich variance estimator to develop a robust genetic association test statistic.

2. Materials and Methods

2.1. Mixture Model Accounting for Covariates

Let $\mathbf{w}$ be a covariate vector. Let y be a random variable indicating the case-control status of an individual such that $y = 1$ if a subject is in the case group and $y = 0$, otherwise. Let $\mathbf{z} = (z_0, z_1, z_2)$ denote an unobservable latent genotype vector, where $\sum_{g=0}^{2} z_g = 1$ and $z_g = 1$ if and only if the genotype is equal to g. Let x and v denote the minor allele read count and the read depth, respectively. The probability function is given by

$$\begin{aligned} p(y, x, v, \mathbf{w}) &= \sum_{\mathbf{z}} p(y, x, v, \mathbf{w}, \mathbf{z}) \\ &= \sum_{\mathbf{z}} p(x|v, \mathbf{w}, \mathbf{z}, y) p(y|v, \mathbf{w}, \mathbf{z}) p(\mathbf{z}|\mathbf{w}, v) p(\mathbf{w}, v) \\ &= p(\mathbf{w}, v) \sum_{\mathbf{z}} p(x|v, \mathbf{z}, y) p(y|\mathbf{z}, \mathbf{w}) p(\mathbf{z}|\mathbf{w}). \end{aligned} \tag{1}$$

If the probability function of the read count x does not depend on the phenotype y, that is, $p(x|v, \mathbf{z}, y) = p(x|v, \mathbf{z})$, then it is called a non-differential error model. We apply a binary logit model to the case-control phenotype response variable y that is the same model for Cochran–Armitage trend test when perfect genotypes are available:

$$p(y|\mathbf{z}, \mathbf{w}) = \frac{e^{y(\beta \mathbf{s}^T \mathbf{z} + \beta_w^T \mathbf{w})}}{1 + e^{\beta_0 + \beta \mathbf{s}^T \mathbf{z} + \beta_w^T \mathbf{w}}}. \tag{2}$$

We assume a binomial error model to the allele read counts as in previous research [13,16,20,21,23]:

$$p(x|v,\mathbf{z},y) = \binom{v}{x}\left(\mathbf{u}_\epsilon^T\mathbf{z}\right)^x\left(1-\mathbf{u}_\epsilon^T\mathbf{z}\right)^{v-x}, \quad (3)$$

where $\mathbf{u}_\epsilon = (\epsilon, 1/2, 1-\epsilon)^T$. For a differential error model, we can use $\mathbf{u}_\epsilon = y(\epsilon_1, 1/2, 1-\epsilon_1)^T + (1-y)(\epsilon_0, 1/2, 1-\epsilon_0)^T$. When perfect genotypes are available, we do not need the conditional probability of the genotype $\mathbf{z}$ given covariates $\mathbf{w}$ to perform genetic association tests since the logistic regression model is a conditional model given the genotypes and covariates. In this work, we assume a multinomial logit model for the latent genotype given the covariates as follows:

$$p(\mathbf{z}|\mathbf{w}) = \frac{\sum_{g=0}^{2} z_g e^{\gamma_g^T\mathbf{w}}}{\sum_{m=0}^{2} e^{\gamma_m^T\mathbf{w}}}, \quad (4)$$

where $\gamma_0 = (0,0,0)^T$ to remove over-parametrization. Other statistical models without the assumptions of a multinomial logit model may also be used for the relationship between covariates and latent genotypes, where we do not know the true model.

The likelihood function L and the log-likelihood function ℓ are written as

$$\begin{aligned} L &= \prod_{k=1}^{N}\left[\sum_{\mathbf{z}_k} p(y_k|\mathbf{z}_k,\mathbf{w}_k)p(x_k|v_k,\mathbf{z}_k,y_k)p(\mathbf{z}_k|\mathbf{w}_k)p(\mathbf{w}_k,v_k)\right] \\ &= \prod_{k=1}^{N}\sum_{i=0}^{2}\left\{\left(\frac{e^{y_k(\beta s_i+\beta_w^T\mathbf{w}_k)}}{1+e^{\beta s_i+\beta_w^T\mathbf{w}_k}}\right)\left(\binom{v_k}{x_k}(u_{\epsilon i})^{x_k}(1-u_{\epsilon i})^{v_k-x_k}\right)\left(\frac{e^{\gamma_i^T\mathbf{w}_k}}{\sum_{m=0}^{2} e^{\gamma_m^T\mathbf{w}_k}}\right)p(\mathbf{w}_k,v_k)\right\}, \quad (5) \end{aligned}$$

$$\begin{aligned} \ell &= \sum_{k=1}^{N}\log\left[\sum_{i=0}^{2}\left\{\left(\frac{e^{y_k(\beta s_i+\beta_w^T\mathbf{w}_k)}}{1+e^{\beta s_i+\beta_w^T\mathbf{w}_k}}\right)\left((u_{\epsilon i})^{x_k}(1-u_{\epsilon i})^{v_k-x_k}\right)\left(\frac{e^{\gamma_i^T\mathbf{w}_k}}{\sum_{m=0}^{2} e^{\gamma_m^T\mathbf{w}_k}}\right)\right\}\right] \\ &+ \sum_{k=1}^{N}\log\binom{v_k}{x_k}p(\mathbf{w}_k,v_k). \quad (6) \end{aligned}$$

The error parameter ϵ is commonly small and hence the estimate of ϵ is often equal to zero. The zero estimate of the error parameter results in a divergent information matrix. It prevents us from calculating Rao's score test statistic. In order to overcome this issue, we include a beta density penalty term to prevent from zero estimate of the error parameter. The penalized log-likelihood function is given by

$$\ell_p = \ell + C\log\left[\epsilon^{a_\epsilon}(1-\epsilon)^{b_\epsilon}\right]. \quad (7)$$

During this work, we choose $C = 1$ as in [24,25]. The penalized complete-data likelihood function is given by

$$L_C = \prod_{k=1}^{N}\prod_{i=0}^{2}\left[\frac{e^{y_k(\beta s_i+\beta_w^T\mathbf{w}_k)}}{1+e^{\beta s_i+\beta_w^T\mathbf{w}_k}}\times\binom{v_k}{x_k}(u_{\epsilon i})^{x_k}(1-u_{\epsilon i})^{v_k-x_k}\epsilon^{\frac{a_\epsilon}{N}}(1-\epsilon)^{\frac{b_\epsilon}{N}}\times\frac{e^{\gamma_i^T\mathbf{w}_k}}{\sum_{m=0}^{2} e^{\gamma_m^T\mathbf{w}_k}}\right]^{z_{ik}} \quad (8)$$

The complete data log-likelihood function is written as

$$\begin{aligned} \ell_C &= \sum_{k=1}^{N}\sum_{i=0}^{2} z_{ik}\left[y_k(\beta s_i+\beta_w^T\mathbf{w}_k)-\log\left(1+e^{\beta s_i+\beta_w^T\mathbf{w}_k}\right)\right] \\ &+ \sum_{k=1}^{N}\sum_{i=0}^{2} z_{ik}\left[x_k\log(u_{\epsilon i})+(v_k-x_k)\log(1-u_{\epsilon i})\right]+a_\epsilon\log\epsilon+b_\epsilon\log(1-\epsilon) \quad (9) \\ &+ \sum_{k=1}^{N}\sum_{i=0}^{2} z_{ik}\left[\gamma_i^T\mathbf{w}_k-\log\left(\sum_{m=0}^{2} e^{\gamma_m^T\mathbf{w}_k}\right)\right]. \end{aligned}$$

2.2. Derivation of EM Algorithm under H_0

We apply the Expectation–Maximization (EM) algorithm [26] to estimate the parameters in our mixture model. Given data and the (r)-th step estimated parameters, the $(r+1)$-th E-step of the EM algorithm is written as

$$\begin{aligned} Q^{(r+1)} &= \sum_{k=1}^{N}\sum_{i=0}^{2}\tau_{ik}^{(r)}\left[y_k(\beta s_i+\beta_w^T\mathbf{w}_k)-\log\left(1+e^{\beta s_i+\beta_w^T\mathbf{w}_k}\right)\right] \\ &+\sum_{k=1}^{N}\sum_{i=0}^{2}\tau_{ik}^{(r)}\left[x_k\log(u_{\epsilon i})+(v_k-x_k)\log(1-u_{\epsilon i})\right]+a_\epsilon\log\epsilon+b_\epsilon\log(1-\epsilon) \\ &+\sum_{k=1}^{N}\sum_{i=0}^{2}\tau_{ik}^{(r)}\left[\gamma_i^T\mathbf{w}_k-\log\left(\sum_{m=0}^{2}e^{\gamma_m^T\mathbf{w}_k}\right)\right], \end{aligned} \tag{10}$$

where

$$\tau_{ik}^{(r)} = \frac{\left(\frac{e^{y_k(\beta^{(r)}s_i+\beta_w^{(r)T}\mathbf{w}_k)}}{1+e^{\beta^{(r)}s_i+\beta_w^{(r)T}\mathbf{w}_k}}\right)\left((u_{\epsilon i}^{(r)})^{x_k}(1-u_{\epsilon i}^{(r)})^{v_k-x_k}\right)\left(\frac{e^{\gamma_i^{(r)T}\mathbf{w}_k}}{\sum_{m=0}^{2}e^{\gamma_m^{(r)T}\mathbf{w}_k}}\right)}{\sum_{g=0}^{2}\left[\left(\frac{e^{y_k(\beta^{(r)}s_g+\beta_w^{(r)T}\mathbf{w}_k)}}{1+e^{\beta^{(r)}s_g+\beta_w^{(r)T}\mathbf{w}_k}}\right)\left((u_{\epsilon g}^{(r)})^{x_k}(1-u_{\epsilon g}^{(r)})^{v_k-x_k}\right)\left(\frac{e^{\gamma_g^{(r)T}\mathbf{w}_k}}{\sum_{m=0}^{2}e^{\gamma_m^{(r)T}\mathbf{w}_k}}\right)\right]}. \tag{11}$$

We note that the posterior probability of subject k belonging to genotype class i depends on the all parameters. In M-step, the $(r+1)$-th estimates of the parameters are obtained by maximizing $Q^{(r+1)}$:

$$\frac{\partial Q^{(r+1)}}{\partial\beta} = \sum_{k=1}^{N}\sum_{i=0}^{2}\tau_{ik}^{(r)}s_i\left(y_k-\pi_{ik}\right)=0 \tag{12}$$

$$\frac{\partial Q^{(r+1)}}{\partial\beta_w} = \sum_{k=1}^{N}\sum_{i=0}^{2}\tau_{ik}^{(r)}\mathbf{w}_k\left(y_k-\pi_{ik}\right)=0 \tag{13}$$

$$\frac{\partial Q^{(r+1)}}{\partial\epsilon} = \sum_{k=1}^{N}\left[\tau_{0k}^{(r)}\left(\frac{x_k}{\epsilon}-\frac{v_k-x_k}{1-\epsilon}\right)+\tau_{2k}^{(r)}\left(\frac{v_k-x_k}{\epsilon}-\frac{x_k}{1-\epsilon}\right)\right]+\frac{a_\epsilon}{\epsilon}-\frac{b_\epsilon}{1-\epsilon}=0 \tag{14}$$

$$\frac{\partial Q^{(r+1)}}{\partial\gamma_i} = \sum_{k=1}^{N}\mathbf{w}_k\left(\tau_{ik}^{(r)}-p_{ik}\right)=0, \tag{15}$$

where we use notations $\pi_{ik}=\pi_{ik}(\beta,\beta_w)=\frac{e^{\beta s_i+\beta_w^T\mathbf{w}_k}}{1+e^{\beta s_i+\beta_w^T\mathbf{w}_k}}$ and $p_{ik}=p_{ik}(\gamma_1,\gamma_2)=\frac{e^{\gamma_i^T\mathbf{w}_k}}{\sum_{m=0}^{2}e^{\gamma_m^T\mathbf{w}_k}}$ for simplicity. From Equation (14), we derive a closed form iteration formula to update the allele read error parameter ϵ:

$$\epsilon^{(r+1)} = \frac{\sum_{k=1}^{N}\left[\tau_{0k}^{(r)}x_k+\tau_{2k}^{(r)}(v_k-x_k)\right]+a_\epsilon}{\sum_{k=1}^{N}\left[(\tau_{0k}^{(r)}+\tau_{2k}^{(r)})v_k\right]+a_\epsilon+b_\epsilon}. \tag{16}$$

There is no closed form iteration formulas to update other parameters β,β_w,γ_i. The M-step for β, β_w, and γ can be obtained by the Newton–Raphson method. The Hessian matrix of $Q^{(r+1)}$ is given by

$$\frac{\partial^2 Q^{(r+1)}}{\partial\beta^2} = -\sum_{k=1}^{N}\sum_{i=0}^{2}\tau_{ik}^{(r)}s_i^2\left[\pi_{ik}(1-\pi_{ik})\right] \tag{17}$$

$$\frac{\partial^2 Q^{(r+1)}}{\partial\beta\partial\beta_w} = -\sum_{k=1}^{N}\sum_{i=0}^{2}\tau_{ik}^{(r)}s_i\mathbf{w}_k\left[\pi_{ik}(1-\pi_{ik})\right] \tag{18}$$

$$\frac{\partial^2 Q^{(r+1)}}{\partial \beta_w \partial \beta_w^T} = -\sum_{k=1}^{N}\sum_{i=0}^{2} \tau_{ik}^{(r)} \mathbf{w}_k \mathbf{w}_k^T \left[\pi_{ik}(1-\pi_{ik})\right] \tag{19}$$

$$\frac{\partial^2 Q^{(r+1)}}{\partial \gamma_i \partial \gamma_i^T} = -\sum_{k=1}^{N} \mathbf{w}_k \mathbf{w}_k^T \left[p_{ik}(1-p_{ik})\right] \tag{20}$$

$$\frac{\partial^2 Q^{(r+1)}}{\partial \gamma_i \partial \gamma_j^T} = \sum_{k=1}^{N} \mathbf{w}_k \mathbf{w}_k^T \left[p_{ik} p_{jk}\right] \tag{21}$$

$$\frac{\partial^2 Q^{(r+1)}}{\partial \gamma_i \partial \beta_w^T} = \frac{\partial^2 Q^{(r+1)}}{\partial \gamma_i \partial \beta} = 0 \tag{22}$$

Let $M = \text{diag}\left(\sum_{i=0}^{2} \tau_{ik}\pi_{ik}(1-\pi_{ik})\right)$ be an $N \times N$ diagonal matrix. Let $W = (w_{ik})$ be the $N \times p$ matrix of covariates. Let μ be an $N \times 1$ vector of $\mu_k = \sum_{ik} \tau_{ik}\pi_{ik}$ and Y be an $N \times 1$ vector of y_k. Initially, we set $\beta^{[0]} = \beta^{(r)}$ and update the parameter estimate by

$$\beta^{[t+1]} = \beta^{[t]} + (W^T M W)^{-1} W^T (Y - \mu). \tag{23}$$

Let $D_{11} = \text{diag}(p_{1k}(1-p_{1k}))$, $D_{12} = D_{21} = -\text{diag}(p_{1k}p_{2k})$, and $D_{22} = \text{diag}(p_{2k}(1-p_{2k}))$. Let $\tau_i = (\tau_{ik})$ be the $N \times 1$ vector and $p_i = (p_{ik})$ be the $N \times 1$ vector. Initially, set $\gamma_i^{[0]} = \gamma_i^{(r)}$ and update the parameters γ_i by

$$\begin{pmatrix} \gamma_1^{[t+1]} \\ \gamma_2^{[t+1]} \end{pmatrix} = \begin{pmatrix} \gamma_1^{[t]} \\ \gamma_2^{[t]} \end{pmatrix} + \begin{pmatrix} W^T D_{11} W & W^T D_{12} W \\ W^T D_{21} W & W^T D_{22} W \end{pmatrix}^{-1} \begin{pmatrix} W^T(\tau_1 - p_1) \\ W^T(\tau_2 - p_2) \end{pmatrix}. \tag{24}$$

In order to obtain $\beta_w^{(r+1)}$ and $\gamma_i^{(r+1)}$, we stop the iterations in the M-step for β and γ_i when $||\beta^{[t+1]} - \beta^{[t]}||^2 + ||\gamma_1^{[t+1]} - \gamma_1^{[t]}||^2 + ||\gamma_2^{[t+1]} - \gamma_2^{[t]}||^2 \le tol^2$ or the number of iterations reaches the prespecified maximum number of iterations. In our work, we set $tol = 10^{-6}$ and fix the maximum iteration as 1000.

2.3. Hypothesis Tests of Genetic Association Controlling for Covariates

To test genetic association between the latent genetic variables and the binary response variable while controlling covariates, we employ Rao's score test. There are several advantages for the use of the score test. Cochran-Armitage trend test with perfect genotypes is a score test, and we extend this test to when the genotypes are highly uncertain. The score test requires less computational cost compared to the likelihood ratio test since it requires the parameter estimates only under the null hypothesis of no association. The score function calculated in previous section is given by

$$S = \sum_{k=1}^{N}\sum_{i=0}^{2} \tau_{ik(0)} s_i \left(y_k - \frac{e^{\beta_{w(0)}^T \mathbf{w}_k}}{1 + e^{\beta_{w(0)}^T \mathbf{w}_k}} \right) \tag{25}$$

where the subscript (0) denotes the estimated parameter under the null hypothesis. Another important issue to be considered when we include the covariates in a low-coverage next-generation sequencing genetic association study is a model misspecification of the latent genotypes on the covariates. To overcome this model misspecification problem, we employ the sandwich variance estimator [27]. In this work, we derive a robust generalized score test using the sandwich variance–covariance estimator. In general, one of the difficulties in applying the sandwich estimator in practice is that it requires analytic derivation for the covariance matrix of the proposed model. For simplicity in our derivation of the sandwich variance estimator, θ denotes the vector of all parameters $\theta = (\beta, \beta_w, \gamma, \epsilon)$, and $\phi = (\beta_w, \gamma, \epsilon)$ denotes the parameter vector except β, and hence $\theta = (\beta, \phi)$. The sandwich variance estimator for the score function S under H_0 is given by

$$v_s = V_{\beta\beta} - V_{\beta\phi} J_{\phi\phi}^{-1} J_{\phi\beta} - J_{\beta\phi} J_{\phi\phi}^{-1} V_{\phi\beta} + J_{\beta\phi} J_{\phi\phi}^{-1} V_{\phi\phi} J_{\phi\phi}^{-1} J_{\phi\beta}, \tag{26}$$

where $V = E_{f_0}\left[\frac{\partial \ell}{\partial \theta}\frac{\partial \ell}{\partial \theta}^T\right]$ and $J = -E_{f_0}\left[\frac{\partial^2 \ell}{\partial\theta\partial\theta^T}\right]$ under the unknown true distribution f_0. For simplicity, we may use h_{ik} during derivation of the sandwich variance estimator:

$$h_{gk} = \left(\frac{e^{y_k(\beta s_g + \beta_w^T \mathbf{w}_k)}}{1 + e^{\beta s_g + \beta_w^T \mathbf{w}_k}}\right)\left((u_{\epsilon g})^{x_k}(1 - u_{\epsilon g})^{v_k - x_k}\right)\left(\frac{e^{\gamma_g^T \mathbf{w}_k}}{\sum_{m=0}^{2} e^{\gamma_m^T \mathbf{w}_k}}\right), \tag{27}$$

so that the likelihood function is written as

$$\ell = \sum_{k=1}^{N} \log\left[\sum_{g=0}^{2} h_{gk}\right] + C. \tag{28}$$

The relationship between J and V can be written as

$$\begin{aligned} J &= \frac{1}{N}\sum_{k=1}^{N}\left[\sum_{g=0}^{2}\tau_{gk}\frac{\partial}{\partial\theta}\log h_{gk}\right]\left[\sum_{g=0}^{2}\tau_{gk}\frac{\partial}{\partial\theta^T}\log h_{gk}\right] \\ &\quad -\frac{1}{N}\sum_{k=1}^{N}\sum_{g=0}^{2}\tau_{gk}\left[\left(\frac{\partial}{\partial\theta}\log h_{gk}\right)\left(\frac{\partial}{\partial\theta^T}\log h_{gk}\right) + \frac{\partial^2}{\partial\theta\partial\theta^T}\log h_{gk}\right] \\ &= V - \frac{1}{N}\sum_{k=1}^{N}\sum_{g=0}^{2}\tau_{gk}\left[\left(\frac{\partial}{\partial\theta}\log h_{gk}\right)\left(\frac{\partial}{\partial\theta^T}\log h_{gk}\right) + \frac{\partial^2}{\partial\theta\partial\theta^T}\log h_{gk}\right] \end{aligned} \tag{29}$$

If there is no model misspecification, we have $J = V$ and the robust score test statistic is reduced to Rao's score test statistic. We denote the difference $R = V - J$ so that $R = \frac{1}{N}\sum_{k=1}^{N}\sum_{g=0}^{2}\tau_{gk}\left[\left(\frac{\partial}{\partial\theta}\log h_{gk}\right)\left(\frac{\partial}{\partial\theta^T}\log h_{gk}\right) + \frac{\partial^2}{\partial\theta\partial\theta^T}\log h_{gk}\right]$. The components of $\frac{\partial}{\partial\theta}\log h_{gk}$ are calculated by

$$\frac{\partial}{\partial\beta}\log h_{gk} = s_g[y_k - \pi_k] \tag{30}$$

$$\frac{\partial}{\partial\beta_w}\log h_{gk} = \mathbf{w}_k[y_k - \pi_k] \tag{31}$$

$$\frac{\partial}{\partial\epsilon}\log h_{gk} = \delta_g(0)\left[\frac{X_k}{\epsilon} - \frac{V_k - X_k}{1-\epsilon}\right] + \delta_g(2)\left[\frac{V_k - X_k}{\epsilon} - \frac{X_k}{1-\epsilon}\right] + \frac{a_\epsilon}{N\epsilon} - \frac{b_\epsilon}{N(1-\epsilon)} \tag{32}$$

$$\frac{\partial}{\partial\gamma_i}\log h_{gk} = \mathbf{w}_k\left[I(g=i) - p_{ik}\right], \tag{33}$$

where $\delta_g(i) = 1$ if $g = i$ and $\delta_g(i) = 0$ if $g \neq i$. It is straightforward to calculate V from the above first derivatives. The second term $\frac{\partial^2}{\partial\theta\partial\theta^T}\log h_{gk}$ of R has components as

$$\frac{\partial^2}{\partial\beta^2}\log h_{gk} = -s_g^2\pi_k(1-\pi_k) \tag{34}$$

$$\frac{\partial^2}{\partial\beta_w\partial\beta_w^T}\log h_{gk} = -\mathbf{w}_k\mathbf{w}_k^T\pi_k(1-\pi_k) \tag{35}$$

$$\frac{\partial^2}{\partial\beta_w\partial\beta}\log h_{gk} = -\mathbf{w}_k s_g\pi_k(1-\pi_k) \tag{36}$$

$$\frac{\partial^2}{\partial\gamma_i\partial\gamma_i^T}\log h_{gk} = -\mathbf{w}_k\mathbf{w}_k^T p_{ik}(1-p_{ik}) \tag{37}$$

$$\frac{\partial^2}{\partial\gamma_i\partial\gamma_{3-i}^T}\log h_{gk} = \mathbf{w}_k\mathbf{w}_k^T p_{ik}p_{3-i,k} \tag{38}$$

$$\frac{\partial^2}{\partial\epsilon^2}\log h_{gk} = -\left(\delta_g(0)\left[\frac{X_k}{\epsilon^2}+\frac{V_k-X_k}{(1-\epsilon)^2}\right]+\delta_g(2)\left[\frac{V_k-X_k}{\epsilon^2}+\frac{X_k}{(1-\epsilon)^2}\right] + \frac{a_\epsilon}{N\epsilon^2}+\frac{b_\epsilon}{N(1-\epsilon)^2}\right), \tag{39}$$

where $i = 1$ or 2. All other second derivatives that are not presented are equal to zero. Using these first and second derivatives of $\log h_{gk}$, we can obtain the components of the difference matrix R as follows:

$$R_{\beta\beta} = \frac{1}{N}\sum_{k=1}^{N}\sum_{g=0}^{2}\tau_{gk}s_g^2\left[(y_k-\pi_k)^2-\pi_k(1-\pi_k)\right] \tag{40}$$

$$R_{\beta_w\beta} = \frac{1}{N}\sum_{k=1}^{N}\sum_{g=0}^{2}\tau_{gk}s_g\mathbf{w}_k\left[(y_k-\pi_k)^2-\pi_k(1-\pi_k)\right] \tag{41}$$

$$R_{\beta_w\beta_w} = \frac{1}{N}\sum_{k=1}^{N}\mathbf{w}_k\mathbf{w}_k^T\left[(y_k-\pi_k)^2-\pi_k(1-\pi_k)\right] \tag{42}$$

$$\begin{aligned} R_{\epsilon\epsilon} = {} & \frac{1}{N}\sum_{k=1}^{N}\left(\tau_{0k}\left[\frac{X_k+a_\epsilon/N}{\epsilon}-\frac{V_k-X_k+b_\epsilon/N}{1-\epsilon}\right]^2+\tau_{1k}\left[\frac{a_\epsilon}{N\epsilon}-\frac{b_\epsilon}{N(1-\epsilon)}\right]^2\right. \\ & +\tau_{2k}\left[\frac{V_k-X_k+a_\epsilon/N}{\epsilon}-\frac{X_k+b_\epsilon/N}{1-\epsilon}\right]^2-\tau_{0k}\left[\frac{X_k+a_\epsilon/N}{\epsilon^2}+\frac{V_k-X_k+b_\epsilon/N}{(1-\epsilon)^2}\right] \\ & \left.-\tau_{1k}\left[\frac{a_\epsilon}{N\epsilon^2}+\frac{b_\epsilon}{N(1-\epsilon)^2}\right]-\tau_{2k}\left[\frac{V_k-X_k+a_\epsilon/N}{\epsilon^2}+\frac{X_k+b_\epsilon/N}{(1-\epsilon)^2}\right]\right) \end{aligned} \tag{43}$$

$$\begin{aligned} R_{\beta\epsilon} = {} & \frac{1}{N}\sum_{k=1}^{N}[y_k-\pi_k]\left(\tau_{0k}s_0\left[\frac{X_k+a_\epsilon/N}{\epsilon}-\frac{V_k-X_k+b_\epsilon/N}{1-\epsilon}\right]+\tau_{1k}s_1\left[\frac{a_\epsilon}{N\epsilon}-\frac{b_\epsilon}{N(1-\epsilon)}\right]\right. \\ & \left.+\tau_{2k}s_2\left[\frac{V_k-X_k+a_\epsilon/N}{\epsilon}-\frac{X_k+b_\epsilon/N}{1-\epsilon}\right]\right) \end{aligned} \tag{44}$$

$$\begin{aligned} R_{\beta_w\epsilon} = {} & \frac{1}{N}\sum_{k=1}^{N}\mathbf{w}_k[y_k-\pi_k]\left(\tau_{0k}\left[\frac{X_k+\epsilon/N}{\epsilon}-\frac{V_k-X_k+b_\epsilon/N}{1-\epsilon}\right]+\tau_{1k}\left[\frac{a_\epsilon}{N\epsilon}-\frac{b_\epsilon}{N(1-\epsilon)}\right]\right. \\ & \left.+\tau_{2k}\left[\frac{V_k-X_k+a_\epsilon/N}{\epsilon}-\frac{X_k+b_\epsilon/N}{1-\epsilon}\right]\right) \end{aligned} \tag{45}$$

$$R_{\gamma_i\gamma_i} = \frac{1}{N}\sum_{k=1}^{N}\mathbf{w}_k\mathbf{w}_k^T\left[(\tau_{ik}-p_{ik})(1-2p_{ik})\right] \tag{46}$$

$$R_{\gamma_1\gamma_2} = \frac{1}{N}\sum_{k=1}^{N}\mathbf{w}_k\mathbf{w}_k^T\left[p_{1k}(p_{2k}-\tau_{2k})+p_{2k}(p_{1k}-\tau_{1k})\right] \tag{47}$$

$$R_{\gamma_i\beta} = \frac{1}{N}\sum_{k=1}^{N}\mathbf{w}_k(y_k-\pi_k)\left[\tau_{ik}s_1-p_{ik}\sum_{g=0}^{2}\tau_{gk}s_g\right] \tag{48}$$

$$R_{\gamma_i\beta_w} = \frac{1}{N}\sum_{k=1}^{N}\mathbf{w}_k\mathbf{w}_k^T(y_k-\pi_k)\left[\tau_{ik}-p_{ik}\right] \tag{49}$$

$$\begin{aligned} R_{\gamma_1\epsilon} = {} & \frac{1}{N}\sum_{k=1}^{N}\mathbf{w}_k\left(-p_{1k}\tau_{0k}\left[\frac{X_k+a_\epsilon/N}{\epsilon}-\frac{V_k-X_k+b_\epsilon/N}{1-\epsilon}\right]\right. \\ & +(1-p_{1k})\tau_{1k}\left[\frac{a_\epsilon}{N\epsilon}-\frac{b_\epsilon}{N(1-\epsilon)}\right] \\ & \left.-p_{1k}\tau_{2k}\left[\frac{V_k-X_k+a_\epsilon/N}{\epsilon}-\frac{X_k+b_\epsilon/N}{1-\epsilon}\right]\right) \end{aligned} \tag{50}$$

$$
\begin{aligned}
R_{\gamma 2\epsilon} &= \frac{1}{N}\sum_{k=1}^{N}\mathbf{w}_k\left(-p_{2k}\tau_{0k}\left[\frac{X_k+a_\epsilon/N}{\epsilon}-\frac{V_k-X_k+b_\epsilon/N}{1-\epsilon}\right]-p_{2k}\tau_{1k}\left[\frac{a_\epsilon}{N\epsilon}-\frac{b_\epsilon}{N(1-\epsilon)}\right]\right.\\
&\left.+(1-p_{2k})\tau_{2k}\left[\frac{V_k-X_k+\epsilon/N}{\epsilon}-\frac{X_k+b_\epsilon/N}{1-\epsilon}\right]\right) \qquad (51)
\end{aligned}
$$

Therefore, our proposed robust score test statistic Z_R can be written as

$$
Z_R = \frac{S}{\sqrt{Nv_s}}, \qquad (52)
$$

which asymptotically has a standard normal distribution under H_0.

Another common approach to obtain p-values is to use Monte Carlo permutation method based on the score vector or function. However, the Monte Carlo permutation p-value calculation given a very small Bonferroni's corrected level of significance needs high computational expenses since it requires at least 10^7 or 10^8 permuted resamples. In this work, we employ the asymptotic permutation p-value calculation. The score function is given by

$$
\begin{aligned}
S &= \sum_{k=1}^{N}\sum_{i=0}^{2}\tau_{ik(0)}s_i\left(y_k-\frac{e^{\beta_{w(0)}^T\mathbf{w}_k}}{1+e^{\beta_{w(0)}^T\mathbf{w}_k}}\right)\\
&= \sum_{k=1}^{N} r_k e_k \qquad (53)
\end{aligned}
$$

where the subscript (0) denotes the estimated parameter under the null hypothesis. We define a score $r_k = \sum_{i=0}^{2}\tau_{ik(0)}s_i$ associated with subject k and the kth residual $e_k = \left(y_k-\frac{e^{\beta_{w(0)}^T\mathbf{w}_k}}{1+e^{\beta_{w(0)}^T\mathbf{w}_k}}\right)$. We can permute the residuals e_k's to calculate the permutation p-value for adjusting covariate effects. The asymptotic permutation test statistic Z_{AP} for a large sample size is given by

$$
Z_{AP} = \frac{S-N\cdot\bar{r}\cdot\bar{e}}{\sqrt{\frac{1}{N-1}\left[\sum_{i=1}^{N}e_i^2-N(\bar{e})^2\right]\left[\sum_{i=1}^{N}r_i^2-N(\bar{r})^2\right]}} \qquad (54)
$$

where $\bar{r}=\frac{1}{N}\sum_{i=1}^{N}r_i$ and $\bar{e}=\frac{1}{N}\sum_{i=1}^{N}e_i$. The simple linear rank test statistic Z_{AP} asymptotically has a standard normal distribution under the null hypothesis [28].

3. Results

3.1. Simulation Study

In this section, we simulate data from the following process:

$$
P(Y=1|w)f(w) = \sum_{i=0}^{2}P(Y=1|G=i,w)P(G=i)f(w) \qquad (55)
$$

For simplicity, we assume genetic relative risk $R_i = \frac{P(Y=1|G=i,w)}{P(Y=1|G=0,w)}$, for $i = 1,2$, does not depend on the covariate W. We assume that the genotype frequency $\pi_i = P(G=i)$ satisfies Hardy–Weinberg equilibrium (HWE), so that $P(G=0)=p^2, P(G=1)=2pq$, and $P(G=2)=q^2$, where q is the minor allele frequency. Then, the prevalence is given by

$$
\begin{aligned}
\phi &= \int P(Y=1|w)f(w)dw\\
&= \int\left[p^2f(w|G=0)+2pqR_1f(w|G=1)+q^2R_2f(w|G=2)\right]P(Y=1|G=0,w)dw \qquad (56)
\end{aligned}
$$

We consider two scenarios when generating covariates w: (1) $f(w|G = i)$ is equal to a standard normal $N(0,1)$ for all $i = 0, 1, 2$, called by a single normal, and (2) $f(w|G = i)$ has a normal distribution with mean μ_i and standard deviation $\sigma = 1$, we call this a normal mixture. For the single normal model,

$$\phi = \left[p^2 + 2pqR_1 + q^2R_2\right] \int P(Y = 1|G = 0, w)f(w)dw \quad (57)$$

We finally assume $P(Y = 1|G = 0, w) = \frac{e^{\alpha+\beta_w w}}{1+e^{\alpha+\beta_w w}}$. During the simulation study, we compute α by numerical integration given prevalence ϕ and other parameters.

3.1.1. Simulation Study for Null Distribution

To evaluate the type I error rate of the proposed test statistic, we perform simulations with 5000 replicates per each parameter setting. We fixed the proportion of cases as 0.5. The parameter settings that we consider are:

(i) Prevalence (ϕ): 0.1, 0.3
(ii) Coverage (v): 4, 30
(iii) Minor allele frequency (q): 0.05, 0.3
(iv) Total sample size (n): 500, 1000, 1500
(v) Covariate (w_1): single normal or normal mixture with mean $\mu = (0, \frac{1}{2}, \frac{1}{2})$ given genotype $(0,1,2)$
(vi) Regression coefficient β_w: 0, 1

We consider prevalence $\phi = 0.3$ that may be large in a genetic association study. It is chosen to reflect pharmacogenomics data that we use in the real data analysis.

Figure 1 shows boxplots of the null simulations. The permutation method appears to have more variability of the empirical rejection rates over different configurations and to have the smaller empirical rejection rates compared to the proposed robust score test based on the sandwich variance estimator. When the sample size was small as 500 and the coverage was 4×, the permutation-based test had less than 2.5% rejection rate though the desired value is 5%. The smallest empirical rejection rate for the proposed robust test was greater than 3.5%, and it appears the empirical rejection rates become closer to 5% as the sample size increases. If the coverage is 30× or higher, then the estimated posterior probabilities in our approach are close to zero-or-one and most inferred genotypes are quite clear. When the coverage was 30×, our proposed test seems to well control the type I error rates regardless of other parameter settings as expected. Table 1 shows the empirical rejection rates under the null settings by combining our simulation results for the lower level of significance.

Table 1. Empirical rejection rates under null settings for level $1 \times 10^{-2}, 1 \times 10^{-3}, 1 \times 10^{-4}$, and 1×10^{-5}.

Method (cvrg)	1×10^{-2}	1×10^{-3}	1×10^{-4}	1×10^{-5}
Permutation (4×)	7.13×10^{-3}	6.14×10^{-4}	4.44×10^{-5}	0
Permutation (30×)	7.73×10^{-3}	7.08×10^{-4}	6.11×10^{-5}	5.56×10^{-6}
Sandwich (4×)	8.34×10^{-3}	6.89×10^{-4}	4.44×10^{-5}	5.56×10^{-6}
Sandwich (30×)	1.02×10^{-2}	1.01×10^{-3}	8.75×10^{-5}	8.33×10^{-6}

3.1.2. Simulation Study for Statistical Power

We used the same parameter settings as in the null simulation study. Additionally, we set multiplicative genetic relative risks vector $(1, 1.5, 1.5^2)$ in the alternative parameter configurations. In the alternative simulations, we calculated empirical rejection rates under Bonferroni corrected level of significance, that is, 5×10^{-8}. Figure 2 shows the boxplots of empirical power under various alternative settings. We removed the results when the sample size was 500 or the minor allele frequency was 0.05 since all the rejection rates were small in Figure 2. It appears interesting that the power of the proposed test when the coverage was 4× and the sample size was 1500 is higher than the power of the test when the coverage was 30× and the sample size was 1000. If the two design costs are similar, then the low-coverage with more samples seems more effective than the high-coverage with less samples.

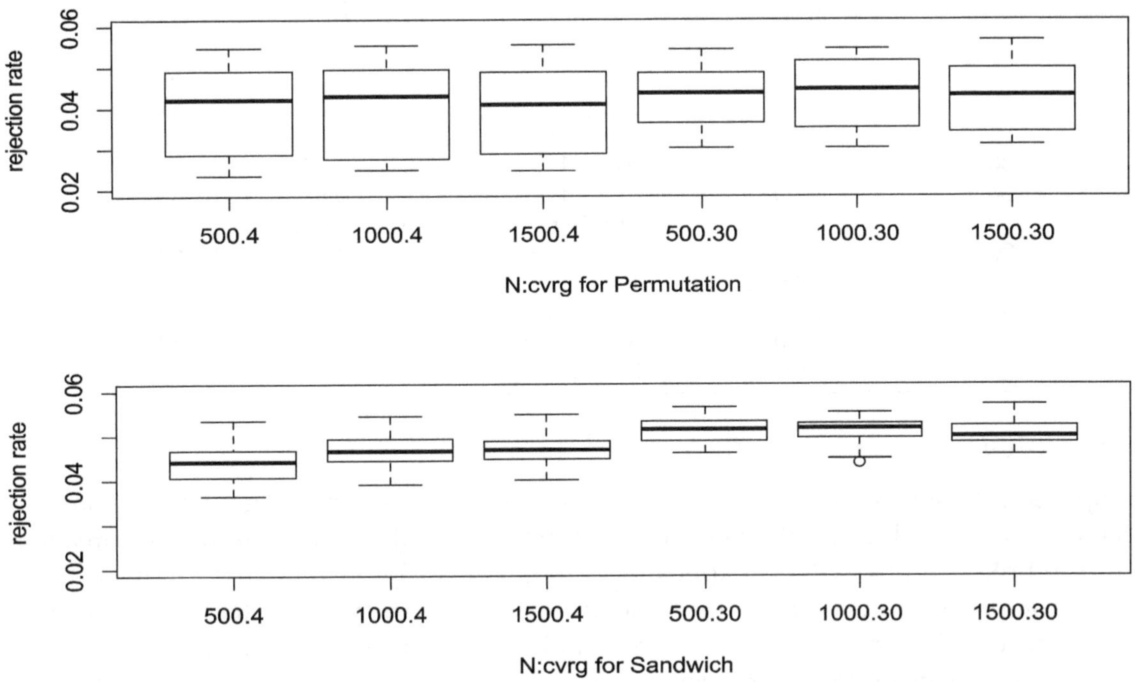

Figure 1. Boxplot of the empirical rejection rates under the null hypothesis.

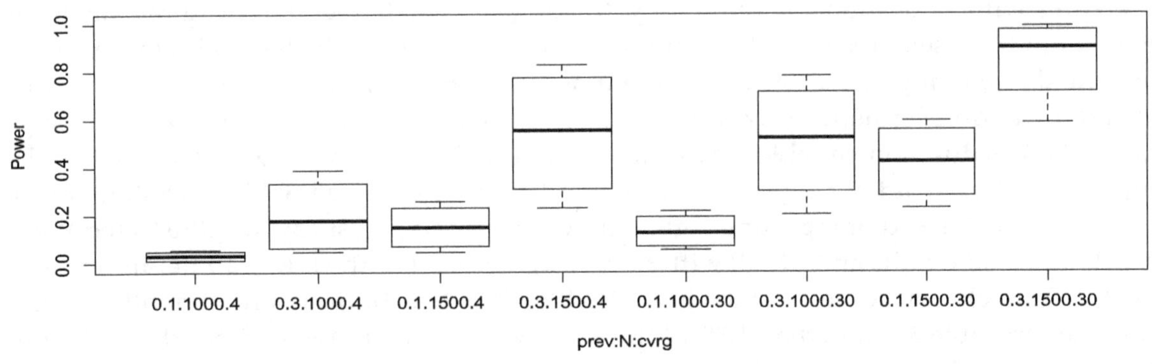

Figure 2. Boxplots of statistical power of the proposed robust test under the alternative settings. The level of significance was set as 5×10^{-8}. The notation 0.1.1000.4 represents prevalence 0.1, total sample size 1000, and coverage 4×.

Table 2 summarizes statistical power of our proposed method and a naive approach. The naive approach uses uncertain genotypes by the maximum posterior probability classification rule [29]. The standard logistic regression was applied to the uncertain genotypes. As expected, the proposed robust method shows higher power than the naive approach when the sequencing coverage is as low as 4×. When the sequencing coverage is high as 30×, two approaches show similar performance in terms of statistical power.

Table 2. Empirical rejection rates under alternative hypothesis. The level of significance was set as 5×10^{-8}.

Coverage	Total Sample Size	Covariate	β_w	Naive	Proposed
4	1000	Normal mixture	0	0.102	0.113
4	1000	Normal mixture	1	0.233	0.261
4	1000	Single normal	0	0.190	0.277
4	1000	Single normal	1	0.269	0.374
4	1500	Normal mixture	0	0.398	0.429
4	1500	Normal mixture	1	0.657	0.701
4	1500	Single normal	0	0.626	0.741
4	1500	Single normal	1	0.736	0.840
30	1000	Normal mixture	0	0.384	0.355
30	1000	Normal mixture	1	0.617	0.603
30	1000	Single normal	0	0.622	0.637
30	1000	Single normal	1	0.734	0.760
30	1500	Normal mixture	0	0.792	0.761
30	1500	Normal mixture	1	0.959	0.954
30	1500	Single normal	0	0.933	0.939
30	1500	Single normal	1	0.978	0.978

3.2. Real Data Analysis

The proposed robust generalized score test was applied to the pharmacogenomics data consisting of 400 epilepsy patients [22]. The data were collected from several epilepsy clinics in Korea and were genotyped for whole-exomes by NGS experiments [30]. All study participants followed the criteria in [31] if the participants had drug-resistant (case group) or drug-responsive (control group) epilepsy. We defined the drug resistance as the occurrence of at least four unprovoked seizures during the past one year at the time of recruitment, with trials of two or more appropriate antiepileptic drugs (AEDs) at maximal tolerated doses. Patients who underwent surgical treatment for drug-resistant epilepsy were classified as having drug-resistant epilepsy, regardless of the surgical outcome. We excluded some patients from the study if they were frequently in poor compliance with AED therapy and had reported seizures with a questionable semiology. In addition, we defined the drug responsiveness as complete freedom from seizures for at least one year up to the date of the last follow-up visit.

We included two non-genetic covariates in our association analysis. The two covariates were age of patient and duration of epileptic seizures. The age variable was definitely independent of genetic information, whereas duration variable may be associated with genetic variables. Due to the relatively small sample size 400, we did not expect to find a significantly associated SNP controlling for the two covariates. Therefore, instead of reporting a genome-wide association study, we illustrated the results of a SNP with low read depths and a SNP with high read depths. For the low read depths example, we selected a SNP from chromosome 1, which is *rs*3811406. The distribution of read depths for the SNP was summarized in Table 3. More than 10% of the sample had five or less read depths and more than 30% of the sample had 10 or less read depths at the SNP. When applying our proposed mixture-based association test, the test statistic value was $z_R = 2.864$ and the p-value was $p = 4.183 \times 10^{-3}$, while the standard logistic regression analysis using pooled genotype calls had $z = 2.601$ and the p-value $p = 9.30 \times 10^{-3}$ that was more than twice the p-value of the proposed robust test.

Table 3. Distribution of read depths at *rs*3811406.

Read Depth v	$v \leq 5$	$5 < v \leq 10$	$10 < v \leq 30$	$v > 30$	Total
Frequency	43	86	95	176	400
Proportion	0.1075	0.215	0.2375	0.44	1

In addition, we applied our proposed test to SNP *rs4915154* at which all patients had 13× or higher read depths and 85% patients had 25× or higher read depths. For this SNP, the proposed robust test statistic was $z_R = 2.940$ with p-value $= 3.28 \times 10^{-3}$ and the multiple logistic regression with the pooled genotype calls reported $z = 2.963$ with p-value $= 3.05 \times 10^{-3}$. The two results were quite close, as expected, due to high read depths at the SNP.

4. Discussion and Conclusions

In the present study, we developed the mixture-based genetic association tests adjusting the effects of non-genetic covariates in low-coverage NGS data. In order to construct a robust test statistic under model misspecification, we derived the sandwich variance estimator of the mixture model. The proposed test statistic is calculated from allele read counts and read depths instead of inferred genotypes so that we can apply this association test to low-coverage NGS data controlling for non-genetic covariates without external imputation or elimination of uncertain genotypes. Another important issue that we addressed in the present study is that the proposed test takes account of potential dependence between latent genotypes and the non-genetic covariates. Regarding computational cost, our proposed method is efficient because it is a generalized score test that uses the estimates of the parameters only under the null hypothesis of no association. When the sequencing depth is 4×, it takes around 1.2 s for sample size 500, 4 s for sample size 1000, and 9 s for sample size 1500 to simulate a dataset and to calculate both test statistics Z_{AP} and Z_R. When the sequencing depth is 30×, it takes approximately 0.13 s for sample size 500, 0.3 s for sample size 1000, and 0.53 s for sample size 1500. Time for these computations is measured based on a single core work of a 3.5 GHz Intel Xeon processor. As illustrated in the real data analysis section, the read depth is not a fixed constant. Therefore, the computational time for real data is usually less than that for the coverage 4× simulation setting. We used statistical software R, which is known to be slow. It would be computationally beneficial to run our proposed methods in other faster program languages for a high-dimensional genome-wide association study.

We applied the penalized likelihood method to avoid singularity of information matrix when calculating the proposed score test statistic. Therefore, the penalty term is not necessary for a non-zero estimate of the error parameter. During our work, we fixed the degree of penalization $C = 1$, $a_\epsilon = 0.01$, and $b_\epsilon = 0.99$ that implies 1% of allele read error as prior information. This parameter choice does not affect the proposed test statistic much since the likelihood function is merely changed when the sample size is greater than 500. It may be of interest to find optimal values for the parameters of the penalty term.

The simulation study confirms that the type I error rates of the proposed test are well controlled under the various parameter settings. The proposed robust test appears to perform better than the permutation based approach. Simulation results indicate that coverage 4× with sample size 1500 shows higher power as compared to coverage 30× with sample size 1000. Our method can be applied to an NGS experimental design by simulations to select coverage and sample size given a fixed amount of budget.

We presented a real data example in which the proposed test and multiple logistic regression results are similar to one another if the sequencing depth is high, whereas the test results may differ when the sequencing depth is low. This might have been caused because the proposed test is an extension of the multiple logistic regression with the unobserved latent genotype predictor. If the sequencing depth is high enough to call accurate genotypes, then our probability model becomes identical to the probability model of the multiple logistic regression. It would be more beneficial to compare with the previous methods by evaluating our proposed methods using a larger sized public dataset.

In this work, we focused on a single variant association test while controlling covariates. By adopting a multivariate mixture model, the proposed method can be extended to the multi-variant genetic association test including covariates. We can also extend the present method to differential genotype misclassifications.

Author Contributions: Conceptualization, W.K.; methodology, J.Y.L. and W.K.; software, J.Y.L. and W.K.; formal analysis, J.Y.L. and W.K.; data curation, M.-K.K. and W.K.; writing—original draft preparation, J.Y.L. and W.K.; writing—review and editing, J.Y.L., M.-K.K., and W.K.; project administration, W.K.; funding acquisition, M.-K.K. and W.K. All authors have read and agreed to the published version of the manuscript.

Abbreviations

The following abbreviations are used in this manuscript:

EM	Expectation–Maximization
GWAS	Genome-wide association study
HWE	Hardy–Weinberg equilibrium
maf	Minor allele frequency
NGS	Next-generation sequence
SNP	Single nucleotide polymorphism
TDT	Transmission disequilibrium test

References

1. Wu, M.C.; Lee, S.; Cai, T.; Li, Y.; Boehnke, M.; Lin, X. Rare-variant association testing for sequencing data with the sequence kernel association test. *Am. J. Hum. Genet.* **2011**, *89*, 82–93. [CrossRef]
2. Cirulli, E.T.; White, S.; Read, R.W.; Elhanan, G.; Metcalf, W.J.; Tanudjaja, F.; Fath, D.M.; Sandoval, E.; Isaksson, M.; Schlauch, K.A.; et al. Genome-wide rare variant analysis for thousands of phenotypes in over 70,000 exomes from two cohorts. *Nat. Commun.* **2020**, *11*, 542. [CrossRef]
3. Lakiotaki, K.; Kanterakis, A.; Kartsaki, E.; Katsila, T.; Patrinos, G.P.; Potamias, G. Exploring public genomics data for population pharmacogenomics. *PLoS ONE* **2017**, *12*, e0182138. [CrossRef] [PubMed]
4. Patrinos, G.P.; Giannopoulou, E.; Katsila, T.; Tsermpini, E.E.; Mitropoulou, C. Integrating next-generation sequencing in the clinical pharmacogenomics workflow. *Front. Pharmacol.* **2019**, *10*, 384.
5. Celesti, F.; Celesti, A.; Wan, J.; Villari, M. Why Deep Learning Is Changing the Way to Approach NGS Data Processing: A Review. *IEEE Rev. Biomed. Eng.* **2018**, *11*, 68–76. [CrossRef]
6. Le, N.Q.K.; Yapp, E.K.Y.; Nagasundaram, N.; Chua, M.C.H.; Yeh, H.Y. Computational identification of vesicular transport proteins from sequences using deep gated recurrent units architecture. *Comput. Struct. Biotechnol. J.* **2019**, *17*, 1245–1254. [CrossRef]
7. Tripathi, R.; Sharma, P.; Chakraborty, P.; Varadwaj, P.K. Next-generation sequencing revolution through big data analytics. *Front. Life Sci.* **2016**, *9*, 119–149. [CrossRef]
8. Cirillo, D.; Valencia, A. Big data analytics for personalized medicine. *Curr. Opin. Biotechnol.* **2019**, *58*, 161–167. [CrossRef]
9. Sims, D.; Sudbery, I.; Ilott, N.E.; Heger, A.; Ponting, C.P. Sequencing depth and coverage: Key considerations in genomic analyses. *Nat. Rev. Genet.* **2014**, *15*, 121–132. [CrossRef]
10. Song, K.; Li, L.; Zhang, G. Coverage recommendation for genotyping analysis of highly heterologous species using next-generation sequencing technology. *Sci. Rep.* **2016**, *6*, 35736. [CrossRef]
11. Gordon, D.; Finch, S.J.; Nothnagel, M.; Ott, J. Power and sample size calculations for case-control genetic association tests when errors are present: Application to single nucleotide polymorphisms. *Hum. Hered.* **2002**, *54*, 22–33. [CrossRef]
12. Ahn, K.; Haynes, C.; Kim, W.; Fleur, R.S.; Gordon, D.; Finch, S.J. The effects of SNP genotyping errors on the power of the Cochran-Armitage linear trend test for case/control association studies. *Ann. Hum. Genet.* **2007**, *71*, 249–261. [CrossRef]

13. Kim, W.; Londono, D.; Zhou, L.; Xing, J.; Nato, A.Q.; Musolf, A.; Matise, T.C.; Finch, S.J.; Gordon, D. Single-variant and multi-variant trend tests for genetic association with next-generation sequencing that are robust to sequencing error. *Hum. Hered.* **2012**, *74*, 172–183. [CrossRef]
14. Hou, L.; Sun, N.; Mane, S.; Sayward, F.; Rajeevan, N.; Cheung, K.; Cho, K.; Pyarajan, S.; Aslan, M.; Miller, P. Impact of genotyping errors on statistical power of association tests in genomic analyses: A case study. *Genet. Epidemiol.* **2017**, *41*, 152–162. [CrossRef]
15. Consortium, .G.P. An integrated map of genetic variation from 1092 human genomes. *Nature* **2012**, *491*, 56.
16. Le, S.Q.; Durbin, R. SNP detection and genotyping from low-coverage sequencing data on multiple diploid samples. *Genome Res.* **2011**, *21*, 952–960. [CrossRef]
17. Li, Y.; Sidore, C.; Kang, H.M.; Boehnke, M.; Abecasis, G.R. Low-coverage sequencing: Implications for design of complex trait association studies. *Genome Res.* **2011**, *21*, 940–951. [CrossRef]
18. Kim, W.; Gordon, D.; Sebat, J.; Kenny, Q.Y.; Finch, S.J. Computing power and sample size for case-control association studies with copy number polymorphism: Application of mixture-based likelihood ratio test. *PLoS ONE* **2008**, *3*, e3475. [CrossRef]
19. Barnes, C.; Plagnol, V.; Fitzgerald, T.; Redon, R.; Marchini, J.; Clayton, D.; Hurles, M.E. A robust statistical method for case-control association testing with copy number variation. *Nat. Genet.* **2008**, *40*, 1245. [CrossRef]
20. Kim, S.Y.; Li, Y.; Guo, Y.; Li, R.; Holmkvist, J.; Hansen, T.; Pedersen, O.; Wang, J.; Nielsen, R. Design of association studies with pooled or un-pooled next-generation sequencing data. *Genet. Epidemiol.* **2010**, *34*, 479–491. [CrossRef]
21. Gordon, D.; Finch, S.J.; De La Vega, F. A new expectation-maximization statistical test for case-control association studies considering rare variants obtained by high-throughput sequencing. *Hum. Hered.* **2011**, *71*, 113–125. [CrossRef]
22. Kim, W.; Kim, Y.H. Genetic association tests when a nuisance parameter is not identifiable under no association. *Commun. Stat. Appl. Methods* **2017**, *24*, 663–671. [CrossRef]
23. Kim, W. Transmission Disequilibrium Tests Based on Read Counts for Low-Coverage Next,-Generation Sequence Data. *Hum. Hered.* **2015**, *80*, 36–49. [CrossRef]
24. Chen, H.; Chen, J.; Kalbfleisch, J.D. A modified likelihood ratio test for homogeneity in finite mixture models. *J. R. Stat. Soc. Ser. B* **2001**, *63*, 19–29. [CrossRef]
25. Zhou, H.; Pan, W. Binomial mixture model-based association tests under genetic heterogeneity. *Ann. Hum. Genet.* **2009**, *73*, 614–630. [CrossRef]
26. Dempster, A.P.; Laird, N.M.; Rubin, D.B. Maximum likelihood from incomplete data via the EM algorithm. *J. R. Stat. Soc. Ser. B* **1977**, *39*, 1–22.
27. White, H. Maximum Likelihood Estimation of Misspecified Models. *Econometrica* **1982**, *50*, 1–25. [CrossRef]
28. Sidak, Z.; Sen, P.K.; Hajek, J. *Theory of Rank Tests*; Academic Press: San Diego, CA, USA, 1999.
29. Anderson, T.W. *An Introduction to Multivariate Statistical Analysis*; Wiley: New York, NY, USA, 1962.
30. Kang, K.W.; Kim, W.; Cho, Y.W.; Lee, S.K.; Jung, K.Y.; Shin, W.; Kim, D.W.; Kim, W.J.; Lee, H.W.; Kim, W. Genetic characteristics of non-familial epilepsy. *PeerJ* **2019**, *7*, e8278. [CrossRef]
31. Kim, M.-K.K.; Moore, J.H.; Kim, J.K.; Cho, K.H.; Cho, Y.W.; Kim, Y.S.; Lee, M.C.; Kim, Y.O.; Shin, M.H. Evidence for epistatic interactions in antiepileptic drug resistance. *J. Hum. Genet.* **2011**, *56*, 71–76. [CrossRef]

14

Application of Functional Data Analysis and FTIR-ATR Spectroscopy to Discriminate Wine Spirits Ageing Technologies

Ofélia Anjos [1,2,3,*], Miguel Martínez Comesaña [4], Ilda Caldeira [5,6], Soraia Inês Pedro [1], Pablo Eguía Oller [4] and Sara Canas [5,6]

1 Instituto Politécnico de Castelo Branco, Escola Superior Agrária, 6001-909 Castelo Branco, Portugal; soraia_p1@hotmail.com
2 CEF, Instituto Superior de Agronomia, Universidade de Lisboa, 1349-017 Lisboa, Portugal
3 CBPBI, Centro de Biotecnologia de Plantas da Beira Interior, 6001-909 Castelo Branco, Portugal
4 Department of Mechanical Engineering, Heat Engines and Fluid Mechanics, Industrial Engineering School, University of Vigo, Maxwell s/n, 36310 Vigo, Spain; migmartinez@uvigo.es (M.M.C.); peguia@uvigo.es (P.E.O.)
5 INIAV, INIAV-Dois Portos, Quinta da Almoínha, 2565-191 Dois Portos, Portugal; ilda.caldeira@iniav.pt (I.C.); sara.canas@iniav.pt (S.C.)
6 MED—MediterraneanInstitute for Agriculture, Environment and Development, Instituto de formação avançada, Universidade de Évora, Pólo da Mitra, Ap. 94, 7006-554 Évora, Portugal
* Correspondence: ofelia@ipcb.pt.

Abstract: Fourier transform infrared spectroscopy (FTIR) with Attenuated Total Reflection (ATR) combined with functional data analysis (FDA) was applied to differentiate aged wine spirits according to the ageing technology (traditional using 250 L wooden barrels versus alternative using micro-oxygenation and wood staves applied in 1000 L stainless steel tanks), the wood species used (chestnut and oak), and the ageing time (6, 12, and 18 months). For this purpose, several features of the wine spirits were examined: chromatic characteristics resulting from the CIELab method, total phenolic index, concentrations of furfural, ellagic acid, vanillin, and coniferaldehyde, and total content of low molecular weight phenolic compounds determined by HPLC. FDA applied to spectral data highlighted the differentiation between all groups of samples, confirming the differentiation observed with the analytical parameters measured. All samples in the test set were differentiated and correctly assigned to the aged wine spirits by FDA. The FTIR-ATR spectroscopy combined with FDA is a powerful methodology to discriminate wine spirits resulting from different ageing technologies.

Keywords: FTIR-ATR; FDA; vector analysis; wine spirit; ageing technology; micro-oxygenation; wood; oak; chestnut; ageing time

1. Introduction

The contact of wine distillate with wood is recognised as a pivotal step of the aged wine spirit production, during which its quality increases and sensory fullness can be reached. Scientific evidence exists on the key role of several physicochemical phenomena, particularly the extraction and oxidation reactions involving the wood-derived compounds of low molecular weight, on the chemical changes (quantitative and qualitative aspects of the beverage's volatile and non-volatile composition) and sensory changes (colour, aroma, and taste) occurred [1–3]. Besides, they mainly depend on the ageing technology, the kind of wood used (oak and chestnut), and the length of the ageing process [1,4,5].

Despite the high-quality spirits attained by the traditional ageing technology, using wooden barrels, this expensive and lengthy process led to the search and study of sustainable alternative technologies. Research has been focused on a technology based on adding wood pieces to the distillate kept in stainless steel tanks, as for other alcoholic beverages [6,7]. Recently, the micro-oxygenation technique, reproducing the oxygen transfer that occurs in the wooden barrel, was applied to optimise this ageing technology. Promising outcomes based on the phenolic composition and colour of the wine spirits in the beginning of ageing were attained [8]. Nevertheless, a comprehensive approach is needed towards the full/robust characterisation of this novel technology, exploring the data acquired over the ageing period through different analytical and statistical methodologies.

Spectroscopic techniques are very useful for food and beverages quality evaluation because they require minimal or no sample preparation (absence of extraction, reactions with some other chemical species, treatment with a chelating agent, masking, sub-sampling, or other manipulation), they are rapid and non-destructive (causing no physicochemical changes during the analysis), and they can be used to simultaneously assess several parameters of a sample. In recent years, significant developments related to the applicability of vibrational spectroscopy combined with multivariate data analysis has been made to give the rapid quantification of several compounds in different matrices or to discriminate different groups of samples. Concerning the alcoholic beverages, the studies were more centred on other drinks than wine spirits. FTIR-ATR (Fourier transform infrared spectroscopy with Attenuated Total Reflection) was used to evaluate different parameter in alcoholic beverages, such as the determination of important quality parameters of beers [9], determination of ethanol content in liquors [10], quantitative analysis of methanol (an adulterant in alcoholic beverages) [11], analysis of ethanol and methanol content in traditional fruit spirits [12], and the authentication of whisky according to its botanical origin [13]. Actually, the studies on wine spirits with FTIR-ATR are scarce. One of them was carried out by Anjos et al. [14], applying the FTIR-ATR methodology to predict the alcoholic strength, methanol, acetaldehyde, and fusel alcohols contents of grape-derived spirits.

In this paper, a new application of functional data analysis (FDA) in the field of quality evaluation using spectroscopic techniques is presented. In recent years, FDA has been used in numerous investigations to analyse processes in continuous time; some examples are energy efficiency [15], medical research [16], econometrics [17], optimisation problems [18], industrial processes [19], environmental research [20,21], and food science [22]. In all these works, FDA showed its usefulness for the study of functions, defined in a specific interval, without missing the correlation between the observations.

FDA allows the analysis of the entire curves, which represent individual observations of the sample under study, detecting different behaviours throughout the curves [15,23]. Ordoñez et al. [23], in a similar context that this paper, showed significant reasons to analyse the sample with curves instead of individual observations. In addition, the contrast of similarity has been carried out from a vectorial approach, but summarising the curves with a single value; in this case, the mean. This is necessary because the curves, although representing individual data, are formed by a set of observations correlated to each other. Martínez et al. [20] explained how this correlation is not taken into account from a vectorial approach. Furthermore, from this point of view, to contrast the similarity between the samples, a classical analysis of variance (ANOVA) [24,25] and anon-parametric Kruskal's test [26] have been performed. Although this is a simpler approach, it contributes to highlight the strengths of FDA. On the one hand, Martínez et al. [15] presented how different conclusions can be reached for each approach because of the biased sample used in the vector analysis (mean of the curves) or different detection of outliers. On the other hand, among others, Sancho et al. [19] demonstrated that FDA presents more realistic and accurate results, showing how and at which specific part of the curve the groups have different spectrometric behaviours.

The aim of this research is to contrast the similarity between the samples obtained with FTIR-ATR spectrometry for wine spirits aged by different ageing technologies (traditional and alternative), with different kinds of wood (chestnut and oak) and over the ageing period (6, 12, and 18 months), using functional data analysis of variance (FANOVA).

Thus, the second section presents, on the one hand, the data used in this study and, on the other hand, the specific methodology applied to obtain the results. The third section presents the results of the comparison and analysis of the differences between the wine spirit samples. Then, in the fourth section, a discussion is carried out about the results and the information obtained with them. Finally, in Section 5, the conclusions of this work are presented.

2. Materials and Methods

2.1. Samples

A total of 30 samples of wine spirit (produced at industrial scale) aged with Limousinoak (*Quercus robur* L.) and Portuguese chestnut (*Castanea sativa* Mill.) for 6, 12, and 18 months of ageing time were analysed (Figure 1), covering the following categories:

- TL—aged in 250 L new oak wooden barrels;
- TC—aged in 250 L new chestnut wooden barrels;
- AL—aged in 1000 L stainless steel tanks with oak wood staves and micro-oxygenation;
- AC—aged in 1000 L stainless steel tanks with chestnut wood staves and micro-oxygenation.

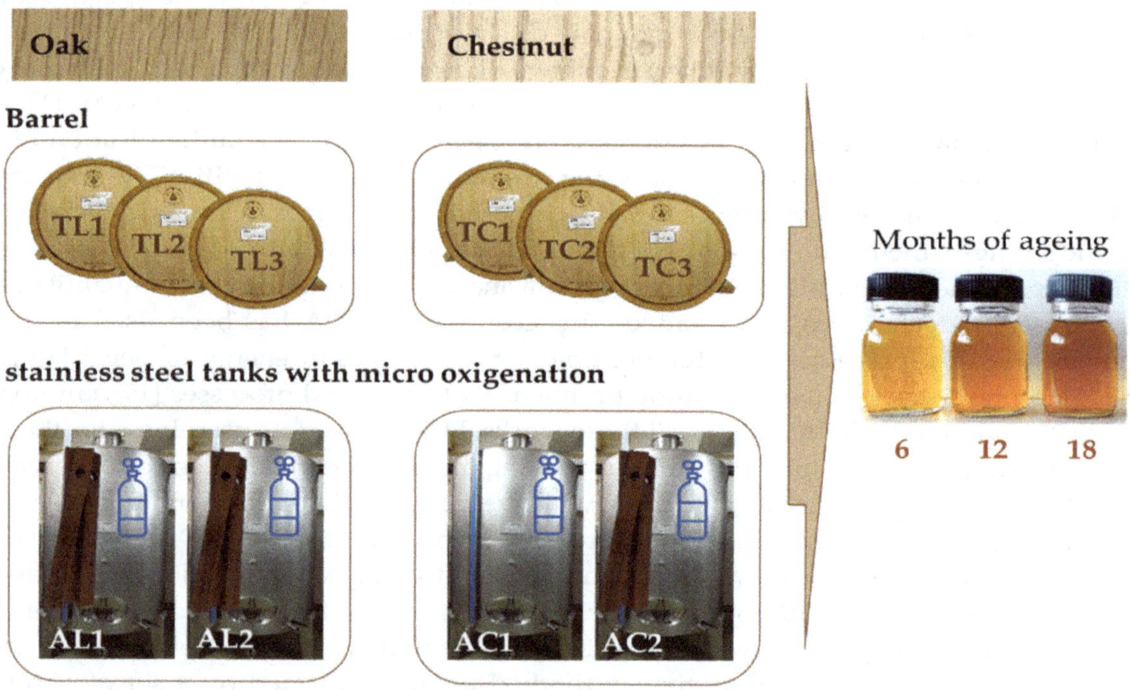

Figure 1. Scheme of the essay. 1, 2, and 3 indicate the essay replicates.

From each sample, four independent samples were taken for spectral analysis, totalising 120 spectra analysed.

The wooden barrels and the stainless-steel tanks were placed in the cellar in similar environmental conditions. The same wine distillate (alcohol strength, 77.4 *v/v*; pH, 5.44; total acidity, as acetic acid, 0.13 g/hL of absolute ethanol; volatile acidity, as acetic acid, 0.11 g/hL of absolute ethanol) produced by the Adega Cooperativa da Lourinhã, Portugal, was used. The wood pieces and barrels were manufactured by J.M. Gonçalves cooperage (Palaçoulo, Portugal) with the medium plus toasting level, as described by Canas et al. [8].

To characterise the wine spirits aged by different technologies, with different wood species, through the ageing time, several analytical determinations were performed for all the samples: CIELAB colour parameters (lightness, saturation, and chromaticity coordinates), total phenolic index, and low molecular weight compounds contents, according to the methodologies described below. All the analyses were done in duplicate.

2.2. *Analytical Procedures*

2.2.1. Analysis of Chromatic Characteristics

The chromatic characteristics of the wine spirits—lightness (L *), varying between 100% (fully transparent) and 0% (fully opaque); saturation (C *); chromaticity coordinates (a * and b *), of which the coordinate a * varies between green (a * < 0) and red hues (a * > 0), and the coordinate b * varies between blue (b * < 0) and yellow hues (b * > 0)—were analysed according to the CIELab method. The analysis was performed in a Varian Cary 100 Bio spectrophotometer (Santa Clara, CA, USA) with a 10 mm glass cell, considering a D65 illuminant and a 10° standard observer. The transmittance measurement was made every 10 nm from 380 to 770 nm. The analysis was performed in duplicate.

2.2.2. Determination of the Total Phenolic Index

The total phenolic index (TPI) of the wine spirits was analysed as described by Cetó et al. [27]: dilution of the samples with ethanol/water 77:23 *v*/*v*; absorbance measurement at 280 nm, using a Varian Cary 100 Bio spectrophotometer (Santa Clara, CA, USA) with a 10 mm quartz cell; calculation of the total phenolic index by multiplying the measured absorbance by the dilution factor. The analysis was performed in duplicate.

2.2.3. Analysis of Low Molecular Weight Compounds

The phenolic and furanic compounds, namelyphenolic acids (gallic acid, vanillic acid, syringic acid and ellagic acid), phenolic aldehydes (vanillin, syringaldehyde, coniferaldehyde, and sinapaldehyde), coumarins (umbelliferone and scopoletin), and furanic aldehydes (5-hydroxymethylfurfural, 5-methylfurfural, and furfural) were quantified by liquid chromatography according to the method proposed by Canas et al. [28]. All compounds were quantified to calculate the total content of low molecular weight compounds. Only some compounds—furfural, ellagic acid, vanillin, and coniferaldehyde—were considered in the discussion of individual compounds contents based on their highly discriminant power.

The chromatographic analysis was carried out in a HPLC Lachrom Merck Hitachi system (Merck, Darmstadt, Germany) composed of a quaternary pump L-7100, a column oven L-7350 equipped with a 250 mm × 4 mm i.d. LiChrospher RP 18 (5 μm) column (Merck, Darmstadt, Germany), a UV–Vis detector L-7400, a fluorescence detector L-7480 (connected to the UV–Vis detector), and an autosampler L-7250. The HSM D-7000 software (Merck, Darmstadt, Germany) was used for the management, acquisition, and treatment of the data. The following chromatographic conditions were used: (i) binary gradient consisting of solvent A, water: formic acid (98:2 *v*/*v*) and solvent B, methanol:water:formic acid (70:28:2 *v*/*v*/*v*) as follows: 0% isocratic B in 3 min, linear gradient from 0% to 40% B in 22 min, from 40% to 60% B in 18 min, 60% isocratic B in 12 min, linear gradient from 60% to 80% B in 5 min, 80% isocratic B in 5 min; (ii) column temperature of 40 °C; (iii) flow rate of 1 mL/min. Phenolic acids and furanic aldehydes were detected at 280 nm, phenolic aldehydes were detected at 320 nm, and coumarins were detected at 325 nm (excitation)/454 nm (emission).

Samples were added with an internal standard (20 mg/L of 4-hydroxybenzaldehyde), filtered through 0.45 μm membrane (Titan, Scientific Resources Ltd., Gloucester, UK) and analysed by the direct injection of 20 μL. The identification of chromatographic peaks was made through comparison of their retention time and UV-Vis spectra with those commercial standards. The assessment of chromatographic purity of the peaks and their UV–Vis spectra (200–400 nm) were performed in Waters system composed of a photodiode-array detector (Waters 996), in the same chromatographic conditions and managed by 'Millennium 2010' software (Waters, Milford, NA, USA). The analysis was performed in duplicate. The quantification of each compound was based on a calibration curve made with the corresponding commercial standard.

2.2.4. Spectroscopic Analyses

Spectra were acquired by the Fourier transform infrared spectroscopic method with platinum Attenuated Total Reflectance (FTIR-ATR) with a Bruker spectrometer (Alpha, Bruker Optic GmbH, Ettlingen, Germany) using a diamond crystal. Four spectra per sample were obtained with 128 scans per spectrum at a spectral resolution of 8 cm^{-1} in the range of 4000 to 450 cm^{-1}.

The FTIR-ATR used was equipped with a flow-through cell with controlled temperature. The cleaning of the cell was done by the injection of water in the flow-through cell, and the background was also measured with distilled water.

2.3. *Statistical Analysis*

2.3.1. Statistical Treatment of Analytical Data

A two-way analysis of variance (ANOVA) was carried out to examine the influence of the two fixed factors—ageing technology (two levels: alternative versus traditional) and kind of wood (two levels: Limousin oak versus chestnut)—on the chromatic characteristics, total phenolic index, and low molecular weight compounds' contents of the wine spirits after 6, 12, and 18 months of ageing. For each significant factor or interaction, the variance percentage was calculated. Fisher's least significant difference (LSD) test was used to compare the average values observed for each sample group. All the calculations were carried out using Statistica vs. 5 (Statsoft Inc., Tulsa, OK, USA).

2.3.2. Functional Analysis

The similarity between the different samples of ageing technology and the different samples of ageing time was contrasted. This analysis was carried out from a vectorial and functional approach. Vectorially, the tests used were the classical ANOVA [25], comparing the mean levels of the groups, and Kruskal's non-parametric test [26], which studies whether the observations of each group come from the same distribution. In addition, from the functional approach, the functional ANOVA (FANOVA) was performed.

Functional Data Analysis (FDA)

The analyses from a functional approach study functions, based on sets of observations, were defined in a determined interval I = [a,b]. One of its strengths is its structure of infinite dimensions that allows to extend the possibilities of data analysis [29,30]. A random variable, measured at a set of discrete points $\{t_g\}_{g=1}^{G} \in [a,b]$, has to take values in metric or semi-metric spaces to be considered functional. Thus, functional data take the form of a matrix with n rows, one for each individual studied, and G columns representing the points of evaluation of the functions [31,32].

Smoothing is the most used process to convert discrete observations into continuous functions, $x(t),\ t \in \mathcal{X} \subset \mathcal{F}$; where $\mathcal{F}$ is the functional space. Specifically, assuming that the functions are observed with error, a functional basis expansion can be adopted as follows:

$$x(t) = \sum_{w=1}^{W} c_w \phi_w(t) \tag{1}$$

where $\{c_w\}_{w=1}^{W}$ is the w-th basis coefficients, $\{\phi_w(t)\}_{w=1}^{W}$ is the w-th basis function, and W is the number of basis functions under consideration [29,32]. Thus, the basis functions used in this work were splines [33] due to their specific properties such as the possibility of generating large basis sets easily or their flexibility [34]. On the other hand, to select the number of bases of each sample, the determination coefficient R^2 was taken into account. The number of bases is the minimum number at which R^2

stops improving significantly or surpasses the value of 0.99 (see Martínez et al. [15]). Moreover, the smoothing process involves solving the following problem:

$$\min_{x \in F} \sum_{g=1}^{G} \{z_g - x(t_g)\}^2 + \lambda \Gamma(x) \tag{2}$$

where $z_g = x(t_g) + \epsilon_g$ is the value obtained by evaluating x at point t_g with ϵ_g being a random noise with zero mean, λ is a parameter controlling the intensity of regularisations, and Γ is a parameter that makes it costly to reach complex solutions. Then, the basis coefficients can be expressed as the solution of the smoothing process as follows [29,35]:

$$c = \left(\mathbf{\Phi}^t \mathbf{\Phi} + \lambda \mathbf{R}\right)^{-1} \mathbf{\Phi}^T \mathbf{z} \tag{3}$$

being $\mathbf{\Phi}$ a GxW matrix formed by $\Phi_{gw} = \phi_w(t_g)$ and $\mathbf{R}$ being a WxW matrix of the elements $\mathbf{R}_{wg} = \int_T D^2\phi_w(t) D^2\phi_g(t)dt$ where $D^n\phi_w(t)$ is the nth-order differential operator of ϕ_w.

Functional Depths

The depth concept, in classical multivariate statistics, was used to measure the centrality of a point $x \in \mathbb{R}^d$ within a data set. The points nearest to the centre obtain a higher depth value [36]. With a functional approach, depths measure the centrality of a curve in relation to the other curves of the sample $x_1, \ldots, x_n$, coming from a stochastic process $\mathcal{X}(\cdot)$ evaluated at a specific interval $[a,b] \in \mathbb{R}$ [37,38].

Although there exist different functional types of depths (Fraiman-Muniz [37], h-modal [39] or Random Projections [38]), the most widely used is the h-modal depth due to its better performance in the correct detection of outliers [36]. Therefore, the functional mode of the sample will be the curve most densely surrounded by other curves. The functional depth of a certain curve in a specific sample is calculated as follows:

$$MD_n(x_i, h) = \sum_{w=1}^{n} K\left(\frac{\|x_i - x_w\|}{h}\right) \tag{4}$$

where $\|\cdot\|$ is the norm in a functional space, $K : \mathbb{R}^+ \to \mathbb{R}^+$ is a kernel function, and h represents the bandwidth parameter [39].

Functional depths, which are considered as a measure of dispersion, are essential in the detection of outliers. In any data analysis, the identification of these atypical data is crucial because they could affect the subsequent estimations. In addition, examining them is important to discover the causes that give these observations a different behaviour from the rest. Besides, in functional analyses, it is even more important because it is possible that individually, the values that form the curve are not outliers in a vectorial way but, from a functional point of view, the entire curve could be [36,40]. Martínez et al. [29] explained in detail how to detect functional outliers within a functional sample.

Functional ANOVA (FANOVA)

Functional data Analysis of Variance, similar to the vector version, contrasts the distance between the mean levels of the factor variables. The aim of this contrast is to find out if the set of functions studied are statistically distinguishable [41]. There will also be Q independent samples $X_{gj}(t)$, $j = 1, \ldots, n_g$; $t \in I = [a,b]$. But these samples are extracted from $\mathcal{L}^2(l)$ processes $X_g(t)$, $g = 1, \ldots, Q$ and their mean function is $E(X_g(t)) = m_g(t)$ [42,43]. If the functional sample is divided into groups like $\{\mathcal{X}_j, \mathcal{A}_j\}_{j=1}^{n} \in \mathcal{F}$ x$\mathcal{A} = \{1, \ldots, A\}$, being $\mathcal{A}$ the factor variable, the hypothesis contrast has the following form:

$$\begin{cases} H_0 : \overline{X}_1(t) = \overline{X}_2(t) = \ldots = \overline{X}_A(t) \\ H_1 : \exists h, e \;\; s.t.\; \overline{X}_h(t) \neq \overline{X}_e(t) \end{cases} \tag{5}$$

The model for the j-th observation belonging to the g-th group has the following form [41]:

$$X_{jg}(t) = \overline{X}(t) + \alpha_g(t) + \epsilon_{jg}(t) \tag{6}$$

where $X_{jg}(t)$ is the functional value ofgroup g, $\alpha_g(t)$ is the effect of being part of a determined group and $\epsilon_{jg}(t)$ represents the unexplained variability for the i-th observation of group g.Furthermore, the model in Equation (6) can be represented in its matrix notation:

$$\mathbf{X}(t) = \mathbf{Z}\boldsymbol{\gamma}(t) + \epsilon(t) \tag{7}$$

being $\mathbf{X}(t)$ a N-dimensional vector, $\boldsymbol{\gamma}(t) = (\overline{X}(t), \alpha_1(t), \ldots, \alpha_Q(t))^T$ a (Q + 1)-dimensional vector, $\epsilon(t)$ a vector of N residual functions, and $\mathbf{Z}$ the design matrix with dimension (N, Q + 1).

Thus, to assure the indentification of the functional effects $\alpha_g(t)$, the sum to zero constraint is introduced [41,44]:

$$\sum_{g=2}^{Q+1} \boldsymbol{\gamma}(t) = 0, \ \forall t \tag{8}$$

The parameter vector $\boldsymbol{\gamma}(t)$ in Equation (7) can be estimated minimising the standard least squares:

$$LMSSE(\boldsymbol{\gamma}) = \int [\mathbf{X}(t) - \mathbf{Z}\boldsymbol{\gamma}(t)]^T [\mathbf{X}(t) - \mathbf{Z}\boldsymbol{\gamma}(t)] dt$$

subject to the constraint (8) [41,44].

Regarding the contrast in Equation (5), most tests are based on F test statistic [32,45]:

$$F_n(t) = \frac{SSR_n(t)/(Q-1)}{SSE_n(t)/(n-Q)}$$

where

$$SSR_n(t) = \sum_{g=1}^{Q} n_g (\overline{X}_g(t) - \overline{X}(t))^2$$

$$SSE_n(t) = \sum_{g=1}^{Q} \sum_{j=1}^{n_g} (X_{gj}(t) - \overline{X}_g(t))^2$$

represents the variations between groups and within groups, respectively. In addition, for these calculations, the sample mean function $\overline{X}(t) = (1/n) \sum_{g=1}^{Q} \sum_{j=1}^{n_g} X_{gj}(t)$ and the sample group mean function $\overline{X}_g(t) = (1/n_g) \sum_{j=1}^{n_g} X_{gj}(t)$, $g = 1, \ldots, Q$ were taken into account.

In this work, two specific tests were used to contrast the similarity between samples. On the one hand, the F test with the reduced bias estimation method (FB) [46]. This test uses both point variations between groups and variations within groups. Specifically, the test statistic has the form:

$$F_n = \frac{\int_Q SSR_n(t) dt/(Q-1)}{\int_Q SSE_n(t)/(n-Q)} \tag{9}$$

The distribution of this statistic is approximated by $F_{(Q-1)k,(n-Q)k}$, where k is estimated by the bias-reduced method. The p-value taken into account comes from $P(F_{(Q-1)k,(n-Q)k} > F_n)$ [45,46].

On other hand, a permutation test based on a representation of the base function (FP) was used. This test is based on the basis representation procedure presented by Górecki and Smaga [47]. The functional observations are be represented by a finite number of basis functions $\varphi_m \in \mathcal{L}^2(I)$, $m = 1, \ldots, K$ as follows:

$$X_{gj}(t) \approx \sum_{m=1}^{K} c_{gjm}\varphi_m(t),\ t \in I \tag{10}$$

where c_{gjm} are random variables with a significantly large K. Moreover, this test uses the following approximation of Equation (9) [45]:

$$\frac{(a-b)/(Q-1)}{(c-a)/(n-Q)} \tag{11}$$

where

$$a = \sum_{g=1}^{Q} \frac{1}{n_g} 1_{n_g}^T \mathbf{C}_g^T \mathbf{J}_\varphi \mathbf{C}_g 1_{n_g}$$

$$b = \frac{1}{n} \sum_{g=1}^{Q} \sum_{j=1}^{Q} 1_{n_g}^T \mathbf{C}_g^T \mathbf{J}_\varphi \mathbf{C}_j 1_{n_j}$$

$$c = \sum_{g=1}^{Q} trace(\mathbf{C}_g^T \mathbf{J}_\varphi \mathbf{C}_g)$$

being 1_a a vector of ones with dimension $ax1$, $\mathbf{C}_g = (c_{gjm})_{j=1,\ldots,n_g; m=1,\ldots,K}$, $i = 1, \ldots, Q$, and $\mathbf{J}_\varphi := \int_I \boldsymbol{\varphi}(t)\boldsymbol{\varphi}^T(t)dt$ is the matrix of cross-products with dimensions KxK based on $\boldsymbol{\varphi}(\mathbf{t}) = (\varphi_1(t), \ldots, \varphi_K(t))^T$.

3. Results

The two-way ANOVA results (Table 1) show that the ageing technology and the kind of wood had a highly significant effect on the colour and total phenolic content acquired by the wine spirits during the ageing process (after 6, 12, and 18 months). Among these factors, greater influence was exerted by the ageing technology (higher percentage of the variance explained) on the chromatic characteristics, while a similar weight of both factors was observed in the phenolic content. Regardless of the sampling time, the wine spirits aged with micro-oxygenation and chestnut wood staves (AC) exhibited a significantly lower value of lightness (L *) and significantly higher values of saturation (C *) and chromaticity coordinates (a * and b *) than the others. This set of chromatic characteristics reflects a more evolved colour, since lower L * and higher C * correspond to a more intense/darker colour, and the combination of higher a * and b * (yellow and red hues, respectively) is associated with a greater intensity of amber or orange hue, which made these spirits look older than the others. The colour of wine spirits from Limousin wooden barrels (TL) was on the opposite side, while the wine spirits aged with micro-oxygenation and Limousin wood staves (AL) and those aged in chestnut wooden barrels (TC) presented intermediate characteristics.

Table 1. Effect of the ageing technology and kind of wood on the chromatic characteristics and total phenolic index acquired by the wine spirits after 6, 12, and 18 months of ageing. AC: Alternative Chestnut, AL: Alternative Oak, TC: Traditional Chestnut, TL: Traditional Oak.

Ageing Months	Code	L *(%)	A *	B *	C *	TPI
6	TC	85.41 ± 1.41 b	3.37 ± 1.08 b	50.96 ± 2.68 b	51.08 ± 2.74 b	24.94 ± 1.98 b
	TL	93.73 ± 0.42 c	−1.25 ± 0.17 a	26.76 ± 2.21 a	26.79 ± 2.20 a	11.99 ± 1.18 a
	AC	77.14 ± 1.26 a	11.79 ± 1.22 c	70.00 ± 1.59 c	70.99 ± 1.77 c	47.79 ± 3.80 c
	AL	87.55 ± 0.23 b	1.75 ± 0.14 b	46.24 ± 0.14 b	46.27 ± 0.15 b	26.88 ± 0.87 b
Variance origin	Technology (S)	61.9 ***	52.7 ***	60.2 ***	60.2 ***	42.1 ***
	Wood (W)	36.7 ***	31.9 ***	38.8 ***	38.8 ***	52.4 ***
	SxW	NS	14.0 **	NS	NS	4.2 *
	Residual	1.5	1.3	1.0	1.0	1.3
12	TC	79.11 ± 1.38 b	9.89 ± 1.31 b	69.41 ± 1.64 b	70.11 ± 1.80 b	37.90 ± 1.94 b
	TL	90.62 ± 1.04 c	−0.49 ± 0.65 a	40.66 ± 3.4 a	40.67 ± 3.40 a	17.55 ± 1.81 a
	AC	65.58 ± 1.97 a	25.63 ± 1.78 c	87.25 ± 0.56 c	90.95 ± 1.04 c	65.72 ± 1.37 c
	AL	79.17 ± 0.60 b	10.38 ± 0.53 b	70.87 ± 0.61 b	71.63 ± 0.68 b	37.86 ± 0.23 b
Variance origin	Technology (S)	49.2 ***	49.9 ***	49.4 ***	50.7 ***	48.6 ***
	Wood (W)	49.7 ***	46.3 ***	43.5 ***	44.9 ***	48.8 ***
	SxW	NS	3.1 *	6.2 ***	3.6 *	2.2 *
	Residual	1.1	0.8	0.9	0.8	0.5
18	TC	77.33 ± 1.25 b	12.06 ± 1.24 b	73.97 ± 1.22 b	74.95 ± 1.40 b	40.98 ± 2.46 b
	TL	89.61 ± 1.17 c	−0.02 ± 0.84 a	44.35 ± 3.74 a	44.36 ± 3.74 a	18.15 ± 1.63 a
	AC	62.29 ± 1.94 a	28.97 ± 1.62 c	89.27 ± 0.12 c	93.86 ± 0.61 c	71.60 ± 2.50 c
	AL	76.59 ± 0.52 b	13.19 ± 0.45 b	75.87 ± 0.44 b	77.01 ± 0.51 b	40.55 ± 1.25 b
Variance origin	Technology (S)	52.2 ***	52.9***	47.7 ***	50.0 ***	47.9 ***
	Wood (W)	46.9 ***	45.2***	40.2 ***	42.3 ***	49.5 ***
	SxW	NS	1.4*	11.1 **	6.8 ***	2.1 *
	Residual	0.9	0.6	0.9	0.9	0.6

L *—lightness; a *, b *—chromaticity coordinates; C *—saturation; TPI—total phenolic index. For each sampling time (6, 12 and 18 months), mean values with the same letter in a column are not statistically different. NS, $p > 0.05$; * $0.01 < p < 0.05$; ** $0.001 < p < 0.01$; *** $p < 0.001$.

In a previous work, this research team have already identified furfural, ellagic acid, vanillin and coniferaldehyde as markers of the ageing technology and the kind of wood used [8]. The results obtained for these compounds are shown in Table 2. It should be stressed that most of the low molecular weight compounds contents were mainly dependent on the ageing technology (higher percentage of the variance explained), as aforementioned for the chromatic characteristics and TPI, but ellagic acid content was closely related to the kind of wood. Indeed, significantly higher levels of furfural, vanillin, and coniferaldehyde were found in the wine spirits aged by the alternative technology; the first two were higher in the modality comprising chestnut wood staves (AC), whereas the latter was higher in the modality comprising Limousin wood staves (AL). Slight differences were found between the wine spirits from chestnut barrels (TC) and Limousin oak barrels (TL). Regarding ellagic acid, higher content was promoted by the chestnut wood (AC and TC), especially in the alternative technology. The contact with Limousin oak together with micro-oxygenation (AL) induced a slightly lower content of this phenolic acid, and a weak performance was showed by the Limousin oak barrels (TC).

Besides the differences between the two ageing technologies and the two kinds of wood, the results also reveal a remarkable role of the ageing time, as indicated by the percentage of variance observed for each factor in ANOVA (Table 2). Regardless of the ageing modality, there was a gradual decrease of lightness and a gradual increase of saturation and chromatic coordinates that correspond to the colour development over the ageing process.

Table 2. Effect of the ageing technology and kind of wood on the contents of low molecular weight compounds (mg/L absolute ethanol) of the wine spirits after 6, 12, and 18 months of ageing.

Ageing Months	Code	Furfural	Ellagic Acid	Vanillin	Coniferaldehyde	sumHPLC
6	TC	38.31 ± 6.90 ᵃ	8.12 ± 1.41 ᵇ	2.03 ± 0.01 ᵇ	6.17 ± 0.63 ᵃ	163.10 ± 24.08 ᵇ
	TL	31.73 ± 6.38 ᵃ	3.43 ± 0.20 ᵃ	1.49 ± 0.08 ᵃ	5.21 ± 0.04 ᵃ	78.99 ± 9.71 ᵃ
	AC	127.05 ± 5.07 ᶜ	15.35 ± 0.38 ᶜ	4.62 ± 0.20 ᵈ	10.60 ± 0.65 ᵇ	296.05 ± 15.91 ᶜ
	AL	87.74 ± 4.11 ᵇ	6.28 ± 0.70 ᵃ,ᵇ	3.26 ± 0.16 ᶜ	12.20 ± 0.52 ᵇ	195.23 ± 3.04 ᵇ
Variance origin	Technology(S)	82.7 ***	30.4 ***	79.1 ***	98.5 ***	63.1 ***
	Wood(W)	8.1 **	56.9 ***	15.1 ***	NS	34.6 ***
	SxW	8.0 ***	10.8 ***	5.3 **	NS	NS
	Residual	1.2	2.0	0.5	1.5	2.3
12	TC	35.85 ± 6.03 ᵃ	12.86 ± 1.16 ᵇ	3.61 ± 0.22 ᵇ	7.00 ± 0.57 ᵃ	231.68 ± 26.34 ᵇ
	TL	31.36 ± 5.80 ᵃ	5.64 ± 0.34 ᵃ	2.66 ± 0.23 ᵃ	6.49 ± 0.58 ᵃ	95.09 ± 13.67 ᵃ
	AC	131.17 ± 4.91 ᶜ	24.89 ± 1.26 ᶜ	8.68 ± 0.02 ᵈ	13.97 ± 0.17 ᵇ	369.24 ± 8.57 ᵈ
	AL	96.08 ± 1.93 ᵇ	11.94 ± 0.88 ᵇ	6.77 ± 0.09 ᶜ	19.61 ± 0.41 ᶜ	275.32 ± 4.56 ᶜ
Variance origin	Technology(S)	87.9 ***	41.3 ***	89.2 ***	79.7 ***	64.5 ***
	Wood(W)	5.2 **	50.0 ***	8.6 ***	5.1 ***	33.8 ***
	SxW	6.1 ***	7.8 ***	1.8 ***	14.8 ***	NS
	Residual	0.8	0.9	0.3	0.4	1.7
18	TC	36.55 ± 7.28 ᵃ	15.48 ± 1.41 ᵇ	4.43 ± 0.34 ᵇ	6.85 ± 0.75 ᵃ	271.85 ± 38.84 ᵇ
	TL	31.63 ± 6.26 ᵃ	6.81 ± 0.49 ᵃ	3.13 ± 0.23 ᵃ	6.41 ± 0.60 ᵃ	104.10 ± 16.10 ᵃ
	AC	113.35 ± 4.27 ᶜ	28.17 ± 1.15 ᶜ	8.61 ± 0.07 ᵈ	11.16 ± 0.15 ᵇ	347.46 ± 4.33 ᶜ
	AL	86.72 ± 0.21 ᵇ	13.81 ± 0.23 ᵇ	7.49 ± 0.24 ᶜ	17.93 ± 0.07 ᶜ	275.63 ± 4.79 ᵇ
Variance origin	Technology(S)	89.4 ***	39.1 ***	92.0 ***	63.3 ***	43.6 ***
	Wood(W)	4.8 ***	53.7 ***	7.2 ***	10.0 ***	40.9 ***
	SxW	4.3 *	6.2 ***	NS	26.0 ***	11.9 *
	Residual	1.5	0.8	0.7	0.7	3.6

Furf—furfural; Ellag—ellagic acid; Vanil—vanillin; Cofde—coniferaldehyde; sumHPLC—total content of low molecular weight compounds determined by HPLC. For each sampling time (6, 12, and 18 months), mean values with the same letter in a column are not. NS, $p > 0.05$; * $0.01 < p < 0.05$; ** $0.001 < p < 0.01$; *** $p < 0.001$.

The levels of significance shown in Table 3 reveal that the variation of chromatic characteristics, phenolic content (TPI and sumHPLC), and most of individual compounds contents were mostly significant between 6 months and 12 months. Significant differences between the three sampling times marked the wine spirits aged by the alternative technology with Limousin staves. In general, the L * parameter decreased over the ageing time, while the other parameters (a *, b *, and C *) increased. Furfural, ellagic acid, vanillin, coniferaldehyde and total content of low molecular weight compounds increased. Similar results were observed in the first months of ageing time of this kind of wine spirit [8].

Table 3. Differences observed for each samples group (technology vs. wood) during the ageing time.

	TC			TL			AC			AL		
	6	12	18	6	12	18	6	12	18	6	12	18
L *(%)	b	a	a	b	a	a	b	a	a	c	b	a
a *	a	b	b	a	a	a	a	b	b	a	b	c
b *	a	b	c	a	b	b	a	b	b	a	b	c
C *	a	b	c	a	b	b	a	b	b	a	b	c
TPI	a	b	b	a	b	b	a	b	b	a	b	b
Furf	a	a	a	a	a	a	a	ab	b	a	ab	b
Ellag	a	b	b	a	b	c	a	b	b	a	b	c
Vanil	a	b	c	a	b	c	a	b	b	a	b	c
Cofde	a	a	a	a	b	b	a	b	a	a	b	c
sumHPLC	a	b	b	a	a	a	a	b	b	a	b	b

The spectra obtained for wine spirit are similar to those reported by others authors [28].

The representative absorbance spectra of wine spirit samples studied is plotted in Figure 2; their spectral information is in accordance with previous reports of other authors [48]. The IR region from 2990 to 3626 cm^{-1} has a very strong influence due to water present in the samples [49]. Nevertheless, for these analyses, the background was measured with water.

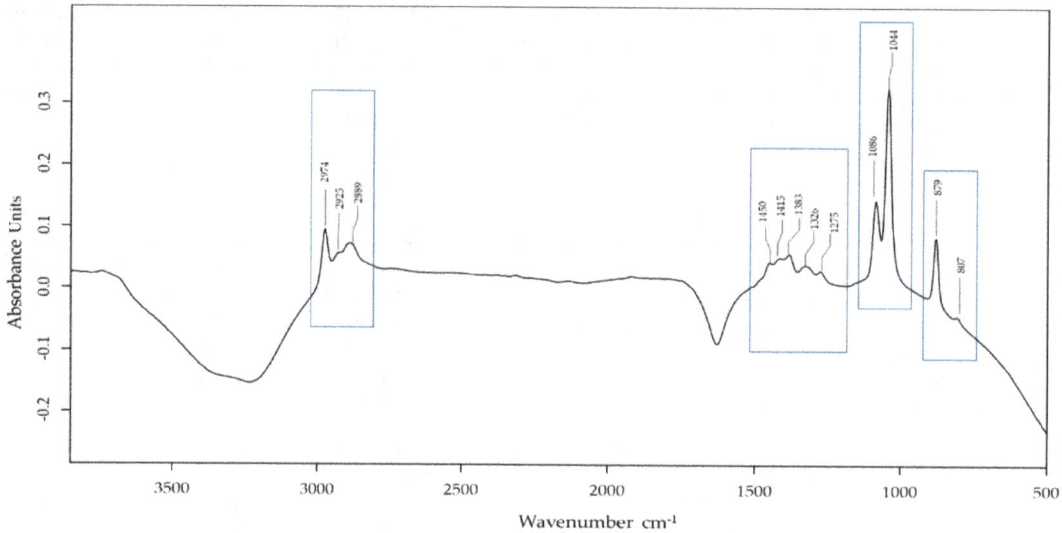

Figure 2. Fourier transform infrared spectroscopy–Attenuated Total Reflection (FTIR-ATR) absorbance spectra of wine spirit samples.

There is IR information in the regions from 3000 to 2900 cm^{-1} due to the O–H stretching of alcohols and C–H stretching of CH_3 and CH_2, and consequently related to the presence of ethanol and methanol in the alcoholic beverages [49].

Regarding the region from 1500 to 860 cm^{-1}, it corresponds to C–C and C–O vibrations in volatile compounds [12,14].

The small peak at 1450 cm^{-1} was assigned to C–OH bending deformation, and the peak at 1275 was assigned to C–O stretching in the acid molecules [11,50].

According to Stuart (2004), the region from 1300 to 840 cm^{-1} shows other absorption bands assigned to the C=O and C=C groups present in furanic compounds. The highest peaks at 1086 and 1044 cm^{-1} are ascribed to the C–O stretch absorption bands, which are important regions for ethanol and methanol identification and quantification respectively, and C–C absorption bands, which are related to ethanol and some organic compounds such as sugars, phenols, alcohols, and esters [14,49,51].

The peak at 879 cm^{-1} could be ascribed to out-of-plane C–H bending of aromatic compounds [10], and to CH–OH, C–C, C–O, and C–H bond stretching due to water, sugars, and phenolic compounds [51].

According to these regions of the FTIR-ATR absorption spectra (with baseline correction) of the wine spirit, a mathematical analysis has been performed to determine the differences between the groups studied. On the one hand, a functional ANOVA (FANOVA) model has been considered using two different tests: FP, the permutation test based on a representation of the base function (Equation (9)); and FB, the F test with the reduced bias estimation method (Equation (11)). On the other hand, a vector analysis based on the classical Analysis of Variance and the non-parametric Kruskal test have been carried out. Although all the areas of the curves have been analysed following the same process, only two were plotted each time to show the differences between the samples (space problem).

First, the different groups based on the ageing technology and kind of wood (AC, AL, TC, and TL) were tested. The contrast is different for each ageing time (18 months, 12 months, and 6 months). With an 18-month period, the hypothesis of similarity between all groups is rejected in all areas of the whole curve and from both points of view (Table 4). The samples of wine spirits aged by the alternative technology, on average, have higher absorbance units than the aged by the traditional one. Within the alternative technology, the wine spirit aged in oak wood always showed higher absorbance units than the one aged in chestnut wood (Figure 3). With an ageing period of 12 months, the similarity hypothesis is also rejected in all areas by the two analyses (Table 4). Even so, in this case, the differences between alternative and traditional ageing technologies are small. The spectrometric curves can hardly be differentiated in the functional part of Figure 4, and the p-values obtained are higher (Table 4). With an ageing time of 6 months, significant reasons were found to reject the similarity hypothesis in all areas through functional and vector analysis (Table 4). With this sample, the difference is more significant than with the 12-month sample but less than with the 18-month sample. Moreover, it can be seen that the wine spirit that gets higher absorbance units in the alternative sample is the one aged in oak, but with the traditional sample, it is the one aged in chestnut (Figure 5). These results are in accordance with those previously observed for the colour and analytical parameters (Tables 1 and 2).

Table 4. Numerical results of the similarity contrast between the groups AC, AL, TC, and TL, depending on the ageing time of the wine spirit samples. Functional results (FDA) are based on FANOVA using two different tests (FP: permutation test based on a representation of the base function, FB: F test with a reduced bias estimation method). In addition, the results of the ANOVA and Kruskal's test representing the vectorial results (VA) are shown. All the results are p-values based on a 5% significance level.

TEST\SAMPLE			3050–2750 cm^{-1}	1525–120 cm^{-1}	1150–960 cm^{-1}	910–750 cm^{-1}
				18 months		
FDA	**FANOVA**	**FP**	≈0	≈0	≈0	≈0
		FB	≈0	≈0	≈0	≈0
VA	**ANOVA**		≈0	≈0	≈0	≈0
	Kruskal		≈0	≈0	≈0	≈0
				12 months		
FDA	**FANOVA**	**FP**	0.001	≈0	≈0	≈0
		FB	0.003	≈0	1.31×10^{-5}	≈0
VA	**ANOVA**		0.007	≈0	0.003	≈0
	Kruskal		0.008	2.23×10^{-6}	0.003	≈0
				6 months		
FDA	**FANOVA**	**FP**	≈0	≈0	≈0	≈0
		FB	≈0	≈0	≈0	1×10^{-4}
VA	**ANOVA**		≈0	≈0	≈0	≈0
	Kruskal		≈0	≈0	≈0	≈0

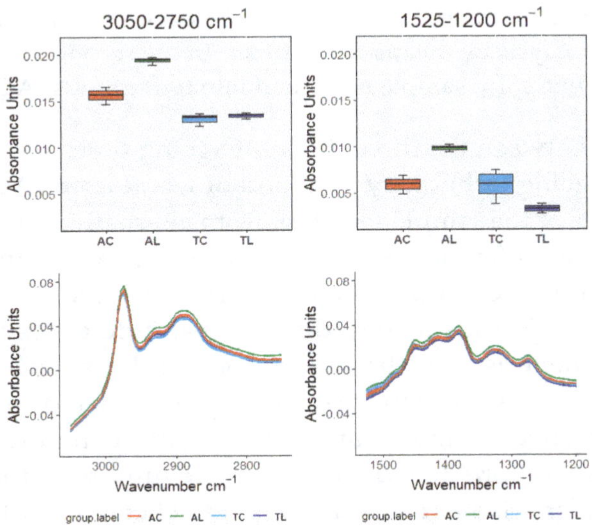

Figure 3. Plots of two of the four significant areas of the curves with an ageing time of 18 months. In the first row, vectorial analysis by means of boxplots. In the second row, functional data analysis (FDA) through curves in the studied interval. The wine spirit sample is divided into four groups (AC, AL, TC, and TL).

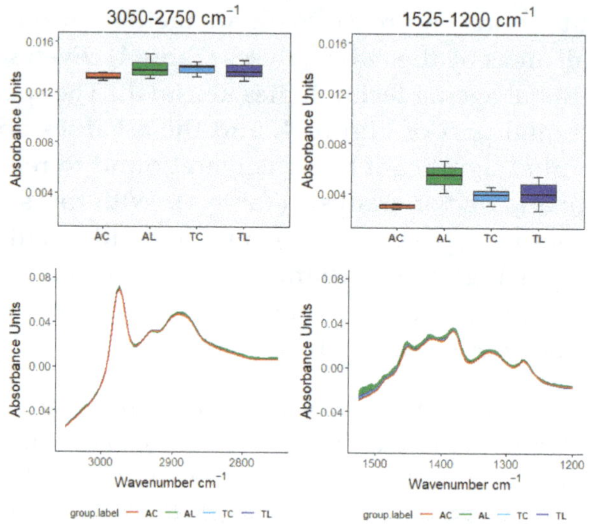

Figure 4. Plots of two of the four significant areas of the curves with an ageing time of 12 months. In the first row, vectorial analysis by means of boxplots. In the second row, FDA through curves in the studied interval. The wine spirit sample is divided into four groups (AC, AL, TC, and TL).

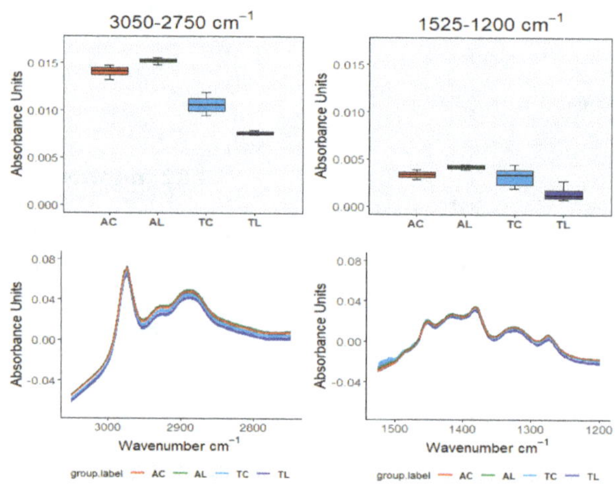

Figure 5. Plots of two of the four significant areas of the curves with an ageing time of 6 months. In the first row, vectorial analysis by means of boxplots. In the second row, FDA through curves in the studied interval. The wine spirit sample is divided into four groups (AC, AL, TC, and TL).

Secondly, the similarity between the three different ageing times was contrasted. This contrast is different depending on the ageing technology and kind of wood used (AC, AL, TC, and TL). Figure 6 shows the boxplots and curves of the sample resulting from alternative technology with chestnut wood. It can be seen that there is little difference between the three ageing times. Especially at the spectral region of 1150–960 cm^{-1} (second d column of Figure 6), in which the similarity hypothesis is rejected in the vector analysis. Instead, FDA is able to detect these differences between the curve samples (Table 5). This region is characteristic of the absorption bands assigned to C=O and C=C groups existing in furanic compounds, C–O stretch absorption bands related to ethanol and methanol, and C–C absorption bands also related to ethanol and some organic compounds such as sugars, phenols, alcohols, and esters previously reported, and all of them are important to differentiate the samples in this study. They are mainly identified at the peaks of the 1044 cm^{-1} and 1086 cm^{-1}, which are chiefly related to the presence of ethanol and methanol but also related to some organic compounds such as sugars, phenols, alcohols, and esters existing in minor concentration in the beverages. In addition, similarity in the other areas of the entire curve within the sample resulting from alternative technology with chestnut wood is rejected. In the case of the alternative technology, the wine spirits aged with

Limousin oak wood are very similar to those aged with chestnut wood but with more distance between the three ageing times (Figure 7), as observed in the chemical analysis. The similarity hypothesis is rejected in all areas and from both points of view (Table 5). Figure 8 shows the differentiation within the samples resulting from chestnut barrels. In this case, the two areas drawn from the whole curve are closer, but the differences between the three ageing times are greater. The hypothesis of similarity between the samples is rejected in all areas and from the vectorial and functional approach (Table 5). Finally, regarding the wine spirits aged in Limousin oak barrels, the 18 and 12-month samples show higher absorbance units than the 6-month sample (Figure 9). The spectrometric curves of the functional graph can be easily distinguished. There are significant reasons to reject the similarity between the three samples in all areas of the full curve from the two analyses (Table 5).

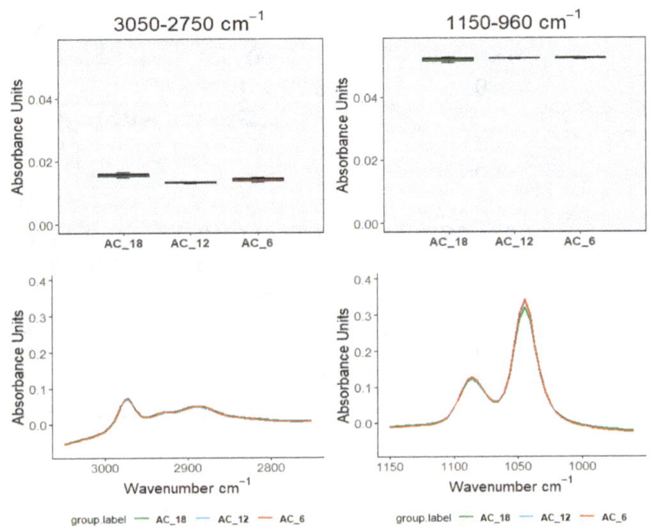

Figure 6. Plots of two of the four significant areas of the Alternative Chestnut (AC) curves. In the first row, vectorial analysis by means of boxplots. In the second row, FDA through curves in the studied interval. The wine spirit sample is divided into three groups depending on the ageing time (18, 12, and 6 months of ageing).

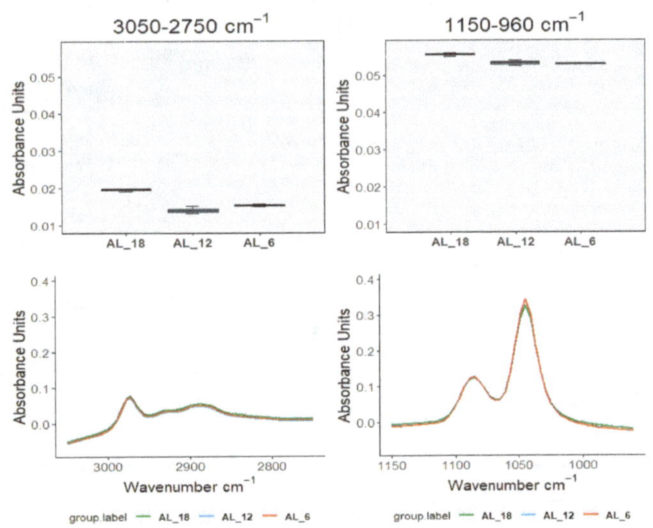

Figure 7. Plots of two of the four significant areas of the Alternative Oak (AL) curves. In the first row, vectorial analysis by means of boxplots. In the second row, FDA through curves in the studied interval. The wine spirit sample is divided into three groups depending on the ageing time (18, 12, and 6 months of ageing).

Table 5. Results of the similarity contrast between the three ageing times (18 months, 12 months, and 6 months), depending on the ageing technology. Functional results (FDA) are based on functional ANOVA (FANOVA) using two different tests (FP: permutation test based on a representation of the base function, FB: F test with a reduced bias estimation method). In addition, the results of the ANOVA and Kruskal's test representing the vectorial results (VA) are shown. All the results are p-values based on a 5% significance level.

TEST\SAMPLE			3050–2750 cm^{-1}	1525–120 cm^{-1}	1150–960 cm^{-1}	910–750 cm^{-1}
			Groups within AC			
FDA	**FANOVA**	**FP**	≈ 0	≈ 0	≈ 0	≈ 0
		FB	≈ 0	≈ 0	≈ 0	≈ 0
VA	**ANOVA**		≈ 0	≈ 0	0.032	≈ 0
	Kruskal		≈ 0	6.72×10^{-5}	0.214	1×10^{-4}
			Groups within AL			
FDA	**FANOVA**	**FP**	≈ 0	≈ 0	≈ 0	≈ 0
		FB	≈ 0	≈ 0	≈ 0	≈ 0
VA	**ANOVA**		≈ 0	≈ 0	≈ 0	≈ 0
	Kruskal		2.07e-06	4.14×10^{-6}	3.34×10^{-5}	3.71×10^{-6}
			Groups within TC			
FDA	**FANOVA**	**FP**	≈ 0	≈ 0	≈ 0	≈ 0
		FB	≈ 0	≈ 0	≈ 0	≈ 0
VA	**ANOVA**		≈ 0	≈ 0	≈ 0	≈ 0
	Kruskal		≈ 0	7.32×10^{-6}	≈ 0	≈ 0
			Groups within TL			
FDA	**FANOVA**	**FP**	≈ 0	≈ 0	≈ 0	≈ 0
		FB	≈ 0	≈ 0	≈ 0	≈ 0
VA	**ANOVA**		≈ 0	≈ 0	≈ 0	≈ 0
	Kruskal		1×10^{-4}	1.95×10^{-6}	≈ 0	≈ 0

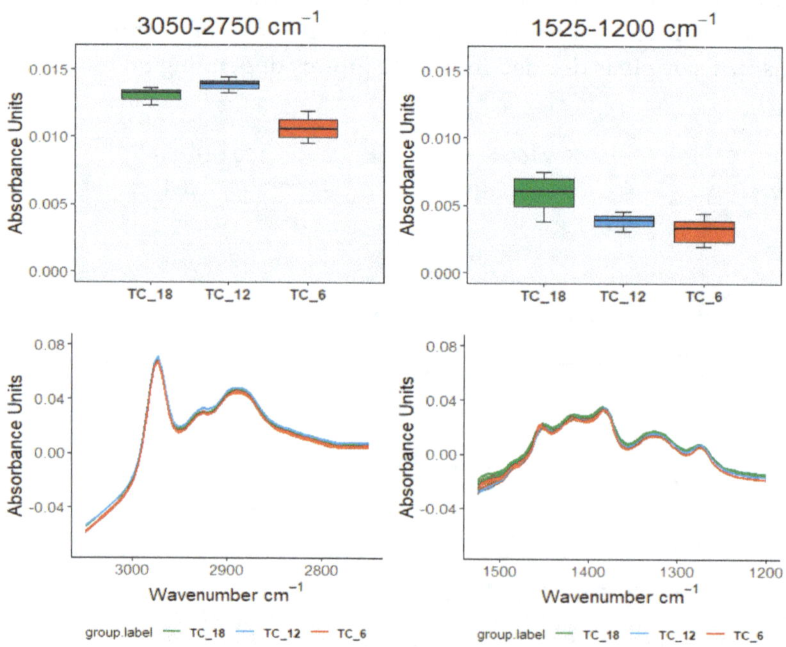

Figure 8. Plots of two of the four significant areas of the Traditional Chestnut (TC) curves. In the first row, vectorial analysis by means of boxplots. In the second row, FDA through curves in the studied interval. The wine spirit sample is divided into three groups depending on the ageing time (18, 12, and 6 months of ageing).

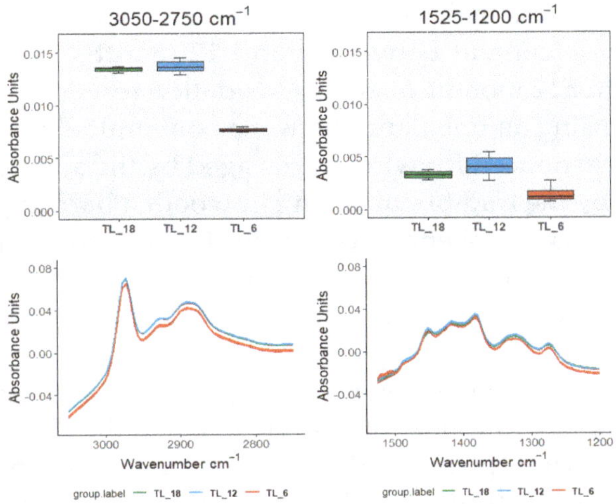

Figure 9. Plots of two of the four significant areas of the Traditional Oak (TL) curves. In the first row, vectorial analysis by means of boxplots. In the second row, FDA through curves in the studied interval. The wine spirit sample is divided into three groups depending on the ageing time (18, 12, and 6 months of ageing).

4. Discussion

The colour and total phenolic content acquired by the wine spirits during the ageing process (Table 1) are in line with the wine spirits' features observed in the first six months of ageing [8], and they are also in accordance with those of recent ageing studies of brandy [2,3]. The colour acquisition has been assigned to extraction, oxidation, and other physicochemical phenomena occurring during ageing [8,52]. Hence, more marked color development in the wine spirits aged by the alternative technology, and with chestnut wood, may have resulted from the direct application of oxygen [8] and/or from more oxygen transferred by this kind of wood due to its higher porosity [53], which may have favored such phenomena. Regarding the total phenolic content (TPI), a consistent behavior with the chromatic characteristics was found, reinforcing previous results [8]: there is a correlation between the enrichment in phenolic compounds and the colour acquired, which are induced by the ageing technology and the kind of wood.

Accordingly, the total content of low molecular weight compounds (phenolic acids, phenolic aldehydes, coumarins, and furanic aldehydes) concentrations determined by HPLC (Table 2) acted as distinctive feature of the aged wine spirits, which can be ordered as follows: AC > AL ~ TC > TL. Therefore, the pattern reported for the first six months of ageing [8] prevailed until the end of the ageing process (18 months).

Low molecular weight compounds, such as furfural, ellagic acid, vanillin, and coniferaldehyde, as aforementioned, stood out as markers of the ageing technology (Table 2). Similar results for these wood-derived phenolic compounds [1,54] were obtained in the beginning of ageing [8]. Their higher accumulation in wine spirits resulting from the alternative technology has been ascribed to a faster extraction from the wood, as described by other authors for the ageing of red wine with wood pieces [55,56]; the behavior of coniferaldehyde is explained by its sensitivity towards oxidation.

The evolution observed for the colour parameters over the ageing process was supported by the progressive increase of the total phenolic content (TPI) and particularly the low molecular weight compounds quantified by HPLC (sumHPLC), which reflected the release of the wood-derived compounds into the wine spirit together with a positive balance between their formation/degradation [4] in the liquid medium. The kinetics of individual compounds shows a continuous increment of ellagic acid content, regardless of the assay modality, which may have resulted from the release of this phenolic acid existing in the wood [1] and from the oxidation and hydrolysis of ellagitannins in the wine spirit during ageing [57,58]. The kinetics of vanillin presented a similar pattern, which was likely due to its release from the wood [59], lignin's hydroalcoholysis, and subsequent oxidation reactions that converted coniferaldehyde into vanillin in the wine spirit [4]. Interestingly, there was an increase of coniferaldehyde content between 6 and 12 months followed by a decrease until 18 months. This suggests that after 12 months of ageing, oxidation reactions prevailed over extraction and lignin's hydroalcoholys is causing an unbalance between coniferaldehyde formation and degradation. Similar results were previously noticed for wine spirits aged by the alternative technology but without micro-oxygenation and also by the traditional ageing in wooden barrels [4]. The evolution of furfural content was different from those of phenolic compounds. Indeed, its kinetics was closely related to the ageing technology: there was an increase from 6 to 12 months and then a decrease in wine spirits aged by the alternative technology; the opposite behavior occurred in wine spirits aged by the traditional one (TC and TL).

The statistical analyses performed during the ageing time reveals significant differences between 6 months and 12 months in a higher number of parameters than the observed between 12 months and 18 months. However, for the wine spirits aged by the alternative technology with Limousin staves significant differences between the three sampling times were observed. Regarding the alternative technology, these outcomes suggest that one year is enough/adequate to obtain a high-quality wine spirit aged with micro-oxygenation and chestnut staves as a consequence of higher pool of compounds and specific anatomical properties of this kind of wood compared with the Limousin oak wood [2,60].

The mathematical analysis performed with functional and vectorial approach corroborates the findings about colour parameters, total phenolic content, and low molecular weight compounds. According to the mathematical analysis, two main conclusions have been reached. First, there are significant reasons to reject the similarity between the wine spirits samples differentiated according to the ageing technology used. It was rejected in all areas of the full spectrometric curve for each of the three ageing times tested, both in vector and functional data analysis. Secondly, evidence was found to state that the samples of wine spirits with different ageing time are not similar. The similarity hypothesis between the three ageing times for each of the different technologies (traditional vs. alternative) used was rejected in all the areas of the spectrometric curve analysed.

On the other hand, FDA shows higher consistency than the vector analysis as observed by Martínez et al. [15,20]. In addition, to provide more information and accuracy, FDA can detect significant differences between groups that vector analysis cannot. This can be seen in the sample obtained by alternative technology with chestnut wood, specifically in the spectral region of 1150–960 cm^{-1} of the entire curve. It is attributed to the functional groups present in furanic compounds and also related to ethanol, methanol, and sugars, phenols, and esters existing in the wine spirit [1,4,5,8,14,49,51]. Vector analysis found no evidence to reject the similarity between the samples based on ageing time, but FDA, taking into account all correlated observations measured in the specific area, did.

5. Conclusions

The aged wine spirit composition and quality depend on the raw material, but mainly on the ageing technology, the wood species used in the ageing process, and the ageing time. Monitoring this quality over time with a fast methodology is very important for the industry.

This study showed a remarkable congruence between the analytical determinations (colour, total phenolic index, and low molecular weight compounds) and the FTIR-ATR/mathematical approach in the differentiation of wine spirits aged by the alternative (micro-oxygenation combined with wood pieces in stainless steel tanks) and the traditional (wooden barrels) technologies, using two kinds of wood (chestnut and oak), over the ageing time (6, 12, and 18 months). Additionally, the analytical results suggest that one year is enough/adequate to achieve a high-quality wine spirit aged with micro-oxygenation and chestnut staves.

FTIR-ATR with appropriate chemometric techniques, specifically functional data analysis, and vector analysis proved to be a powerful tool for an easier and faster monitoring of the wine spirit's ageing process. FDA showed higher consistency than the vector analysis, providing more information and accuracy; it detected significant differences between groups that the vector analysis did not detect.

It was clearly demonstrated in the sample resulting from the alternative technology with chestnut wood, specifically in the area 1150–960 cm^{-1} of the entire curve. Vector analysis found no evidence to reject similarity between the time ageing samples, but FDA, taking into account all correlated observation measured in the specific area, found it.

It was also possible to identify the more accurate spectral region to perform a calibration model to be applied by the wine spirit industry.

Author Contributions: Conceptualisation, O.A. and S.C.; methodology, O.A., S.C., M.M.C., P.E.O.; formal analysis, S.C., O.A., I.C., S.I.P.; Statistical analysis: M.M.C., P.E.O., O.A.; investigation, S.C., O.A., I.C., S.I.P.; resources, S.C.; O.A.; writing, O.A., M.M.C., S.C.; supervision, O.A.; project administration, S.C.; funding acquisition, S.C. All authors have read and agreed to the published version of the manuscript.

Acknowledgments: The authors thank João Pedro Catela, Nádia Santos, Manuela Gomes, Eugénia Gomes and Inês Antunes from Adega Cooperativa da Lourinhã, and Tanoaria J.M. Gonçalves for the technical support.

References

1. Canas, S. Phenolic Composition and Related Properties of Aged Wine Spirits: Influence of Barrel Characteristics. A Review. *Beverages* **2017**, *3*, 55. [CrossRef]
2. García-Moreno, M.V.; Sánchez-Guillén, M.M.; Mier, M.R.D.; Delgado-González, M.J.; Rodríguez-Dodero, M.C.; García-Barroso, C.; Guillén-Sánchez, D.A. Use of Alternative Wood for the Ageing of Brandy de Jerez. *Foods* **2020**, *9*, 250. [CrossRef] [PubMed]
3. Schwarz, M.; Rodríguez-Dodero, C.M.; Jurado, S.M.; Puertas, B.G.; Barroso, C.; Guillén, A.D. Analytical Characterization and Sensory Analysis of Distillates of Different Varieties of Grapes Aged by an Accelerated Method. *Foods* **2020**, *9*, 277. [CrossRef] [PubMed]
4. Canas, S.; Caldeira, I.; Belchior, A.P. Extraction/oxidation kinetics of low molecular weight compounds in wine brandy resulting from different ageing technologies. *Food Chem.* **2013**, *138*, 2460–2467. [CrossRef]
5. Caldeira, I.; Santos, R.; Ricardo-da-Silva, J.M.; Anjos, O.; Mira, H.; Belchior, A.P.; Canas, S. Kinetics of odorant compounds in wine brandies aged in different systems. *Food Chem.* **2016**, *211*, 937–946. [CrossRef]
6. Rodríguez-Solana, R.; Rodríguez-Freigedo, S.; Salgado, J.M.; Domínguez, J.M.; Cortés-Diéguez, S. Optimisation of accelerated ageing of grape marc distillate on a micro-scale process using a Box–Benhken design: Influence of oak origin, fragment size and toast level on the composition of the final product. *Aust. J. Grape Wine Res.* **2017**, *23*, 5–14. [CrossRef]
7. Gómez-Plaza, E.; Bautista-Ortín, A.B. Chapter 10—Emerging Technologies for Aging Wines: Use of Chips and Micro-Oxygenation. In *Red Wine Technology*; Morata, A., Ed.; Academic Press: Cambridge, MA, USA, 2019; pp. 149–162. [CrossRef]
8. Canas, S.; Caldeira, I.; Anjos, O.; Belchior, A. Phenolic profile and colour acquired by the wine spirit in the beginning of ageing: Alternative technology using micro-oxygenation vs traditional technology. *LWT Food Sci. Technol.* **2019**, *111*, 260–269. [CrossRef]
9. Llario, R.; Iñón, F.A.; Garrigues, S.; de la Guardia, M. Determination of quality parameters of beers by the use of attenuated total reflectance-Fourier transform infrared spectroscopy. *Talanta* **2006**, *69*, 469–480. [CrossRef]
10. Yadav, P.K.; Sharma, R.M. Classification of illicit liquors based on their geographic origin using Attenuated total reflectance (ATR)—Fourier transform infrared (FT-IR) spectroscopy and chemometrics. *Forensic Sci. Int.* **2019**, *295*, e1–e5. [CrossRef]
11. Nagarajan, R.; Mehrotra, R.; Bajaj, M.M. Quantitative analysis of methanol, an adulterant in alcoholic beverages, using attenuated total reflectance spectroscopy. *J. Sci. Ind. Res.* **2006**, *65*, 416–419.
12. Teodora Emilia, C.; Carmen, S.; Florinela, F.; FloricuÅ£a, R.; Raluca Maria, P.O.P.; Mira, F. Rapid Quantitative Analysis of Ethanol and Prediction of Methanol Content in Traditional Fruit Brandies from Romania, using FTIR Spectroscopy and Chemometrics. *Not. Bot. Horti Agrobot. Cluj-Napoca* **2013**, *41*. [CrossRef]
13. Wiśniewska, P.; Boqué, R.; Borràs, E.; Busto, O.; Wardencki, W.; Namieśnik, J.; Dymerski, T. Authentication of whisky due to its botanical origin and way of production by instrumental analysis and multivariate classification methods. *Spectrochim. Acta. Part A Mol. Biomol. Spectrosc.* **2017**, *173*, 849–853. [CrossRef] [PubMed]
14. Anjos, O.; Santos, A.J.A.; Estevinho, L.M.; Caldeira, I. FTIR–ATR spectroscopy applied to quality control of grape-derived spirits. *Food Chem.* **2016**, *205*, 28–35. [CrossRef] [PubMed]

15. Martínez Comesaña, M.; Martínez Mariño, S.; Eguía Oller, P.; Granada Álvarez, E.; Erkoreka González, A. A Functional Data Analysis for Assessing the Impact of a Retrofitting in the Energy Performance of a Building. *Mathematics* **2020**, *8*, 547. [CrossRef]
16. Martínez, J.; Ordoñez, C.; Matìas, J.M.; Taboada, J. Determining noise in an aggregates plant using functional statistics. *Hum. Ecol. Risk Assess.* **2011**, *17*, 521–533. [CrossRef]
17. Müller, H.-G.; Sen, R.; Stadtmüller, U. Functional data analysis for volatility. *J. Econom.* **2011**, *165*, 233–245. [CrossRef]
18. López, M.; Martínez, J.; Matías, J.M.; Taboada, J.; Vilán, J.A. Shape functional optimization with restrictions boosted with machine learning techniques. *J. Comput. Appl. Math.* **2010**, *234*, 2609–2615. [CrossRef]
19. Sancho, J.; Pastor, J.J.; Martínez, J.; García, M.A. Evaluation of Harmonic Variability in Electrical Power Systems through Statistical Control of Quality and Functional Data Analysis. *Procedia Eng.* **2013**, *63*, 295–302. [CrossRef]
20. Martínez, J.; Pastor, J.; Sancho, J.; McNabola, A.; Martínez, M.; Gallagher, J. A functional data analysis approach for the detection of air pollution episodes and outliers: A case study in Dublin, Ireland. *Mathematics* **2020**, *8*, 225. [CrossRef]
21. Di Battista, T.; Fortuna, F. Functional confidence bands for lichen biodiversity profiles: A case study in Tuscany region (central Italy). *Stat. Anal. Data Min. ASA Data Sci. J.* **2017**, *10*, 21–28. [CrossRef]
22. Ruiz-Bellido, M.A.; Romero-Gil, V.; García-García, P.; Rodríguez-Gómez, F.; Arroyo-López, F.N.; Garrido-Fernández, A. Data on the application of Functional Data Analysis in food fermentations. *Data Brief* **2016**, *9*, 401–412. [CrossRef] [PubMed]
23. Ordoñez, C.; Martínez, J.; Matías, J.M.; Reyes, A.N.; Rodríguez-Pérez, J.R. Functional statistical techniques applied to vine leaf water content determination. *Math. Comput. Model.* **2010**, *52*, 1116–1122. [CrossRef]
24. Crawley, M.J. *The R Book*; John Wiley & Sons: Hoboken, NJ, USA, 2013.
25. Montgomery, D.C. *Design and Analysis of Experiments*; John Wiley & Sons, Inc.: Hoboken, NJ, USA, 2013.
26. Ostertagová, E.; Ostertag, O.; Kovác, J. Methodology and application of the Kruskal-Wallis test. *Appl. Mech. Mater.* **2014**, *611*, 115–120. [CrossRef]
27. Cetó, X.; Gutiérrez, J.M.; Gutiérrez, M.; Céspedes, F.; Capdevila, J.; Mínguez, S.; Jiménez-Jorquera, C.; del Valle, M. Determination of total polyphenol index in wines employing a voltammetric electronic tongue. *Anal. Chim. Acta* **2012**, *732*, 172–179. [CrossRef]
28. Canas, S.; Belchior, A.; Spranger, M.; Sousa, R. High-performance liquid chromatography method for analysis of phenolic acids, phenolic aldehydes, and furanic derivatives in brandies. Development and validation. *J. Sep. Sci.* **2003**, *26*, 496–502. [CrossRef]
29. Martínez, J.; Saavedra, Á.; García, P.J.; Piñeiro, J.I.; Iglesias, C.; Taboada, J.; Sancho, J.; Pastor, J. Air quality parameters outliers detection using functional data analysis in the Langreo urban area (Northern Spain). *Appl. Math. Comput.* **2014**, *241*, 1–10. [CrossRef]
30. Wang, J.L.; Chiou, J.M.; Müller, H.G. Functional Data Analysis. *Annu. Rev. Stat. Appl.* **2016**, *3*, 257–295. [CrossRef]
31. Kramosil, I.; Michálek, J. Fuzzy metrics and statistical metric spaces. *Kybernetika* **1975**, *11*, 336–344.
32. Ramsay, J.O.; Silverman, B.W. *Functional Data Analysis*, 2nd ed.; Springer: New York, NY, USA, 2005.
33. Cormier, E.; Genest, C.; Nešlehová, J.G. Using B-splines for nonparametric inference on bivariate extreme-value copulas. *Extremes* **2014**, *17*, 633–659. [CrossRef]
34. Kwok, W.Y.; Moser, R.D.; Jiménez, J. A Critical Evaluation of the Resolution Properties of B-Spline and Compact Finite Difference Methods. *J. Comput. Phys.* **2001**, *174*, 510–551. [CrossRef]
35. Piñeiro, J.I.; Torres, J.M.; García, P.J.; Alonso, J.R.; Muñiz, C.D.; Taboada, J. Analysis and detection of functional outliers in waterquality parameters from different automated monitoring stationsin the Nalón River Basin (Northern Spain). *Environ. Sci. Pollut. Res. Int.* **2015**, *22*, 387–396. [CrossRef] [PubMed]
36. Febrero, M.; Galeano, P.; Wenceslao, G.M. Outlier detection in functional data by depth measures, with application to identify abnormal NOx levels. *Environmetrics* **2008**, *19*, 331–345. [CrossRef]
37. Fraiman, R.; Muniz, G. Trimmed means for functional data. *TEST* **2001**, *10*, 419–440. [CrossRef]
38. Cuevas, A.; Febrero, M.; Fraiman, R. Robust estimation and classification for functional data via projection-based notions. *Comput. Stat.* **2007**, *22*, 481–496. [CrossRef]
39. Cuevas, A.; Febrero, M.; Fraiman, R. On the use of bootstrap for estimating functions with functional data. *Comput. Stat. Data Anal.* **2006**, *51*, 1063–1074. [CrossRef]

40. Millán-Roures, L.; Epifanio, I.; Martínez, V. Detection of Anomalies in Water Networks by Functional Data Analysis. *Math. Probl. Eng.* **2018**, *2018*, 13. [CrossRef]
41. Maturo, F.; Battista, T.; Fortuna, F. Parametric functional analysis of variance for fish biodiversity assessment. *J. Environ. Inform.* **2016**, *28*, 101–109. [CrossRef]
42. Cuesta-Albertos, J.A.; Febrero, M. A simple multiway ANOVA for functional data. *TEST* **2010**, *19*, 537–557. [CrossRef]
43. Zhang, J.-T. Analysis of Variance for Functional Data. In *A Chapman & Hall Book*; Press, C.R.C., Ed.; Taylor & Francis Group: Abingdon, UK, 2013; p. 412.
44. Aguilera, A.; Fortuna, F.; Escabias, M.; Battista, T. Assessing Social Interest in Burnout Using Google Trends Data. *Soc. Indic. Res.* **2019**, 1–13. [CrossRef]
45. Górecki, T.; Smaga, Ł. fdANOVA: An R software package for analysis of variance for univariate and multivariate functional data. *Comput. Stat.* **2019**, *34*, 571–597. [CrossRef]
46. Zhang, J.-T. Statistical inferences for linear models with functional responses. *Stat. Sin.* **2011**, *21*, 1431–1451. [CrossRef]
47. Górecki, T.; Smaga, Ł. A comparison of tests for the one-way ANOVA problem for functional data. *Comput. Stat.* **2015**, *30*, 987–1010. [CrossRef]
48. Moreira, J.; Santos, L. Spectroscopic interferences in Fourier transform infrared wine analysis. *Anal. Chim. Acta* **2003**, *513*, 263–268. [CrossRef]
49. Shurvell, H.F. Spectra—Structure Correlations in the Mid- and Far-Infrared. In *Handbook of Vibrational Spectroscopy*; John Wiley & Sons: Hoboken, NJ, USA, 2001.
50. Tarantilis, P.A.; Troianou, V.E.; Pappas, C.S.; Kotseridis, Y.S.; Polissiou, M.G. Differentiation of Greek red wines on the basis of grape variety using attenuated total reflectance Fourier transform infrared spectroscopy. *Food Chem.* **2008**, *111*, 192–196. [CrossRef]
51. Öztürk, B.; Yucesoy, D.; Ozen, B. Application of Mid-infrared Spectroscopy for the Measurement of Several Quality Parameters of Alcoholic Beverages, Wineand Raki. *Food Anal. Methods* **2012**, *5*, 1435–1442. [CrossRef]
52. Fujieda, M.; Tanaka, T.; Suwa, Y.; Koshimizu, S.; Kouno, I. Isolation and Structure of Whiskey Polyphenols Produced by Oxidation of Oak Wood Ellagitannins. *J. Agric. Food. Chem.* **2008**, *56*, 7305–7310. [CrossRef]
53. Carvalho, A. Identificação anatómica e caracterização física e mecânica das madeiras utilizadas no fabrico de quartolas para produção de aguardentes velhas de qualidade—Denominação Lourinhã. *Ciência Técnica Vitivinícola* **1998**, *13*, 71–105.
54. Martínez-Gil, A.; del Alamo-Sanza, M.; Sánchez-Gómez, R.; Nevares, I. Alternative Woods in Enology: Characterization of Tannin and Low Molecular Weight Phenol Compounds with Respect to Traditional Oak Woods. A Review. *Molecules* **2020**, *25*, 1474. [CrossRef]
55. Oberholster, A.; Elmendorf, B.L.; Lerno, L.A.; King, E.S.; Heymann, H.; Brenneman, C.E.; Boulton, R.B. Barrel maturation, oak alternatives and micro-oxygenation: Influence on red wine aging and quality. *Food Chem.* **2015**, *173*, 1250–1258. [CrossRef]
56. Laqui-Estaña, J.; López-Solís, R.; Peña-Neira, Á.; Medel-Maraboli, M.; Obreque-Slier, E. Wines in contact with oak wood: The impact of the variety (Carménère and Cabernet Sauvignon), format (barrels, chips and staves), and aging time on the phenolic composition. *J. Sci. Food Agric.* **2019**, *99*, 436–448. [CrossRef]
57. Karvela, E.; Makris, D.P.; Kefalas, P.; Moutounet, M. Extraction of phenolics in liquid model matrices containing oak chips: Kinetics, liquid chromatography-mass spectroscopy characterisation and association with in vitro antiradical activity. *Food Chem.* **2008**, *110*, 263–272. [CrossRef] [PubMed]
58. García-Estévez, I.; Alcalde-Eon, C.; Martínez-Gil, A.M.; Rivas-Gonzalo, J.C.; Escribano-Bailón, M.T.; Nevares, I.; del Alamo-Sanza, M. An Approach to the Study of the Interactions between Ellagitannins and Oxygen during Oak Wood Aging. *J. Agric. Food. Chem.* **2017**, *65*, 6369–6378. [CrossRef] [PubMed]
59. Martínez-Gil, A.; Cadahía, E.; Fernández de Simón, B.; Gutiérrez-Gamboa, G.; Nevares, I.; del Álamo-Sanza, M. Phenolic and volatile compounds in Quercus humboldtii Bonpl. wood: Effect of toasting with respect to oaks traditionally used in cooperage. *J. Sci. Food Agric.* **2019**, *99*, 315–324. [CrossRef] [PubMed]
60. Anjos, O.; Carmona, C.; Caldeira, I.; Canas, S. Variation of Extractable Compounds and Lignin Contents in Wood Fragments Used in the Aging of Wine Brandies. *BioResources* **2013**, *8*, 4484–4496. [CrossRef]

Permissions

All chapters in this book were first published by MDPI; hereby published with permission under the Creative Commons Attribution License or equivalent. Every chapter published in this book has been scrutinized by our experts. Their significance has been extensively debated. The topics covered herein carry significant findings which will fuel the growth of the discipline. They may even be implemented as practical applications or may be referred to as a beginning point for another development.

The contributors of this book come from diverse backgrounds, making this book a truly international effort. This book will bring forth new frontiers with its revolutionizing research information and detailed analysis of the nascent developments around the world.

We would like to thank all the contributing authors for lending their expertise to make the book truly unique. They have played a crucial role in the development of this book. Without their invaluable contributions this book wouldn't have been possible. They have made vital efforts to compile up to date information on the varied aspects of this subject to make this book a valuable addition to the collection of many professionals and students.

This book was conceptualized with the vision of imparting up-to-date information and advanced data in this field. To ensure the same, a matchless editorial board was set up. Every individual on the board went through rigorous rounds of assessment to prove their worth. After which they invested a large part of their time researching and compiling the most relevant data for our readers.

The editorial board has been involved in producing this book since its inception. They have spent rigorous hours researching and exploring the diverse topics which have resulted in the successful publishing of this book. They have passed on their knowledge of decades through this book. To expedite this challenging task, the publisher supported the team at every step. A small team of assistant editors was also appointed to further simplify the editing procedure and attain best results for the readers.

Apart from the editorial board, the designing team has also invested a significant amount of their time in understanding the subject and creating the most relevant covers. They scrutinized every image to scout for the most suitable representation of the subject and create an appropriate cover for the book.

The publishing team has been an ardent support to the editorial, designing and production team. Their endless efforts to recruit the best for this project, has resulted in the accomplishment of this book. They are a veteran in the field of academics and their pool of knowledge is as vast as their experience in printing. Their expertise and guidance has proved useful at every step. Their uncompromising quality standards have made this book an exceptional effort. Their encouragement from time to time has been an inspiration for everyone.

The publisher and the editorial board hope that this book will prove to be a valuable piece of knowledge for researchers, students, practitioners and scholars across the globe.

List of Contributors

Gi-Sung Lee
Department of Children Welfare, Woosuk University, Wanju Jeonbuk 55338, Korea

Ki-Hak Hong
Department of Computer Science, Dongshin University, Naju Jeonnam 58245, Korea

Chang-Kyoon Son
Department of Applied Statistics, Dongguk University, Gyeongju Gyeongbuk 38066, Korea

Javier Martínez Torres
Department of Applied Mathematics I. Telecommunications Engineering School, University of Vigo, 36310 Vigo (Pontevedra), Spain

Jorge Pastor Pérez
Centro de Evaluación, Formación y Calidad de Aragón, 50018 Zaragoza, Spain

Joaquín Sancho Val
Centro Universitario de la Defensa. Academia General Militar, 50090 Zaragoza, Spain

Aonghus McNabola and John Gallagher
Department of Civil, Structural and Environmental Engineering, Trinity College Dublin, University of Dublin, Dublin D02 PN40, Ireland

Miguel Martínez Comesaña
Escuela de Ingeniería Industrial, University of Vigo, 36310 Vigo (Pontevedra), Spain

Javier E. Contreras-Reyes
Departamento de Estadística, Facultad de Ciencias, Universidad del Bío-Bío, Concepción 4081112, Chile

Mohsen Maleki
Department of Statistics, College of Sciences, Shiraz University, Shiraz 71946 85115, Iran

Daniel Devia Cortés
Departamento de Evaluación de Pesquerías, Instituto de Fomento Pesquero, Valparaíso 2361827, Chile

Elías Berriochoa, Alicia Cachafeiro and Alberto Castejón
Departamento de Matemática Aplicada I, Universidad de Vigo, 36201 Vigo, Pontevedra, Spain

José Manuel García-Amor
Departamento de Matemáticas, Instituto E. S. Valle Inclán, 36001 Pontevedra, Spain

Md Showaib Rahman Sarker, Michael Pokojovy and Sangjin Kim
Department of Mathematical Sciences, The University of Texas at El Paso, El Paso, TX 79968, USA

Miguel Martínez Comesaña, Sandra Martínez Mariño, Pablo Eguía Oller and Enrique Granada Álvarez
Department of Mechanical Engineering, Heat Engines and Fluid Mechanics, Industrial Engineering School, University of Vigo, Maxwell s/n, 36310 Vigo, Spain

Aitor Erkoreka González
ENEDI Research Group, Department of Thermal Engineering, University of the Basque Country, 48013 Bilbao, Spain

Abhijeet R Patil
Computational Science, University of Texas at El Paso, El Paso, TX 79968, USA

Ishaq Adeyanju Raji and Muhammad Hisyam Lee
Department of Mathematical Sciences, Universiti Teknologi Malaysia, Skudai 81310, Malaysia

Muhammad Riaz and Nasir Abbas
Department of Mathematics and Statistics, King Fahd University of Petroleum and Minerals, Dhahran 31261, Saudi Arabia

Mu'azu Ramat Abujiya
Preparatory Year Mathematics Program, King Fahd University of Petroleum and Minerals, Dhahran 31261, Saudi Arabia

Hohsuk Noh
Department of Statistics, Sookmyung Women's University, Seoul 04310, Korea

Seong J. Yang
Department of Statistics (Institute of Applied Statistics), Jeonbuk National University, Jeollabuk-do 54896, Korea

Manuel Oviedo-de La Fuente
Department of Statistics, Mathematical Analysis and Optimization, Universidad de Santiago de Compostela, 15782 Santiago de Compostela, Spain

Celestino Ordóñez
Department of Mining Exploitation and Propsecting, Universidad de Oviedo, Escuela Politécnica de Mieres, 33600 Mieres, Spain

Javier Roca-Pardiñas
Department of Statistics and Operation Research, Universidad de Vigo, 36310 Vigo, Spain

Jong-Min Kim
Division of Sciences and Mathematics, University of Minnesota at Morris, Morris, MN 56267, USA

Miguel Flores
MODES Group, Department of Mathematics, Escuela Politécnica Nacional, 170517 Quito, Ecuador

Salvador Naya
MODES Group, CITIC, ITMATI, Department of Mathematics, Escola Politécnica Superior, Campus Industrial, Universidade da Coruña, Mendizábal s/n, 15403 Ferrol, Spain

Paula Raña and Rubén Fernández-Casal
MODES Group, CITIC, Department of Mathematics, Faculty of Computer Science, Campus de Elviña, Universidade da Coruña, 15008 A Coruña, Spain

Sonia Zaragoza
PROTERM Group, Department of Naval and Industrial Engineering, Campus Industrial, Universidade da Coruña, Mendizábal s/n, 15403 Ferrol, Spain

Javier Tarrío-Saavedra
MODES Group, CITIC, Department of Mathematics, Escola Politécnica Superior, Campus Industrial, Universidade da Coruña, Mendizábal s/n, 15403 Ferrol, Spain

Jung Yeon Lee
Department of Psychiatry, New York University School of Medicine, New York, NY 10016, USA

Myeong-Kyu Kim
Department of Neurology, Chonnam National University Medical School, Gwangju 61469, Korea

Wonkuk Kim
Department of Applied Statistics, Chung-Ang University, Seoul 06974, Korea

Ofélia Anjos
Instituto Politécnico de Castelo Branco, Escola Superior Agrária, 6001-909 Castelo Branco, Portugal
CEF, Instituto Superior de Agronomia, Universidade de Lisboa, 1349-017 Lisboa, Portugal
CBPBI, Centro de Biotecnologia de Plantas da Beira Interior, 6001-909 Castelo Branco, Portugal

Soraia Inês Pedro
Instituto Politécnico de Castelo Branco, Escola Superior Agrária, 6001-909 Castelo Branco, Portugal

Miguel Martínez Comesaña and Pablo Eguía Oller
Department of Mechanical Engineering, Heat Engines and Fluid Mechanics, Industrial Engineering School, University of Vigo, Maxwell s/n, 36310 Vigo, Spain

Ilda Caldeira and Sara Canas
INIAV, INIAV-Dois Portos, Quinta da Almoínha, 2565-191 Dois Portos, Portugal
MED – MediterraneanInstitute for Agriculture, Environment and Development, Instituto de formação avançada, Universidade de Évora, Pólo da Mitra, Ap. 94, 7006-554 Évora, Portugal

Index

A

Adaptive Lasso, 66-67, 70, 72, 74-80, 102-103, 108, 120, 171, 184

Ageing Technology, 226-227, 230, 233-236, 238, 240-242

Ageing Time, 226, 228, 230, 234-242

Air Pollution, 17-18, 20, 31-33, 98-99, 167, 244

Allele Read Counts, 212-213, 223

Average Run Length, 20, 125-126, 128, 191, 199

B

Bioinformatics, 80-81, 103, 105, 109-110, 122-124, 170, 182-184

C

Chromatic Characteristics, 226, 229-230, 233-235, 241

Chromaticity, 228-229, 233-234

Colon Cancer, 66-67, 78-80, 176, 178-179, 184

Coniferaldehyde, 226, 229, 234-235, 241-242

Control Chart, 20, 27, 99, 125-130, 135, 139-141, 186-199, 201-204, 206-208, 210

D

Data Depth, 186-190, 192, 194-197, 199, 203-204, 206, 208, 210

Depth Measurement, 187, 190, 203-204, 208

E

Efficiency Comparison, 142-144, 151

Elastic Net, 66-70, 72, 74-80, 102-103, 108, 123-124, 170-171, 180, 184

Ellagic Acid, 226, 229, 234-235, 241

Energy Efficiency, 82, 97, 186-187, 193-194, 197-198, 201, 208, 210, 227

F

Feature Selection, 66, 80, 102-103, 105, 109-110, 114, 116, 118-120, 122-123, 170, 182-184

Functional Analysis, 17, 21, 30, 83, 90-97, 230, 245

Functional Data Analysis, 17-18, 20-21, 23, 32, 34-35, 82-84, 99-100, 158-159, 168-169, 186-187, 210-211, 226-227, 230, 237, 242, 244-245

Functional Outliers, 22, 30-31, 99, 244

Furanic Compounds, 229, 236, 238, 242

Furfural, 226, 229, 234-235, 241-242

G

Gene Expression, 66-67, 78, 80-81, 102-103, 105, 110, 118-119, 122, 170-171, 173, 176, 179-183

Genetic Information, 212, 222

Genotype, 212-213, 219-220, 222-224

Geometric Distribution, 1-2, 15-16

Geometric Mean, 102, 112, 170-171, 175-176

I

Information Gain, 103, 105, 120, 170, 183

Information Matrix, 36, 47, 223

Iterations, 74-75, 78, 109-110, 113, 115-117, 121, 135, 176-177, 179

K

Kullback-leibler Divergence, 36, 46-47, 49

L

Lagrange Interpolation, 50-51, 60, 62, 64-65

Laplacian Matrix, 66-71

Lighting Consumption, 84, 87-88, 90-93, 96-97

Log-likelihood Function, 45, 68, 108, 172

Low-coverage, 212-213, 221, 223, 225

M

Machine Learning, 66, 83, 99, 102-103, 105, 122-123, 168, 170, 187, 209, 212, 244

Median Absolute Deviation, 125-126, 129, 132, 140

Methanol, 227, 229, 236, 238, 242-243

Micro-oxygenation, 226-228, 233-234, 242-243, 245

Minimax Concave Penalty, 102, 109, 123, 171-172, 184

Misclassification, 72, 212

Mixture Model, 212-213, 223, 225

Molecular Weight, 226, 228-230, 234-235, 241-243, 245

N

Nodal Systems, 50-51, 61, 64-65

Nonparametric Control Chart, 186

O

Objective Function, 68-69

Optimization, 81, 98-99, 158, 244

Outlier, 17, 21, 31-32, 34-35, 83, 85-86, 97-98, 125-126, 129-137, 139-141, 150, 158, 168, 187, 190, 194-195, 202-203, 206, 208-211, 245

P

Penalized Logistic Regression, 66-67, 72, 74, 78-81, 103, 122, 171-172, 184

Penalized Regression, 81, 103, 108, 110, 123

Penalty Function, 68, 70-71, 108, 172

Phenolic Index, 226, 228-230, 234, 242

Photolithography, 125-126, 136-137, 140

Pollution Episodes, 17-18, 21, 31-32, 99, 158-159, 161, 163-166, 244

R

Randomization Device, 1-2, 12, 15-16

Redundancy, 102-103, 105, 120, 122, 170, 183
Refurbishment, 82-84, 87-98
Regression Coefficient, 110, 220
Retrofitting, 82-83, 87-91, 93, 95-98, 244

S

Sandwich Variance Estimator, 212-213, 220, 223
Sea Surface Temperature, 36, 43-44, 47
Semiparametric Regression, 142, 144-146, 153, 188
Sensitive Attribute, 1-2, 12, 15
Separation Properties, 50-52
Shewhart Control Chart, 126, 129-130, 135, 139, 210
Standard Deviation, 20, 47, 74-80, 120, 125-128, 137-138, 140, 176-179
Statistical Process Control, 17, 19, 26, 32, 34, 99, 125, 141, 186-187, 209-211
Stochastic Frontier Model, 142-144, 150
Stochastic Process, 85, 158-159, 190, 199-200

U

Uncertainty Region, 158, 160-161
Unit Circle, 50-52, 64-65

V

Vanillin, 226, 229, 234-235, 241-242
Vector Analysis, 95, 226-227, 236-238, 242-243
Vectorial Analysis, 30, 82-83, 86, 90-97, 237-241

W

Wine Spirit, 226, 228, 235-243
Wooden Barrel, 226

www.ingramcontent.com/pod-product-compliance
Lightning Source LLC
La Vergne TN
LVHW082125200926
848484LV00008B/2007